Step-by-Step Solution to Selected Problems in SIGNALS & SYSTEMS

Hamid Saeedi
University of Doha for Science and Technology

Hossein Pishro-Nik
University of Massachusetts Amherst

Published by Kappa Research, LLC.

Printed in the United States of America

ISBN: 978-0-9906372-3-3

Contents

Preface

Courses on continuous and discrete-time signals and systems are a fundamental part of most Electrical Engineering programs. As a result, there are numerous textbooks on this subject, each offering a variety of examples and unsolved problems for assignments. Additionally, there are books available that provide solutions to these problems or similar ones. However, after evaluating many of these resources, we identified a gap: a lack of books that guide students step-by-step through the solutions. Even when such resources exist, the difficulty level of the problems often fails to adequately prepare students for a rigorous final exam on signals and systems. This realization motivated us to co-author this book. In some similar books, solutions frequently skip steps, assuming students are already familiar with the missing information. In this book, we strive to avoid such assumptions and provide detailed, step-by-step solutions.

The book consists of 10 chapters with over 200 problems, each containing multiple questions on related topics. The chapters are organized to align with the sequence found in most signals and systems textbooks, particularly following the structure of the seminal work by Oppenheim & Willsky[1]. At the beginning of each chapter, we offer a concise yet focused review of the topic to minimize the need for students to refer back to the textbook. We primarily adhere to the notations and terminology used in Oppenheim & Willsky's book.

The preparation of this book would not have been possible without the invaluable assistance of two of our students, Mr. Mohamad Khas and Mr. Mohamad Mir Ahmadi. They were involved in every step of the process, and we relied on their perspectives as students to ensure that the solutions are as accessible as possible for those using this book. The second author has previously published a well-received book on probability and stochastic processes. We hope this book will be similarly successful and help students enhance their practical skills in this important subject.

[1]Oppenheim, A. V., and Willsky, A. S., with S. H. Nawab, Signals and Systems, Second Edition, Upper Saddle River, NJ: Prentice Hall, Inc., 1997

Chapter 1

Continuous-time and Discrete-time Signals

1.1 Continuous-time and Discrete-time Signals

Basically, a signal is a function that conveys information to the behavior of a system or nature of the phenomenon. In other words a signal is represented as a function of an independent variable t or n that donates time. In the case of continuous-time signals, the independent variable is continuous, and the signal is donated by $x(t)$. On the other hand, if the independent variable is a discrete variable, then the signal is called a discrete-time signal and are represented by $x[n]$. In a nutshell:

$$x(t) \text{ is continuous-time if } \quad t \in \mathbb{R}, \quad \text{i.e., } t \text{ is a real number}$$

$$x[n] \text{ is discrete-time if } \quad n \in \mathbb{Z}, \quad \text{i.e., } n \text{ is integer}$$

1.1.1 Real and Complex Signals

A signal $x(t)$ is a real signal if its value is a real number. On the other hand it is a complex signal if its value is a complex number. Complex signals can be thought as a pair of real signals that represented by

$$x(t) = x_1(t) + jx_2(t), \tag{1.1}$$

where $x_1(t)$ and $x_2(t)$ are continuous real signals and $j = \sqrt{-1}$.

Note that the above definition also applies to discrete-time signals.

1.1.2 Exponential and Sinusoidal Signals

The continuous-time complex exponential signal is defined as

$$x(t) = e^{jw_0 t}. \tag{1.2}$$

Using Euler's formula, this signal can be written as

$$x(t) = e^{jw_0 t} = \cos w_0 t + j \sin w_0 t. \quad (1.3)$$

A real signal that is closely related to the complex exponential signal is the sinusoidal signal

$$x(t) = A\cos(w_0 t + \theta), \quad (1.4)$$

where A is the amplitude (real), w_0 is the radian frequency in radians per second (real), and θ is the phase angle in radians (real).

Using Euler's formula, then we have

$$x(t) = Ae^{(jw_0 t + \theta)} = A\cos(w_0 t + \theta) + jA\sin(w_0 t + \theta). \quad (1.5)$$

Similarly, the discrete-time complex exponential signal is written as

$$x[n] = e^{j\Omega_0 n} = \cos \Omega_0 n + j \sin \Omega_0 n. \quad (1.6)$$

Moreover, a closely related signal to the complex exponential signal is the sinusoidal signal

$$x[n] = A\cos(\Omega_0 n + \theta), \quad (1.7)$$

where

$$A\cos(\Omega_0 n + \theta) = A\,\mathrm{Re}\{e^{(j\Omega_0 n + \theta)}\}.$$

1.2 Periodic Signals

A continuous-time signal $x(t)$ is said to be periodic with period T if

$$x(t+T) = x(t) \quad \forall t,$$

where T is a positive value. If this condition is satisfied for a given T, it is also satisfied for all integer multiples of T. The fundamental period T_0 of $x(t)$ is the smallest positive value of T for which the above relationship is satisfied.

Analogously, a discrete-time signal is periodic if there is a positive integer N for which

$$x[n+N] = x[n] \quad \forall n.$$

We can readily deduce that

$$x[n+mN] = x[n] \quad \text{for any integer } m.$$

The fundamental period N_0 of $x[n]$ is the smallest positive integer N for which which the above relationship is satisfied.

The fundamental periods of the continuous-time complex exponential and sinusoidal signals are as follows:

$$\left\{ \begin{array}{l} x(t) = \cos w_0 t \\ x(t) = \sin w_0 t \Rightarrow \\ x(t) = e^{jw_0 t} \end{array} \right. \quad T_0 = \frac{2\pi}{w_0}. \quad (1.8)$$

Note that all signals in (1.8) are periodic for any value of w_0.

Analogously, the fundamental periods of the discrete-time complex exponential and sinusoidal signals are as follows:

$$\begin{cases} x[n] = \cos \Omega_0 n \\ x[n] = \sin \Omega_0 n \Rightarrow \qquad N_0 = (\frac{2\pi}{\Omega_0}) \\ x[n] = e^{j\Omega_0 n} \end{cases} \tag{1.9}$$

Note that signals in (1.9) are not periodic for any value of Ω_0. They are periodic only if $\Omega_0/2\pi$ is a rational number.

Tip: If a signal is composed of two or more than two parts, then to obtain its fundamental period, first we must determine the fundamental period of each part and then obtain their Least Common Multiple (LCM).

1.3 Even and Odd Signals

A signal is even if it satisfies the condition

$$x(t) = x(-t), \qquad x[n] = x[-n].$$

On the other hand, a signal is referred to as an odd signal if

$$x(t) = -x(-t), \qquad x[n] = -x[-n].$$

Note that an odd signal must be 0 at $t = 0$ and $n = 0$.

It is noted that any signal can be broken into a sum of two signals, one of which is even and one of which is odd. That is,

$$x_e(t) = \frac{1}{2}[x(t) + x(-t)] \quad \text{even part of } x(t) \tag{1.10}$$

$$x_o(t) = \frac{1}{2}[x(t) - x(-t)] \quad \text{odd part of } x(t). \tag{1.11}$$

Exactly analogous definitions hold for the discrete-time case.

Note that the sum of two even signals or two odd signals is an even signal and that the sum of an even signal and an odd signal is an odd signal.

We can prove the following properties for even and odd signals:

$$x(t) \text{ is odd } \Rightarrow \int_{-a}^{a} x(t)dt = 0,$$

$$x[n] \text{ is odd } \Rightarrow \sum_{-m}^{m} x[n] = 0,$$

$$x(t) \text{ is even } \Rightarrow x(0) = x_e(0),$$

$$x[n] \text{ is even } \Rightarrow x[0] = x_e[0].$$

1.4 Signal Energy and Power

The energy E and the power P of a continuous-time signal $x(t)$ are defined respectively, as

$$E = \int_{-\infty}^{\infty} |x(t)|^2 dt \tag{1.12}$$

$$,P = \lim_{T\to\infty} \frac{1}{2T} \int_{-T}^{T} |x(t)|^2. \tag{1.13}$$

Similarly, for a discrete-time signal $x[n]$, the energy E and the power P of $x[n]$ are defined respectively, as

$$E = \sum_{n=-\infty}^{\infty} |x[n]|^2 \tag{1.14}$$

$$,P = \lim_{N\to\infty} \frac{1}{2N+1} \sum_{n=-N}^{N} |x[n]|^2. \tag{1.15}$$

Tip: Based on Definitions (1.12) to (1.15), to obtain the energy E and the power P of a signal, we can divide the signal into separate parts that do not overlap in time, and determine the energy and the power of each part. Then the total energy is the sum of the energy of each part and the total power is determined as well, i.e., if

$$x(t) = x_1(t) + x_2(t) + x_3(t) + \ldots,$$

where $x_i(t)x_j(t) = 0, \forall i \neq j$, and we have

$$E\{x_1(t)\} = E_1 \quad E\{x_2(t)\} = E_2 \quad E\{x_3(t)\} = E_3 \quad \ldots,$$

$$P\{x_1(t)\} = P_1 \quad P\{x_2(t)\} = P_2 \quad P\{x_3(t)\} = P_3 \quad \ldots,$$

then energy E and power P of $x(t)$ are obtained as

$$E\{x(t)\} = \sum_i E_i = E_1 + E_2 + E_3 + \ldots$$

$$P\{x(t)\} = \sum_i P_i = P_1 + P_2 + P_3 + \ldots$$

Similarly, for a discrete-time signal $x[n]$, the energy E and the power P of $x[n]$ are the sum of the energy and the power of each part.
According to Definitions (1.12) to (1.15), signals are divided into three classes:

1. A signal $x(t)$ (or $x[n]$) is an energy signal if and only if

$$0 < E < \infty, \qquad P = 0.$$

2. A signal $x(t)$ (or $x[n]$) is a power signal if and only if

$$0 < P < \infty, \qquad E = \infty.$$

3. Signals that do not satisfy the above properties are neither energy signals nor power signals.

Tip: A periodic signal is a power signal if its energy content per period is finite. Therefore, to obtain the power, we need to calculate the power only over a period:

$$x(t) \quad \longrightarrow \quad P = \frac{1}{T}\int_0^T |x(t)|^2 dt,$$

$$x[n] \quad \longrightarrow \quad P = \frac{1}{N}\sum_{n=0}^{N-1} |x[n]|^2,$$

where T and N are the fundamental periods of $x(t)$ and $x[n]$, respectively.

1.5 Basic Continuous-Time and Discrete-time Signals

1.5.1 The Continuous-Time Unit Impulse and Unit Step Functions

The *unit impulse* function (or *Dirac delta* function) is one of the most useful, yet controversial, continuous-time signals, which is defined as

$$\delta(t) = \begin{cases} 0 & t \neq 0 \\ \infty & t = 0 \end{cases} \quad \longrightarrow$$

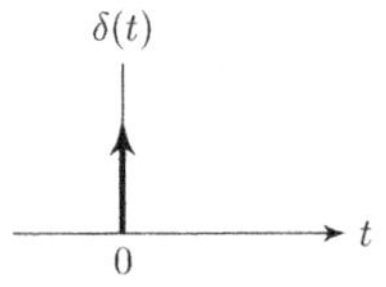

$$\int_{\epsilon^-}^{\epsilon^+} \delta(t)dt = 1.$$

Also the shifted unit impulse function is defined as

$$\delta(t - t_0) = \begin{cases} 0 & t \neq t_0 \\ \infty & t = t_0 \end{cases} \quad \longrightarrow$$

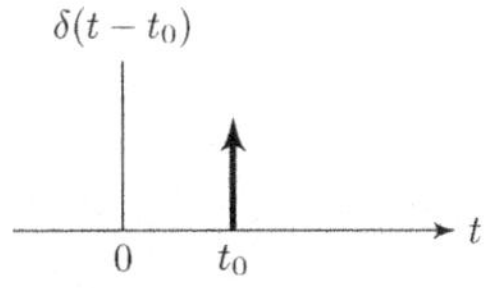

Some properties of $\delta(t)$ are

1. $\delta(at) = \frac{1}{a}\delta(t)$
2. $\delta(-t) = \delta(t)$
3. $x(t)\delta(t) = x(0)\delta(t)$
4. $x(t)\delta(t - t_0) = x(t_0)\delta(t - t_0)$
5. $\int_{0^-}^{0^+} x(t)\delta'(t)dt = -x'(0)$
6. $\delta(f(t)) = \sum_i \frac{\delta(t-t_i)}{|f'(t_i)|}$.

The continuous-time *unit step* function $u(t)$ is defined as

$$u(t) = \begin{cases} 1 & t > 0 \\ 0 & t < 0 \end{cases} \longrightarrow$$

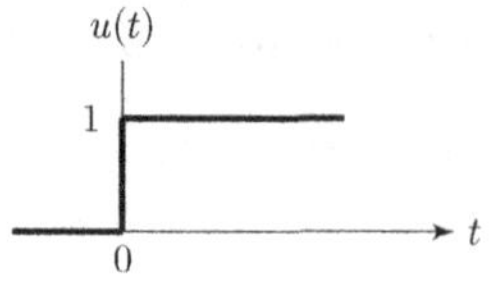

Note that it is discontinuous at $t = 0$ and that the value at $t = 0$ is undefined. Similarly, the shifted unit step function $u(t - t_0)$ is defined as

$$u(t - t_0) = \begin{cases} 1 & t > t_0 \\ 0 & t < t_0 \end{cases} \longrightarrow$$

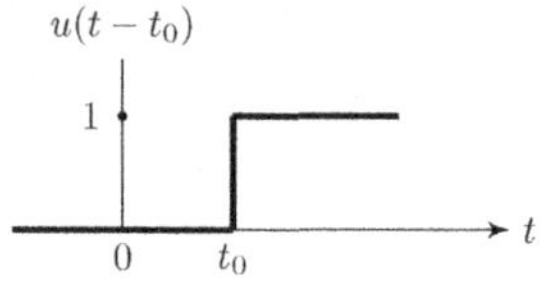

The relationships between the *unit impulse* function $\delta(t)$ and the *unit step* function $u(t)$ are

$$\delta(t) = u'(t) = \frac{du(t)}{dt}, \tag{1.16}$$

$$u(t) = \int_{-\infty}^{t} \delta(t)dt. \tag{1.17}$$

1.5.2 The Discrete-Time Unit Impulse and Unit Step Signals

The discrete-time *unit impulse* signal $\delta[n]$ is defined as

$$\delta[n] = \begin{cases} 0 & n \neq 0 \\ 1 & n = 0 \end{cases} \longrightarrow$$

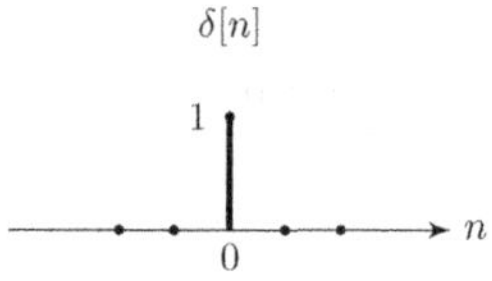

Also the shifted unit impulse signal is defined as

$$\delta[n - k] = \begin{cases} 0 & n \neq k \\ 1 & n = k \end{cases} \longrightarrow$$

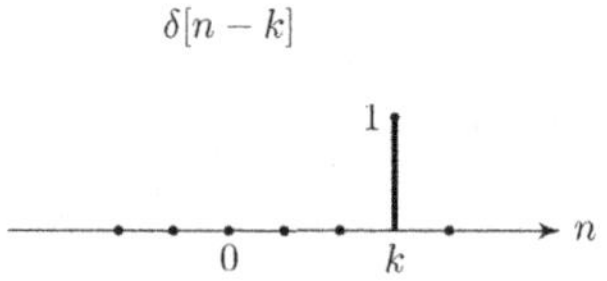

Some properties of $\delta[n]$ are

1. $\delta[an] = \delta[n]$
2. $x[n]\delta[n] = x[0]\delta[n]$
3. $x[n]\delta[n - k] = x[k]\delta[n - k]$.

The *unit step* signal $u[n]$ is defined as

$$u[n] = \begin{cases} 0 & n < 0 \\ 1 & n \geq 0 \end{cases} \longrightarrow$$

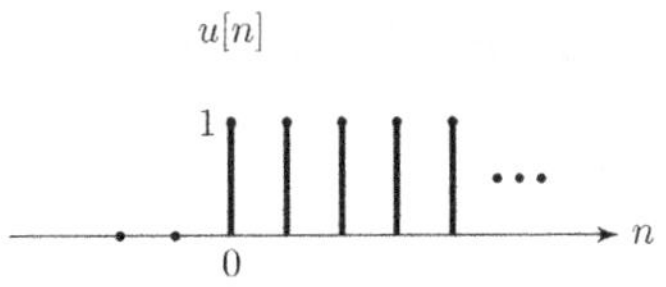

Similarly, the shifted *unit step* signal $u[n-k]$ is defined as

$$u[n-k] = \begin{cases} 0 & n < k \\ 1 & n \geq k \end{cases} \longrightarrow$$

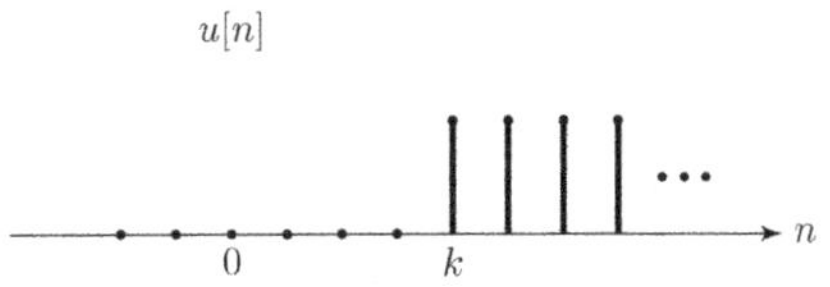

From above definitions, we have

$$\delta[n] = u[n] - u[n-1], \tag{1.18}$$

$$u[n] = \sum_{k=-\infty}^{n} \delta[k] = \sum_{k=0}^{\infty} \delta[n-k]. \tag{1.19}$$

1.6 Time Shifts/Time Scaling

As we know, there are two very important transformations of the independent variable of a signal, *time shifting* and *time scaling.* A general form of transforming the independent variable of a signal $x(t)$ is $x(\alpha t + \beta)$, where α and β are constant numbers. Note that signal $x(t)$ is linearly stretched if $|\alpha| < 1$, linearly compressed if $|\alpha| > 1$, reversed in time if $\alpha < 0$, and shifted in time if β is nonzero. There are two approaches to determine $x(\alpha t + \beta)$ for signal $x(t)$:

1. The first (or systematic) approach is to first determine $f(t) = x(t + \beta)$, in other words to delay or advance $x(t)$ with regards to the value of β, and then to obtain $f(\alpha t) = x(\alpha t + \beta)$ to address time scaling and/or time reversal with regards to α.

2. The second approach is to first determine $f(t) = x(\alpha t)$ and then to perform time shifting on the resulting signal $f(t)$ with regard to β/α, as $f(t + \beta/\alpha) = x(\alpha t + \beta)$.

For a discrete-time signal $x[n]$, time shifting and time reversal are the same as continuous-time. Note that the signal for an integer m, $x[mn]$ is the time-compressed signal of $x[n]$ where some samples of $x[n]$ will be removed. Also signal $x_m[n]$ is the stretched signal of $x[n]$ in accordance with the value of m which is defined as follows:

$$x_m[n] = \begin{cases} x[\frac{n}{m}], & n = km \\ 0 & n \neq km \end{cases}. \tag{1.20}$$

Note that the samples of $x[n]$ are getting far away from each other according to the value of m and the value of empty spots between the samples will be set to zero.

1.7 Some Important Signals

The Unit Ramp function:

$$r(t) = tu(t) = \begin{cases} t & t > 0 \\ 0 & t < 0 \end{cases} \qquad \longrightarrow$$

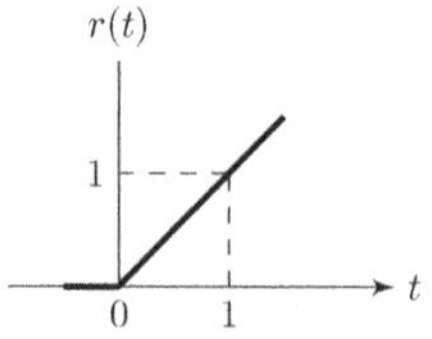

The Rectangular Pulse

$$\prod(\frac{t}{2T}) = \begin{cases} 1 & |t| < T \\ 0 & \text{other} \end{cases} \qquad \longrightarrow$$

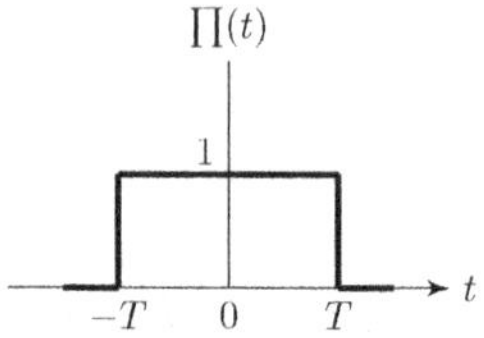

1.8 Problems

1.8.1 Continuous-time Signals

Problem 1. Determine which of the following signals are periodic. If a signal is periodic, determine its fundamental period.

(a) $y(t) = \sin(\frac{5\pi}{4}t)$

(b) $y(t) = e^{j(t+\frac{\pi}{3})}$

(c) $y(t) = \cos(\frac{3}{5}t) + |\sin(\frac{5}{3}t)|$

Solution.

(a) $y(t) = \sin(\frac{5\pi}{4}t)$.
By definition

$$y(t+T_0) = \sin(\frac{5\pi}{4}(t+T_0)) = \sin(\frac{5\pi}{4}t + \frac{5\pi}{4}T_0)$$

$$\Rightarrow \qquad \sin(\frac{5\pi}{4}t + \frac{5\pi}{4}T_0) = \sin(\frac{5\pi}{4}t).$$

Using the trigonometric identity $\sin(t + 2k\pi) = \sin(t)$, we can write

$$\frac{5\pi}{4}T_0 = 2k\pi.$$

The smallest positive T_0 is obtained with $k = 1$, so the fundamental period of $y(t)$ is $T_0 = \frac{8}{5}$.

(b) $y(t) = e^{j(t+\frac{\pi}{3})}$.

By definition

$$y(t+T_0) = e^{j((t+T_0)+\frac{\pi}{3})} = e^{j(t+\frac{\pi}{3})} \times e^{jT_0}$$

$$\Rightarrow \qquad e^{j(t+\frac{\pi}{3})} \times e^{jT_0} = e^{j(t+\frac{\pi}{3})},$$

thus,

$$e^{jT_0} = 1 \qquad \Rightarrow \qquad T_0 = 2k\pi.$$

The smallest positive T_0 is obtained with $k = 1$, so the fundamental period of $y(t)$ is $T_0 = 2\pi$.

(c) $y(t) = \cos(\frac{3}{5}t) + |\sin(\frac{5}{3}t)| = y_1(t) + y_2(t)$.

(I) By definition

$$y_1(t + T_0^1) = \cos(\frac{3}{5}(t + T_0^1)) = \cos(\frac{3}{5}t + \frac{3}{5}T_0^1)$$

$$\Rightarrow \qquad \cos(\frac{3}{5}t + \frac{3}{5}T_0^1) = \cos(\frac{3}{5}t).$$

Using the trigonometric identity $\cos(\phi + 2k\pi) = \cos(\phi)$, we can write

$$\frac{3}{5}T_0^1 = 2k\pi.$$

The smallest positive T_0^1 is obtained with $k = 1$. Thus, the fundamental period of $y_2(t)$ is $T_0^1 = \frac{10\pi}{3}$.

(II) With similar argument, the fundamental period of $\sin(\frac{5}{3}t)$ is $\frac{6\pi}{5}$. However, due to the absolute value operator, the period of $y_2(t) = |\sin(\frac{5}{3}t)|$ is half of $\frac{6\pi}{5}$, i.e., $T_0^2 = \frac{3\pi}{5}$. To see this fact better, in Fig. 1.1, we have plotted $\sin(t)$ and $|\sin(t)|$ and as can be seen, the period of the latter is half of the period of the former.

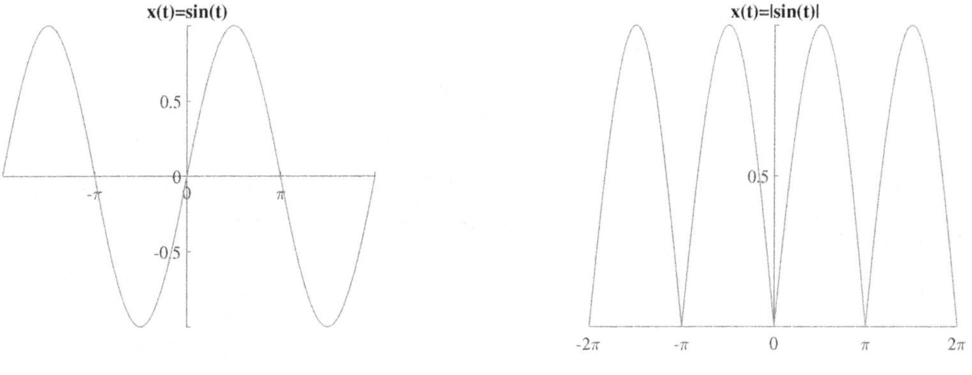

Figure 1.1: Problem 1.

So given that sum of two periodic signals is also periodic and the fundamental period is the Least Common Multiplier of the two fundamental periods, the fundamental period of signal y(t) is:

$$T_0 = \text{LCM}(\frac{3\pi}{5}, \frac{10\pi}{3}) = 30\pi.$$

Problem 2. A continuous-time signal $x(t)$ is shown in Fig. 1.2. Find the fundamental period of $y(t) = x(\cos^2(t))$.

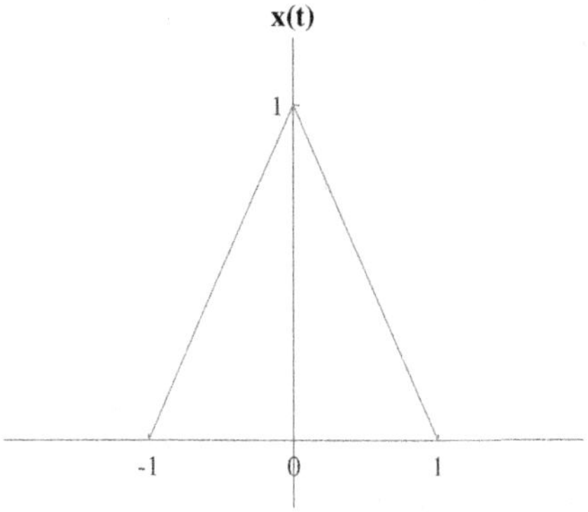

Figure 1.2: Problem 2.

Solution. From Fig. 1.2, we obtain

$$x(t) = (1 - |t|)\big[u(t+1) - u(t-1)\big], \tag{1.21}$$

substituting t with $\cos^2(t)$ in (1.21), we get

$$x(\cos^2(t)) = (1 - |\cos^2(t)|)\big[u(\cos^2(t)+1) - u(\cos^2(t)-1)\big]. \tag{1.22}$$

Also we have

$$\forall\, t \in \mathrm{R}: \quad 0 \le \big(\cos^2(t) + 1\big) \le 2 \quad \Rightarrow \quad u(\cos^2(t)+1) = 1, \tag{1.23}$$

and

$$\forall\, t \in \mathrm{R}: \quad -1 \le \big(\cos^2(t) - 1\big) \le 0 \quad \Rightarrow \quad u(\cos^2(t)-1) = 0. \tag{1.24}$$

By reordering (1.22), we get

$$y(t) = (1 - |\cos^2(t)|)u(\cos^2(t)+1) - (1 - |\cos^2(t)|)u(\cos^2(t)-1) = y_1(t) + y_2(t), \tag{1.25}$$

where

$$y_1(t) = (1 - |\cos^2(t)|)u(\cos^2(t)+1), \tag{1.26}$$

$$y_2(t) = -(1 - |\cos^2(t)|)u(\cos^2(t)-1). \tag{1.27}$$

By substituting (1.23) into (1.26), we get

$$y_1(t) = (1 - |\cos^2(t)|), \tag{1.28}$$

and by substituting (1.24) into (1.27), we obtain

$$y_2(t) = 0. \tag{1.29}$$

Finally by substituting (1.28) and (1.29) into (1.25), we get

$$y(t) = 1 - |\cos^2(t)| = 1 - \cos^2(t) = 1 - \frac{\cos(2t) + 1}{2} \tag{1.30}$$

The period of this signal is simply π.

Problem 3. Sketch the following signals and determine and sketch their even and odd components (Hint : $\prod(t) = u(t + \frac{1}{2}) - u(t - \frac{1}{2})$).

(a) $x(t) = (\frac{t+1}{2})\prod(\frac{t}{2}) + (2 - t)\prod(t - \frac{3}{2})$

(b) $x(t) = 2e^{2t}u(-t) - e^{-2t}u(t)$

(c) $x(t) = \prod(\frac{t-6}{6})$

Solution.

(a) $x(t) = (\frac{t+1}{2})\prod(\frac{t}{2}) + (2 - t)\prod(t - \frac{3}{2})$

$$\begin{aligned} x_e(t) &= \frac{x(t) + x(-t)}{2} \\ &= (\frac{t+1}{4})\prod(\frac{t}{2}) + \frac{2-t}{2}\prod(t - \frac{3}{2}) + (\frac{-t+1}{4})\prod(\frac{-t}{2}) + \frac{2+t}{2}\prod(-t - \frac{3}{2}) \end{aligned}$$

then,

$$x_e(t) = \frac{1}{2}\prod(\frac{t}{2}) + \frac{2-t}{2}\prod(t - \frac{3}{2}) + \frac{2+t}{2}\prod(-t - \frac{3}{2}),$$

and

$$\begin{aligned} x_o(t) &= \frac{x(t) - x(-t)}{2} \\ &= (\frac{t+1}{4})\prod(\frac{t}{2}) + \frac{2-t}{2}\prod(t - \frac{3}{2}) - (\frac{-t+1}{4})\prod(\frac{-t}{2}) - \frac{2+t}{2}\prod(-t - \frac{3}{2}) \end{aligned}$$

then,

$$x_o(t) = \frac{t}{2}\prod(\frac{t}{2}) + \frac{2-t}{2}\prod(t - \frac{3}{2}) - \frac{2+t}{2}\prod(-t - \frac{3}{2}).$$

The signals $x(t), x(-t), x_e(t), x_o(t)$ have been sketched in Fig. 1.3.

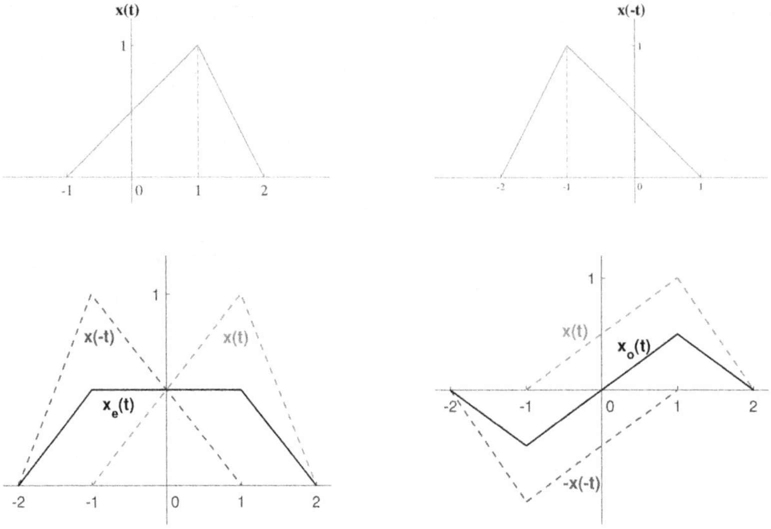

Figure 1.3: Problem 3 - Part (a).

(b) $x(t) = 2e^{2t}u(-t) - e^{-2t}u(t)$

$$x_e(t) = \frac{x(t)+x(-t)}{2} = e^{2t}u(-t) - \frac{1}{2}e^{-2t}u(t) + e^{-2t}u(t) - \frac{1}{2}e^{2t}u(-t)$$

$$\Rightarrow \quad x_e(t) = \frac{1}{2}(e^{2t}u(-t) + e^{-2t}u(t)),$$

and

$$x_o(t) = \frac{x(t)-x(-t)}{2} = e^{2t}u(-t) - \frac{1}{2}e^{-2t}u(t) - e^{-2t}u(t) + \frac{1}{2}e^{2t}u(-t)$$

$$\Rightarrow \quad x_o(t) = \frac{1}{2}(3e^{2t}u(-t) - 3e^{-2t}u(t)).$$

The signals $x(t), x(-t), x_e(t), x_o(t)$ have been sketched in Fig. 1.4.

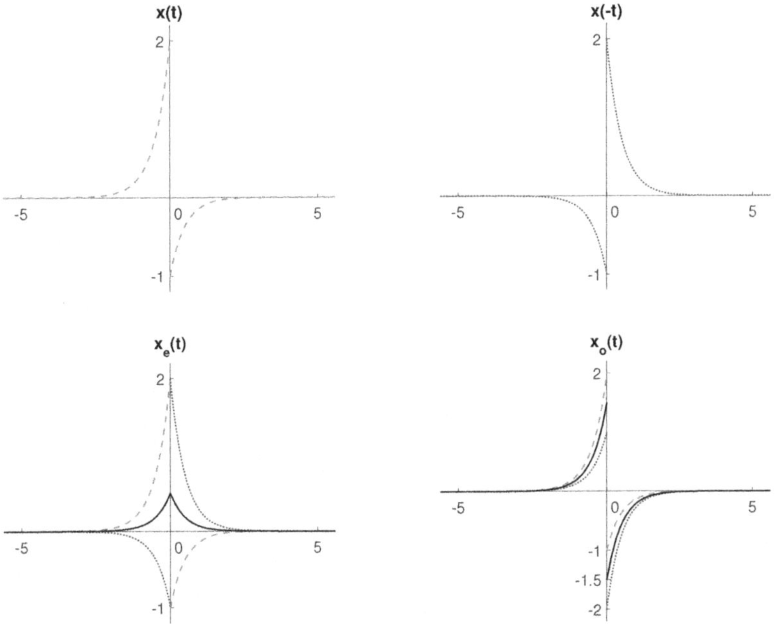

Figure 1.4: Problem 3 - Part (b).

(c) $x(t) = \prod(\frac{t-6}{6})$

$$x_e(t) = \frac{x(t)+x(-t)}{2} = \frac{1}{2}(\prod(\frac{t-6}{6}) + \prod(\frac{-t-6}{6}))$$

and

$$x_o(t) = \frac{x(t)-x(-t)}{2} = \frac{1}{2}(\prod(\frac{t-6}{6}) - \prod(\frac{-t-6}{6})).$$

The signals $x(t), x(-t), x_e(t), x_o(t)$ have been sketched in Fig. 1.5.

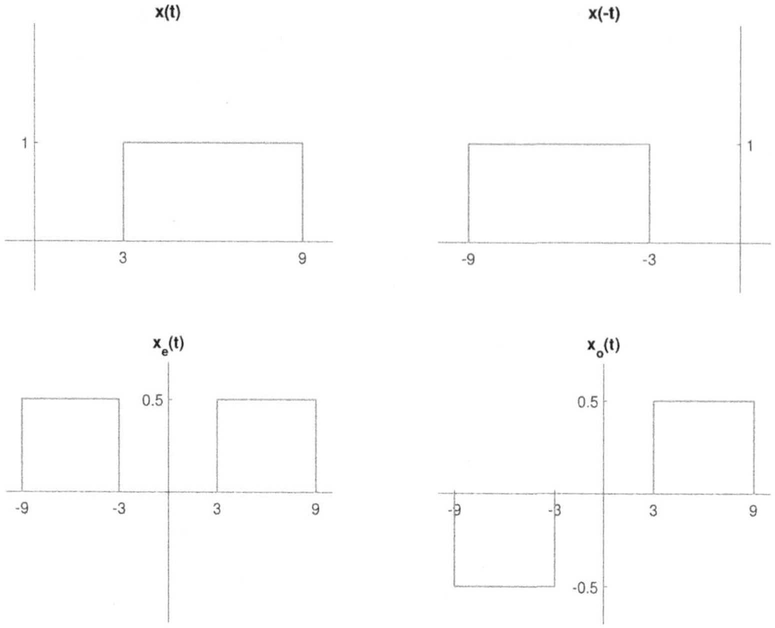

Figure 1.5: Problem 3 - Part (c).

Problem 4. Let $x_o(t)$ be the odd component of the signal shown in Fig. 1.6. Compute the following integral:

$$\int_0^{\infty} x_o(t)dt \;= ?$$

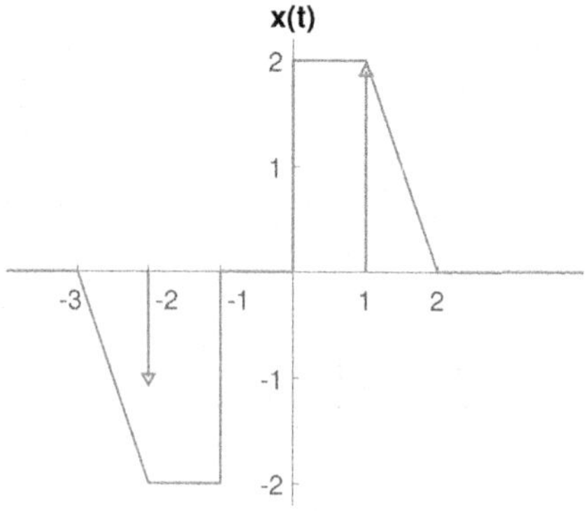

Figure 1.6: Problem 4.

Solution.

We expand $x_o(t)$ based on $x(t)$:

$$\begin{aligned}\int_0^{\infty} x_o(t)dt &= \int_0^{\infty} \frac{x(t)-x(-t)}{2}dt \\ &= \frac{1}{2}\int_0^{\infty} x(t)dt - \frac{1}{2}\int_0^{\infty} x(-t)dt \\ &= \frac{1}{2}(2+3) - \frac{1}{2}(-1-3) = 4.5\end{aligned}$$

Problem 5. A continuous-time signal $x(t)$ is shown in Fig. 1.7. Obtain the second derivative of this signal.

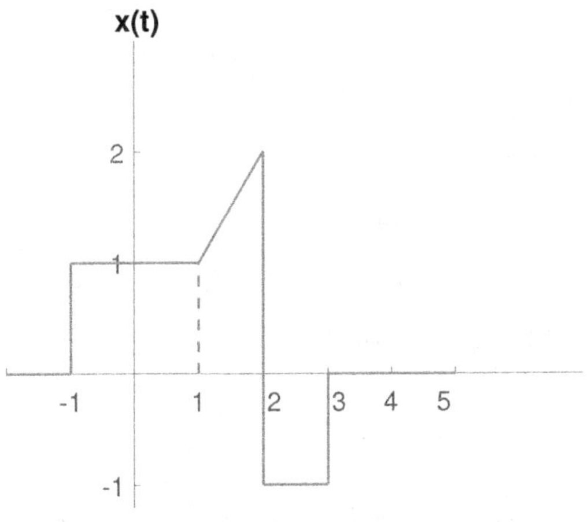

Figure 1.7: Problem 5.

Solution.
We can write $x(t)$ as (see Problem (16)):

$$x(t) = \mathrm{u}(t+1) + \mathrm{r}(t-1) - \mathrm{r}(t-2) - 3\,\mathrm{u}(t-2) + \mathrm{u}(t-3)$$

and according to this, the first derivative of signal $x(t)$ is equal to

$$x^{'}(t) = \delta(t+1) + \mathrm{u}(t-1) - \mathrm{u}(t-2) - 3\delta(t-2) + \delta(t-3).$$

Thus, the second derivative of signal $x(t)$ is equal to

$$x^{''}(t) = \delta^{'}(t+1) + \delta(t-1) - \delta(t-2) - 3\delta^{'}(t-2) + \delta^{'}(t-3).$$

Problem 6. Evaluate the following expressions using the unit impulse and unit step functions properties:

(a) $\int_0^3 \sin(t)\delta(3t-6)dt$

(b) $\int_0^4 2\cos(t-2)\delta^{'}(t-2)dt$

(c) $x(t) = t\delta^{'}(t) + \delta(t)$

(d) $\int_{-\infty}^{\infty} \delta(4-t^2)dt$

(e) $x(t) = \int_{t-1}^{t-4} \tau^2\delta(t-\tau-2)d\tau$

Solution.

(a) $\int_0^3 \sin(t)\delta(3t-6)dt \xrightarrow{\delta(at)=\frac{1}{|a|}\delta(t)} \frac{1}{3}\int_0^3 \sin(t)\delta(t-2)dt = \frac{1}{3}\sin(2).$

(b) Let $A = \int_0^4 2\cos(t-2)\delta^{'}(t-2)dt$. Setting $t' = t-2$, we get

$$A = \int_{-2}^{2} 2\cos(t')\delta^{'}(t')dt'.$$

From Section 1.5.1 we know that

$$\int_{0^-}^{0^+} x(t)\delta^{'}(t)dt = -x^{'}(0). \tag{1.31}$$

Thus, according to (1.31) we can write

$$A = -x^{'}(0),$$

where

$$x(t) = 2\cos(t) \quad \Rightarrow \quad A = -2\sin(0) = 0.$$

(c) $x(t) = t\delta^{'}(t) + \delta(t)$.

$$x(t) = \frac{d}{dt}\left[\int_{-\infty}^{t}(\tau\delta^{'}(\tau) + \delta(\tau))d\tau\right] = \begin{cases} \text{if } \ t \leq 0 & \frac{d}{dt}[\cancelto{0}{\int_{-\infty}^{t}\tau\delta^{'}(\tau)d\tau} + \cancelto{0}{\int_{-\infty}^{t}\delta(\tau)d\tau}] = 0 \\ \text{if } \ t > 0 & \frac{d}{dt}[\cancelto{-1}{\int_{-\infty}^{t}\tau\delta^{'}(\tau)d\tau} + \cancelto{1}{\int_{-\infty}^{t}\delta(\tau)d\tau}] = 0 \end{cases}$$

thus,

$$x(t) = 0.$$

(d) Let $A = \int_{-\infty}^{\infty} \delta(4-t^2)dt$.

From Section 1.5.1, we already have that

$$\delta(f(t)) = \sum_i \frac{\delta(t-t_i)}{|f'(t_i)|}, \tag{1.32}$$

where t_i's are the roots of function $f(t)$. Using (1.32), we get

$$f(t) = (4-t^2) \xrightarrow{t_1=-2, t_2=2} f^{'}(t) = -2t, f^{'}(t_1) = 4, f^{'}(t_2) = -4.$$

Thus

$$A = \int_{-\infty}^{\infty} \delta(4-t^2)dt = \int_{-\infty}^{\infty} \frac{\delta(t-2)}{4} + \frac{\delta(t+2)}{4} = \frac{1}{2}.$$

(e) $x(t) = \int_{t-1}^{t-4} \tau^2\delta(t-\tau-2)d\tau$.

Unit impulse is an even function, therefore: $\delta(t-\tau-2) = \delta(\tau-t+2)$. Thus,

$$x(t) = \int_{t-1}^{t-4} \tau^2\delta(\tau-t+2).$$

The integrand is zero unless $\tau = t-2$. Thus

$$x(t) = \int_{t-1}^{t-4} (t-2)^2\delta(\tau-(t-2))d\tau = -\int_{t-4}^{t-1} (t-2)^2\delta(\tau-(t-2))d\tau$$

$$\rightarrow \quad x(t) = -(t-2)^2\int_{t-4}^{t-1} \delta(\tau-(t-2))d\tau \quad \xrightarrow{t-4<t-2<t-1} \quad x(t) = -(t-2)^2.$$

Problem 7. Determine and sketch signal $y(t) = \int_{-\infty}^{t} x(t)dt$ for the signals shown in the following figures.

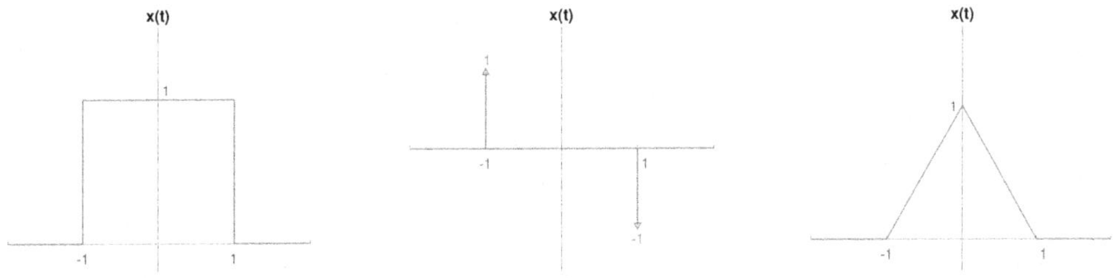

(a) (b) (c)

Figure 1.11: Problem 7.

Solution.

(a) $x(t) = \mathrm{u}(t+1) - \mathrm{u}(t-1)$.

$y(t) = \int_{-\infty}^{t} x(\tau)d\tau = \int_{-\infty}^{t} \mathrm{u}(\tau+1) - \mathrm{u}(\tau-1)d\tau = \mathrm{r}(t+1) - \mathrm{r}(t-1)$.

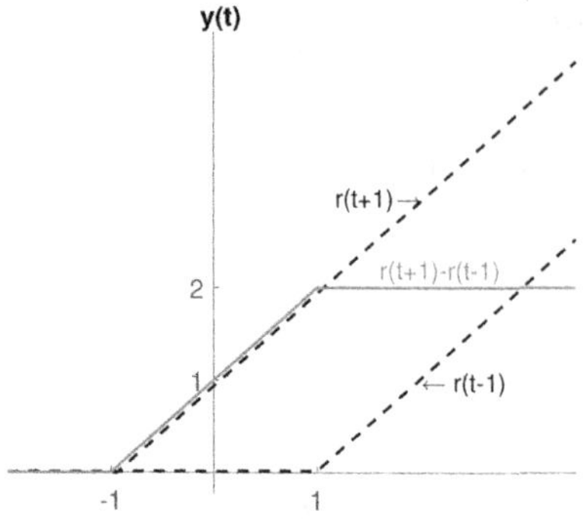

Figure 1.12: Problem 7 - Part (a).

(b) $x(t) = \delta(t+1) - \delta(t-1)$.

$y(t) = \int_{-\infty}^{t} x(\tau)d\tau = \int_{-\infty}^{t} \delta(\tau+1) - \delta(\tau-1)d\tau = \mathrm{u}(t+1) - \mathrm{u}(t-1)$.

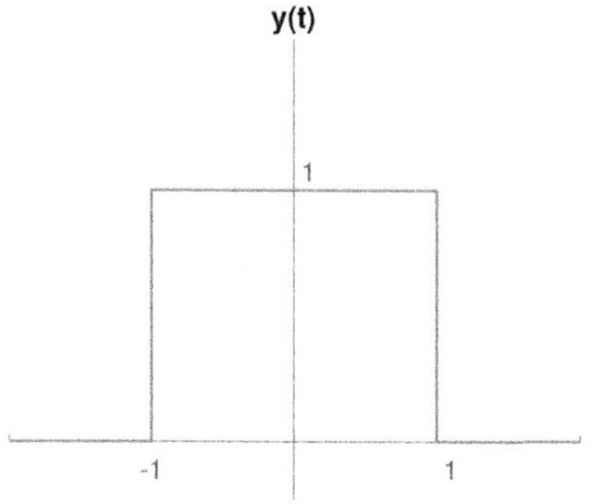

Figure 1.13: Problem 7 - Part (b).

(c) $x(t) = (t+1)\,\mathrm{u}(t+1) - 2t\,\mathrm{u}(t) + (t-1)\,\mathrm{u}(t-1)$.

$$y(t) = \int_{-\infty}^{t} x(\tau)d\tau = \int_{-\infty}^{t} \big((\tau+1)\,\mathrm{u}(\tau+1) - 2\tau\,\mathrm{u}(\tau) + (\tau-1)\,\mathrm{u}(\tau-1)\big)d\tau$$

$$\rightarrow \quad y(t) = \int_{-\infty}^{t} (\tau+1)\,\mathrm{u}(\tau+1)d\tau - 2\int_{-\infty}^{t} \tau\,\mathrm{u}(\tau)d\tau + \int_{-\infty}^{t} (\tau-1)\,\mathrm{u}(\tau-1)d\tau$$

$$\rightarrow \quad y(t) = \mathrm{u}(t+1)\int_{-1}^{t} (\tau+1)d\tau - 2\,\mathrm{u}(t)\int_{0}^{t} \tau d\tau + \mathrm{u}(t-1)\int_{1}^{t} (\tau-1)d\tau$$

$$\rightarrow \quad y(t) = \mathrm{u}(t+1)(\frac{t^2}{2} + t + \frac{1}{2}) - 2\,\mathrm{u}(t)(\frac{t^2}{2}) + \mathrm{u}(t-1)(\frac{t^2}{2} - t + \frac{1}{2}),$$

thus

$$y(t) = \begin{cases} 0, & t \le -1 \\ \frac{t^2}{2} + t + \frac{1}{2}, & -1 < t \le 0 \\ -\frac{t^2}{2} + t + \frac{1}{2}, & 0 < t \le 1 \\ 1 & t > 1 \end{cases}.$$

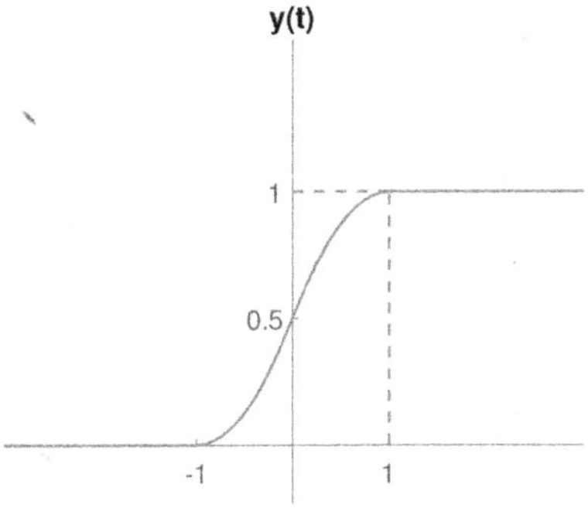

Figure 1.14: Problem 7 - Part (c).

Problem 8. A continuous-time signal x(t) is shown in Fig. 1.15. Sketch and label each of the following signals.

(a) $y(t) = x(t)\,\mathrm{u}(t) + x(-t)\,\mathrm{u}(t)$

(b) $y(t) = x(2t - 1)$

(c) $y(t) = x(-\frac{t}{2} + 1)$

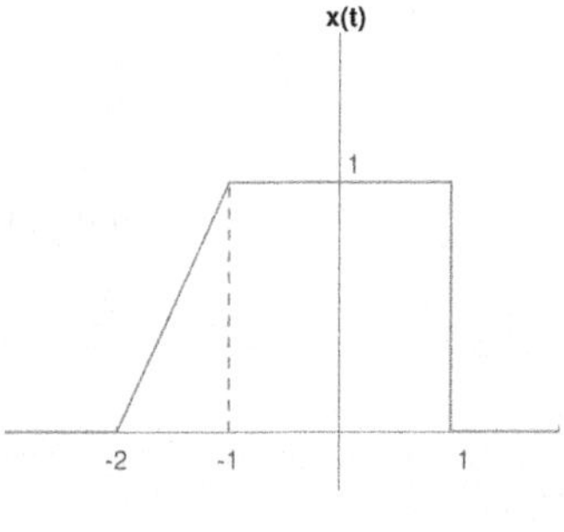

Figure 1.15: Problem 8.

Solution.

(a) $y(t) = x(t)\,\mathrm{u}(t) + x(-t)\,\mathrm{u}(t) = (x(t) + x(-t))\,\mathrm{u}(t)$.

Let

$$\begin{cases} y_1(t) = x(-t) \\ y_2(t) = x(t) + x(-t) \\ y(t) = y_2(t)\,\mathrm{u}(t) \end{cases}$$

Then we obtain $y(t)$ based on the following figures:

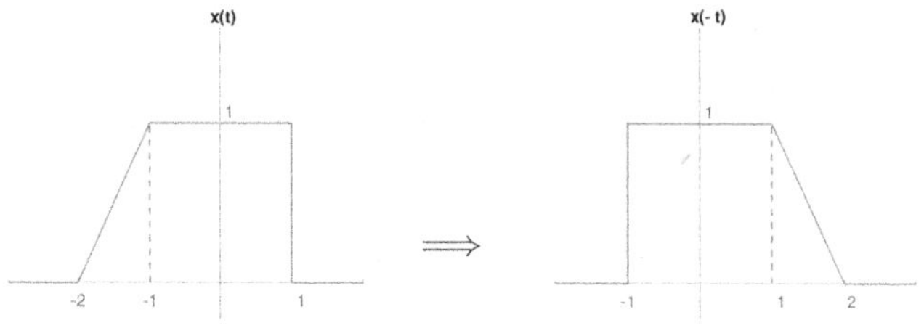

Signal $x(t)$ Signal $y_1(t)$

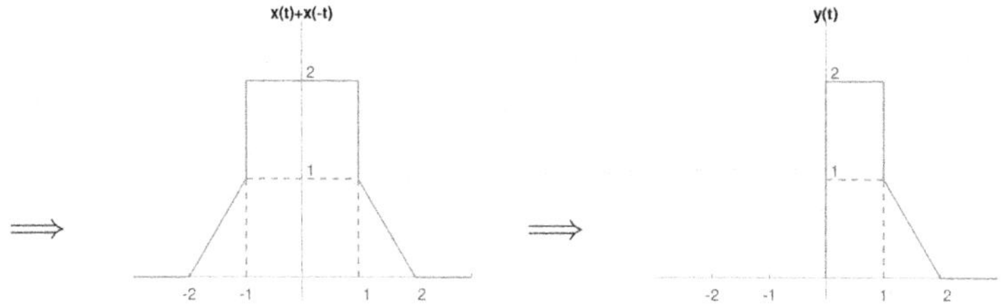

Signal $y_2(t)$ Signal $y(t)$

Figure 1.20: Problem 8 - Part (a).

(b) $y(t) = x(2t - 1)$.

Let

$$\begin{cases} y_1(t) = x(t-1) \\ y(t) = y_1(2t) \end{cases}.$$

Then we obtain $y(t)$ based on the following figures:

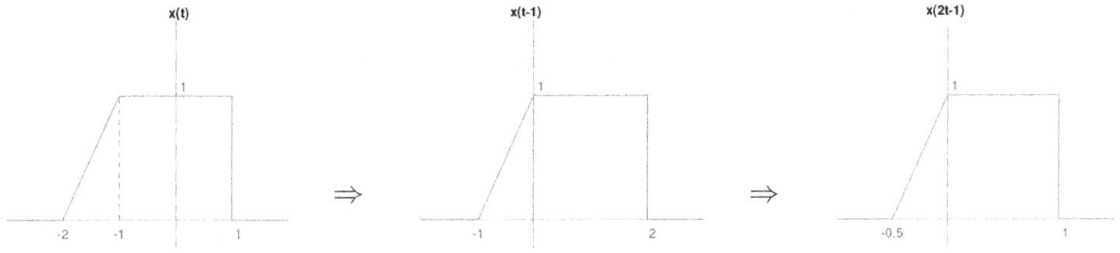

Signal $x(t)$ Signal $y_1(t)$ Signal $y(t)$

Figure 1.24: Problem 8 - Part (b).

(c) $y(t) = x(-\frac{t}{2} + 1)$.

Let

$$\begin{cases} y_1(t) = x(t+1) \\ y_2(t) = y_1(\frac{t}{2}) \\ y(t) = y_2(-t) \end{cases}.$$

Then we obtain $y(t)$ based on the following figures:

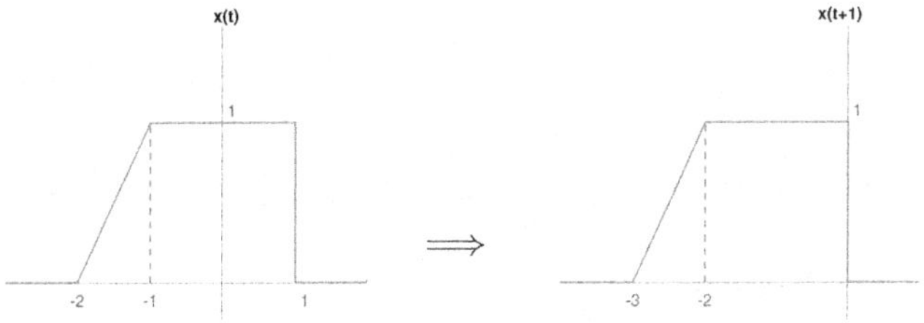

Signal $x(t)$ Signal $y_1(t)$

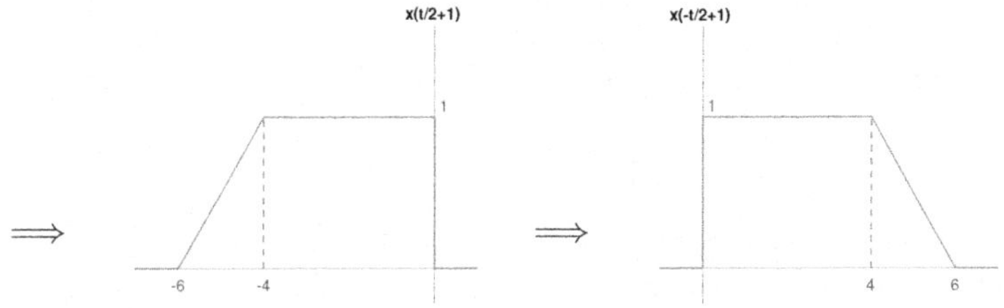

Signal $y_2(t)$ Signal $y(t)$

Figure 1.29: Problem 8 - Part (c).

Problem 9. Calculate the energy and power for the following continuous time signals and determine whether the signals are energy signals, power signals, or neither.

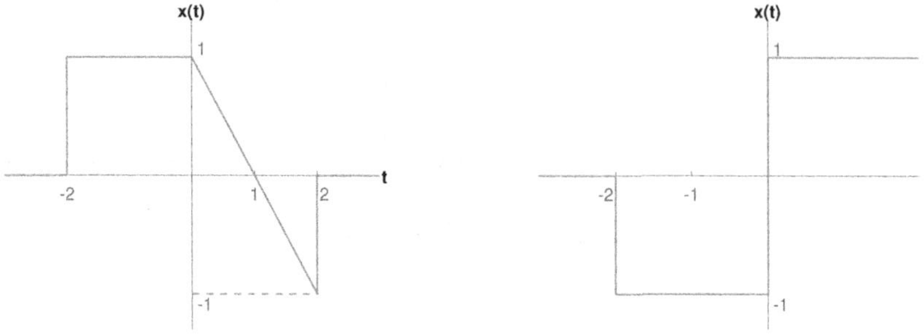

(a) (b)

Figure 1.32: Problem 9.

Solution.

(a) $P = \lim_{T\to\infty} \frac{1}{2T}\int_{-T}^{T} |x(t)|^2 dt = \lim_{T\to\infty} \frac{1}{2T}\int_{-2}^{2} |x(t)|^2 dt = \lim_{T\to\infty} \frac{\int_{-2}^{2} |x(t)|^2 dt}{2T} = 0,$

$E = \int_{-\infty}^{\infty} |x(t)|^2 dt = \int_{-2}^{0} |1|^2 dt + \int_{0}^{2} |1-t|^2 dt = 2 + \frac{2}{3}.$

Thus, $x(t)$ is an energy signal.

(b) $P = \lim_{T\to\infty} \frac{1}{2T}\int_{-T}^{T} |x(t)|^2 dt = \lim_{T\to\infty} \frac{1}{2T}\{\int_{-2}^{0} |-1|^2 dt + \int_{0}^{T} |1|^2 dt\}$

$= \lim_{T\to\infty} \frac{\int_{-2}^{0} |-1|^2 dt}{2T} + \lim_{T\to\infty} \frac{\int_{0}^{T} |1|^2 dt}{2T} = 0 + \lim_{T\to\infty} \frac{T}{2T} = \frac{1}{2},$

$E = \int_{-\infty}^{\infty} |x(t)|^2 dt = \int_{-2}^{0} |-1|^2 dt + \int_{0}^{\infty} |1|^2 dt \to \infty,$

since $0 \leq P < \infty$, $x(t)$ is a power signal.

As can be seen, although the signals are pretty similar, one is an energy signal and the other is a power signal. The reason is that the former is bounded in time (it is non-zero only in a finite interval of time) while the latter is not.

Problem 10. Calculate energy and power for the following signal and determine whether the signal is energy signal, power signal, or neither.

$x(t) = \cos(t) + \cos(\alpha t)$, where $\alpha > 0$

Solution.

$x(t) = \cos(t) + \cos(\alpha t)$.

$$\begin{aligned}
P &= \lim_{T\to\infty} \frac{1}{2T}\int_{-T}^{T} |x(t)|^2 dt \\
&= \lim_{T\to\infty} \frac{1}{2T}\int_{-T}^{T} |\cos(t) + \cos(\alpha t)|^2 dt \\
&= \lim_{T\to\infty} \frac{1}{2T}\int_{-T}^{T} (\cos^2(t) + \cos^2(\alpha t) + 2\cos(t)\cos(\alpha t))dt \\
&= \lim_{T\to\infty} \frac{1}{T}\int_{0}^{T} (\frac{1}{2} + \frac{1}{2}\cos(2t) + \frac{1}{2} + \frac{1}{2}\cos(2\alpha t) + \cos(\alpha t + t)) + \cos(\alpha t - t))dt \\
&= \lim_{T\to\infty} \frac{1}{T}(T + \frac{1}{4}\sin(2T) + \frac{1}{4\alpha}\sin(2\alpha T) + \frac{1}{\alpha+1}\sin(T\alpha + T) + \frac{1}{\alpha-1}\sin(T\alpha - T)) \\
&= 1.
\end{aligned}$$

Thus, $x(t)$ is a power signal.

Problem 11. Consider a continuous-time system with input $x(t)$ and output $y(t)$. The input-output relationship for this system is

$$y(t) = tx(3t) + 2.$$

Determine the output of the system when the input is $x(t-5)$.

Solution.

Let $x_1(t) = x(t-5)$. The system output will be

$$y_1(t) = tx_1(3t) + 2, \tag{1.33}$$

where

$$x_1(3t) = x(3t-5). \tag{1.34}$$

Substituting (1.34) into (1.33), we get

$$y_1(t) = tx(3t-5) + 2.$$

Problem 12. Consider a continuous-time system with input $x(t)$ and output $y(t)$. The input-output relationship for this system is

$$y(t) = \begin{cases} x(t) - x(-t) & x(t) \geq x(-t) \\ x(-t) & x(t) < x(-t) \end{cases}.$$

For the continuous-time signal $x(t)$ shown in Fig. 1.33, determine the output of the system.

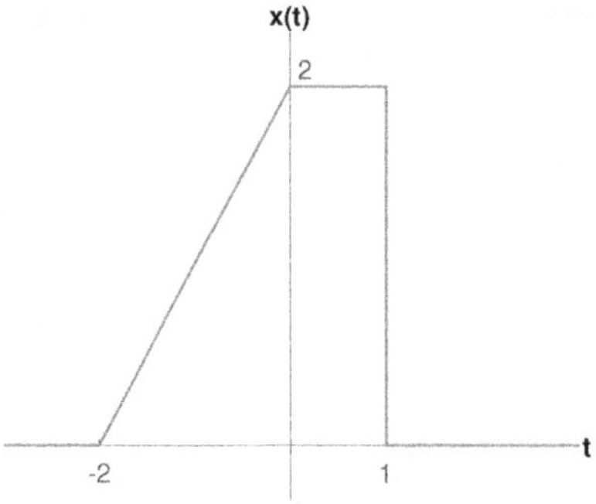

Figure 1.33: Problem 12.

Solution.

First we sketch both $x(t)$ and $x(-t)$ signals on a figure.

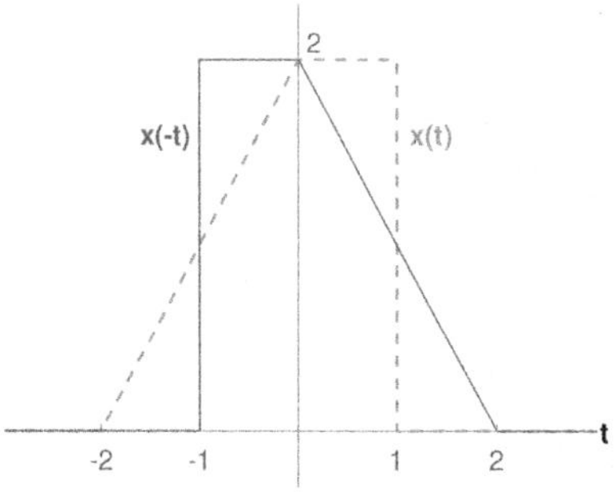

Figure 1.34: Problem 12, $x(t)$ and $x(-t)$.

Then we decompose signal $y(t)$ to following partitions:

1. $t < -2$

   $\Rightarrow$ According to the figure, $x(t) = x(-t) = 0$, so $y(t) = 0$.

2. $-2 \leq t < -1$

   $\Rightarrow$ According to the figure, $x(t) > x(-t)$, so $y(t) = x(t) - x(-t)$.

3. $-1 \leq t < 0$

   $\Rightarrow$ According to the figure, $x(t) < x(-t)$, so $y(t) = x(-t)$.

4. $0 \le t < 1$

$\Rightarrow$ According to the figure, $x(t) > x(-t)$, so $y(t) = x(t) - x(-t)$.

5. $1 \le t < 2$

$\Rightarrow$ According to the figure, $x(t) < x(-t)$, so $y(t) = x(-t)$.

6. $t \ge 2$

$\Rightarrow$ According to the figure, $x(t) = x(-t) = 0$, so $y(t) = 0$.

Now, according to the above description, we sketch the output:

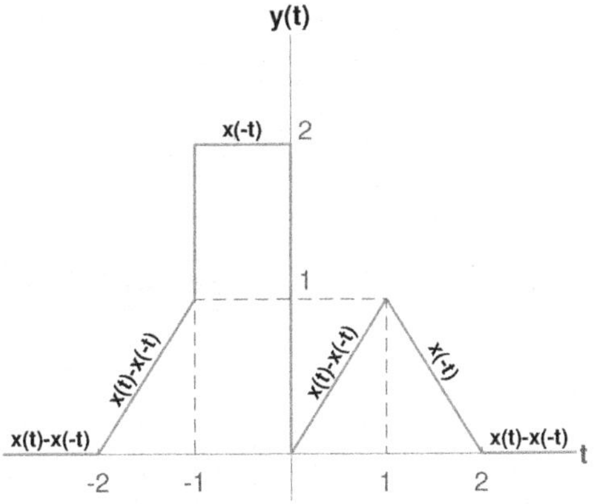

Figure 1.35: Problem 12, output $y(t)$.

1.8.2 Discrete-time Signals

Problem 13. Determine which of the following signals are periodic. If a signal is periodic, determine its fundamental period.

(a) $x[n] = \sin(\frac{7\pi}{3}n)$

(b) $x[n] = \cos(n\pi) + \cos(n)$

(c) $x[n] = e^{j(n+3)}$

(d) $x[n] = \cos(\frac{\pi}{4}n^2)$

Solution.

(a) $x[n] = \sin(\frac{7\pi}{3}n)$.

By definition

$$x[n + N_0] = \sin(\frac{7\pi}{3}(n + N_0)) = \sin(\frac{7\pi}{3}n + \frac{7\pi}{3}N_0)$$

$$\Rightarrow \quad \sin(\frac{7\pi}{3}n + \frac{7\pi}{3}N_0) = \sin(\frac{7\pi}{3}n).$$

Using the trigonometric identity $\sin(\phi + 2k\pi) = \sin(\phi)$, we can write

$$\frac{7\pi}{3}N_0 = 2k\pi,$$

thus, we must have

$$N_0 = \frac{6}{7}k.$$

The smallest positive integer N_0 is obtained with $k = 7$, so the fundamental period of $x[n]$ is $N_0 = 6$.

(b) $x[n] = \cos(n\pi) + \cos(n) = x_1[n] + x_2[n]$.

(I) By definition

$$x_1[n + N_0] = \cos((n + N_0)\pi) = \cos(n\pi + N_0\pi)$$
$$\Rightarrow \qquad \cos(n\pi + N_0\pi) = \cos(n\pi).$$

Using the trigonometric identity $\cos(\phi + 2k\pi) = \cos(\phi)$, we can write

$$N_0\pi = 2k\pi \qquad \Rightarrow \qquad N_0 = 2k.$$

The smallest positive integer N_0 is obtained with $k = 1$, so the fundamental period of $x[n]$ is $N_0 = 2$.

(II) By definition

$$x_2[n + N_0] = \cos(n + N_0)$$
$$\Rightarrow \qquad \cos(n + N_0) = \cos(n).$$

Using the trigonometric identity $\cos(\phi + 2k\pi) = \cos(\phi)$, we can write

$$N_0 = 2k\pi.$$

As can be seen, there is no value of k for which N_0 is an integer number, so $x_2[n]$ is not periodic.

Finally as the second signal is not periodic, $x[n] = x_1[n] + x_2[n]$ is not periodic.

(c) $x[n] = e^{j(n+3)}$.

By definition

$$x[n + N_0] = e^{j((n+N_0)+3)} = e^{j(n+3)} \times e^{jN_0}$$

Setting

$$e^{j(n+3)} \times e^{jN_0} = e^{j(n+3)},$$

we must have

$$e^{jN_0} = 1,$$

which is not possible, except for $N_0 = 0$. As such, the signal is not periodic.

(d) $x[n] = \cos(\frac{\pi}{4}n^2)$.

By definition

$$x[n+N_0] = \cos\left(\frac{\pi}{4}(n+N_0)^2\right) = \cos\left(\frac{\pi}{4}n^2 + \frac{\pi}{2}nN_0 + \frac{\pi}{4}N_0^2\right)$$

$$\Rightarrow \qquad \cos\left(\frac{\pi}{4}n^2 + \frac{\pi}{2}nN_0 + \frac{\pi}{4}N_0^2\right) = \cos(\frac{\pi}{4}n^2).$$

Using the trigonometric identity $\cos(\phi + 2k\pi) = \cos(\phi)$, we can write

$$\frac{\pi}{2}nN_0 + \frac{\pi}{4}N_0^2 = 2k\pi \qquad \Rightarrow \qquad \frac{nN_0}{2} + \frac{N_0^2}{4} = 2k,$$

$$N_0 = 1 \qquad \Rightarrow \qquad \frac{n}{2} + \frac{1}{4} = 2k,$$

$$N_0 = 2 \qquad \Rightarrow \qquad n + 1 = 2k,$$

In the former case, k can not be an integer but in the latter case, k can be. Thus, the fundamental period of $x[n]$ is $N_0 = 2$.

Problem 14. Sketch the following signal and determine and sketch its even and odd components.

$$x[n] = 3\delta[n+4] - 2\delta[n+3] + u[n+2]$$

Solution. $x[n] = 3\delta[n+4] - 2\delta[n+3] + u[n+2]$

$$x_e[n] = \frac{x[n]+x[-n]}{2} = \frac{3}{2}\delta[n+4] - \delta[n+3] + \frac{1}{2}\,\mathrm{u}[n+2] + \frac{3}{2}\delta[-n+4] - \delta[-n+3] + \frac{1}{2}\,\mathrm{u}[-n+2],$$

and

$$x_o[n] = \frac{x[n]-x[-n]}{2} = \frac{3}{2}\delta[n+4] - \delta[n+3] + \frac{1}{2}\,\mathrm{u}[n+2] - \frac{3}{2}\delta[-n+4] + \delta[-n+3] - \frac{1}{2}\,\mathrm{u}[-n+2].$$

Therefore according to the above equations we sketch signals $x[n], x[-n], x_e[n], x_o[n]$.

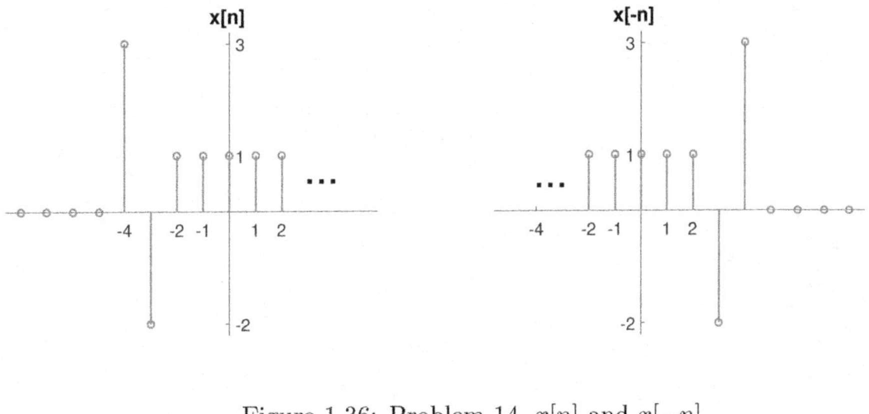

Figure 1.36: Problem 14, $x[n]$ and $x[-n]$.

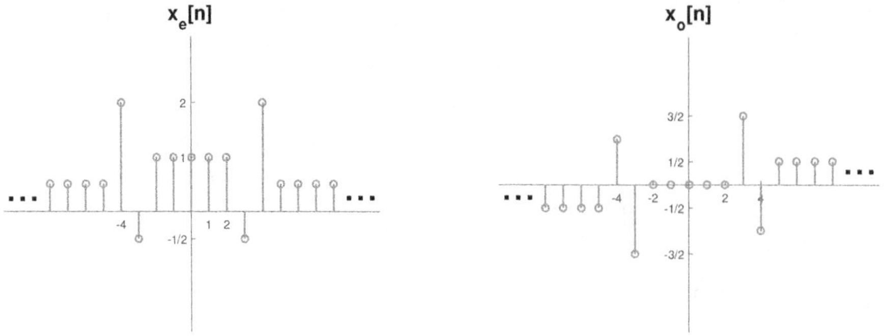

Figure 1.37: Problem 14, the even and odd components of $x[n]$.

Problem 15. If the even component of the signal is shown in the following figure and for $n < 0$, $x[n] = 0$, find $x[n]$.

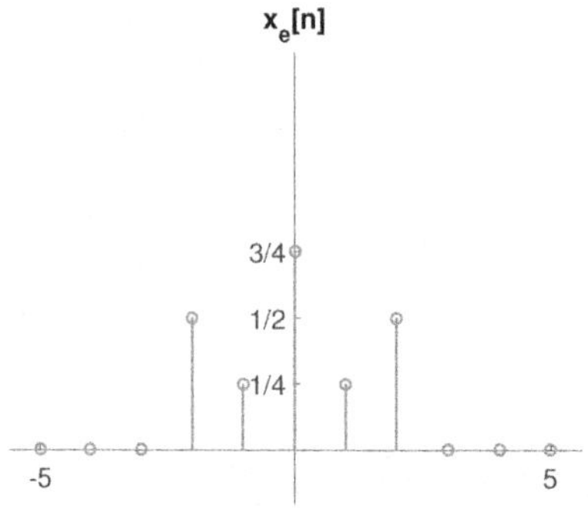

Figure 1.38: Problem 15, the even part of $x[n]$.

Solution.

$$x_e[n] = \frac{x[n] + x[-n]}{2},$$

Then, we have $x_e[0] = \frac{x[0] + x[0]}{2} = \frac{3}{4} \Rightarrow x[0] = \frac{3}{4}$,
$x_e[1] = \frac{x[1] + \cancelto{0}{x[-1]}}{2} = \frac{1}{4} \Rightarrow x[1] = \frac{1}{2}$,
$x_e[2] = \frac{x[2] + \cancelto{0}{x[-2]}}{2} = \frac{1}{2} \Rightarrow x[2] = 1$,
$x_e[3] = \frac{x[3] + \cancelto{0}{x[-3]}}{2} = 0 \Rightarrow x[3] = 0$,
$x_e[4] = \frac{x[4] + \cancelto{0}{x[-4]}}{2} = 0 \Rightarrow x[4] = 0$,
$\Rightarrow \quad \forall \; k \geq 3 \quad x[k] = 0.$

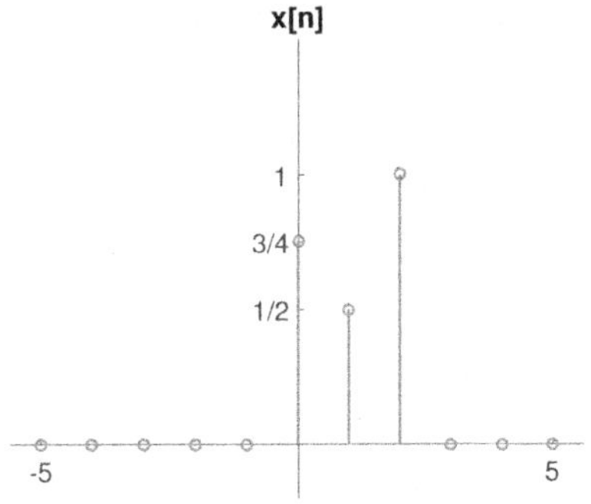

Figure 1.39: Problem 15, $x[n]$.

Problem 16. Write the following signals using basic signals.

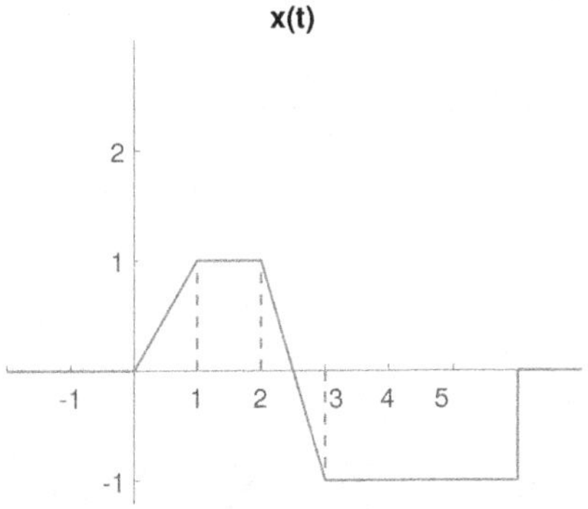

(a)

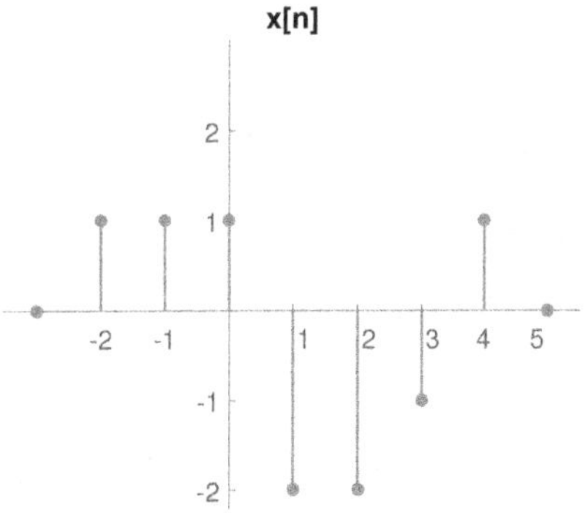

(b)

Solution.

(a) We decompose signal $x(t)$ to the following partitions:

(a) $t < 0$ $\Rightarrow$ $x(t) = 0$.

(b) $0 \leq t < 1$ $\Rightarrow$ We use the unit ramp signal $\mathrm{r}(t)$.

(c) $1 \leq t < 2$ $\Rightarrow$ We use $-\,\mathrm{r}(t-1)$ to cancel the ramp from the previous segment.

(d) $2 \leq t < 3$ $\Rightarrow$ We use $-2\,\mathrm{r}(t-2)$ for this segment.

(e) $3 \leq t < 6$ $\Rightarrow$ We use signal $2\,\mathrm{r}(t-3)$ to cancel the ramp in the previous step.

(f) $t \geq 6$ $\Rightarrow$ We use $\mathrm{u}(t-6)$, to remove the negative constant value inherited from the previous step.

Thus:

$$x(t) = \mathrm{r}(t) - \mathrm{r}(t-1) - 2\,\mathrm{r}(t-2) + 2\,\mathrm{r}(t-3) + \mathrm{u}(t-6).$$

(b) We decompose signal $x[n]$ to the following partitions:

(a) $n \leq -3$ $\Rightarrow$ $x[n] = 0$.

(b) $-2 \leq n \leq 0$ $\Rightarrow$ We use $\mathrm{u}[n+2]$ for this part.

(c) $1 \leq n \leq 2$ $\Rightarrow$ We use $-3\,\mathrm{u}[n-1]$ to remove the positive constant from the last segment and go down to -2.

(d) $n \geq 3$ $\Rightarrow$ We use $2\,\mathrm{u}[n-3]$ to cancel the effect of the negative constant value from the last part. We also use $-\,\delta[n-3]$ and $\delta[n-4]$ to generate an impulses in $n=3$ and $n=4$.

thus,

$$x[n] = \mathrm{u}[n+2] - 3\,\mathrm{u}[n-1] + 2\,\mathrm{u}[n-3] - \delta[n-3] + \delta[n-4].$$

Note: For discrete signals, we can always write them in terms of impulse functions. The aim here is to use this function as little as possible.

Problem 17. Evaluate the following expression using the unit impulse and unit step functions properties.

(a) $x[n] = \sum_{k=-\infty}^{7} \delta[n+3-k]$ (b) $x[n] = \sum_{k=-\infty}^{n} 3^k \delta[4-k]$

Solution.

(a) $x[n] = \sum_{k=-\infty}^{7} \delta[n+3-k]$.

$$x[n] = \sum_{k=-\infty}^{7} \delta[n+3-k] \xrightarrow{3-k=k'} x[n] = \sum_{k'=\infty}^{-4} \delta[n+k'] = \sum_{k'=-4}^{\infty} \delta[n+k'].$$

Unit impulse is an even function, so $\delta[n+k'] = \delta[-n-k']$. Thus,

$$x[n] = \sum_{k'=-4}^{\infty} \delta[-n-k'].$$

We know that $\sum_{k=k_0}^{\infty} \delta[n-k] = \mathrm{u}[k-k_0]$. So

$$\sum_{k'=-4}^{\infty} \delta[n-k'] = \mathrm{u}[n+4].$$

As such,

$$x[n] = \sum_{k'=-4}^{\infty} \delta[-n-k'] = \mathrm{u}[-n+4].$$

(b) $x[n] = \sum_{k=-\infty}^{n} 3^k \delta[4-k]$.

Signal $\delta[k-4]$ is zero except for $k=4$. Therefore we can write

$$x[n] = 3^4 \sum_{k=-\infty}^{n} \delta[4-k] = \begin{cases} 0, & if \;\; n < 4 \\ 3^4, & if \;\; n \geq 4 \end{cases}$$

thus,

$$x[n] = 81\,\mathrm{u}[n-4].$$

Problem 18. A discrete-time signal $x[n]$ is shown in the following figure.

Sketch and label each of the following signals.

(a) $y[n] = x[-2n+1]$

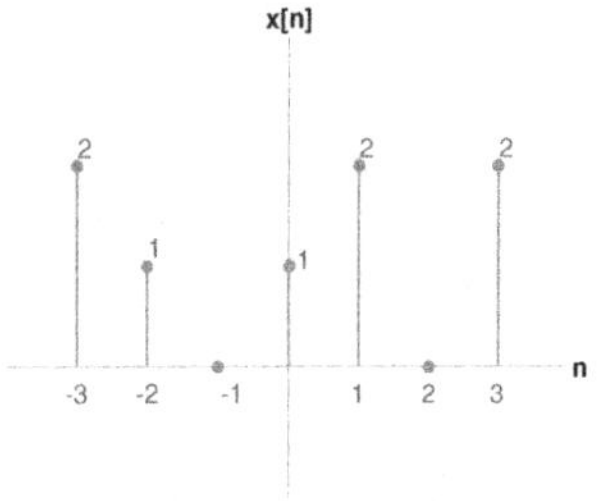

Figure 1.40: Problem 18.

(b) $y[n] = x_2[n+4]$

Solution.

(a) $y[n] = x[-2n+1]$.
For the systems below, if the input is $z[n]$, we have indicated the output and based on that, we can track the changes of $x[n]$:

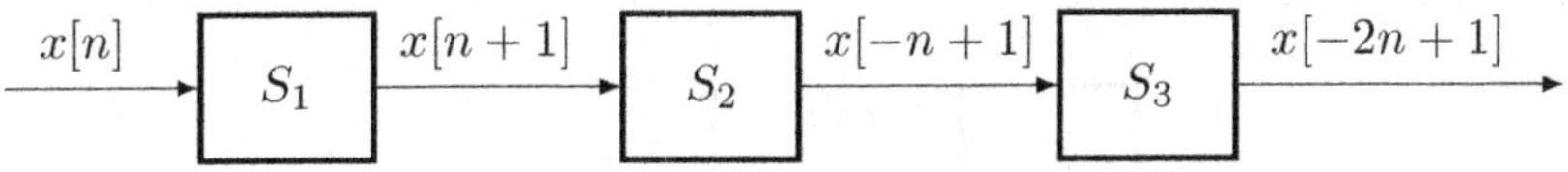

$$\begin{cases} S_1: & z[n] \rightarrow z[n+1] \\ S_2: & z[n] \rightarrow z[-n] \\ S_3: & z[n] \rightarrow z[2n] \end{cases}$$

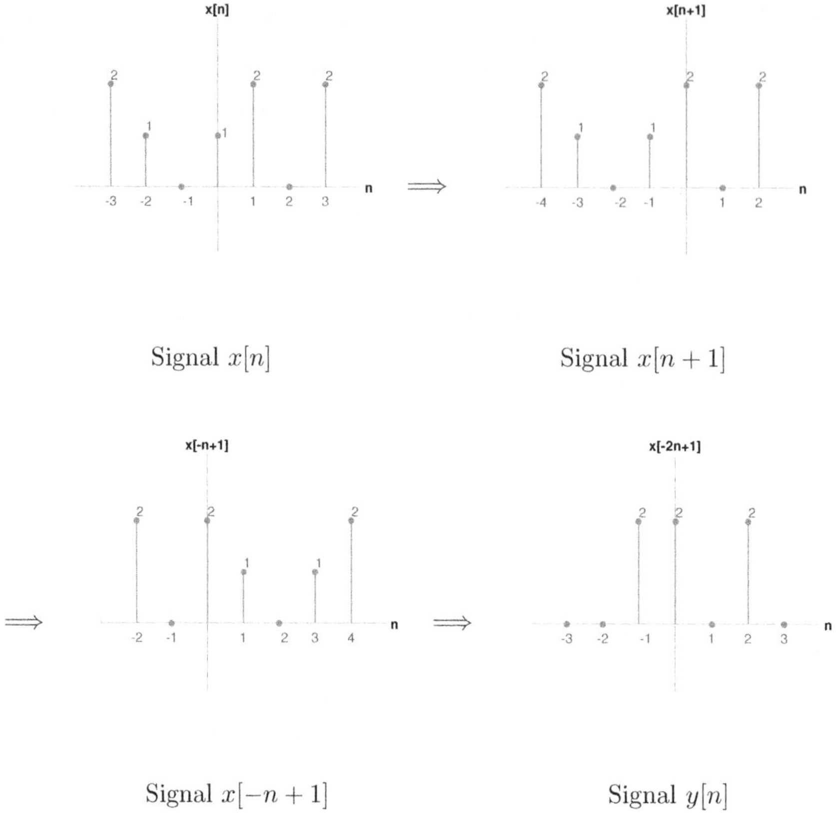

Figure 1.45: Problem 18 - Part (a).

(b) $y[n] = x_2[n+4]$.
For the systems below, if the input is $z[n]$, we have indicated the output and based on that, we can track the changes of $x[n]$:

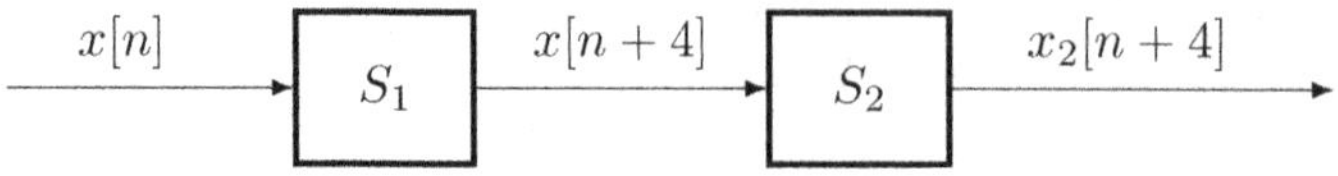

$$\begin{cases} S_1: & z[n] \rightarrow z[n+4] \\ S_2: & z[n] \rightarrow z_2[n] \end{cases}$$

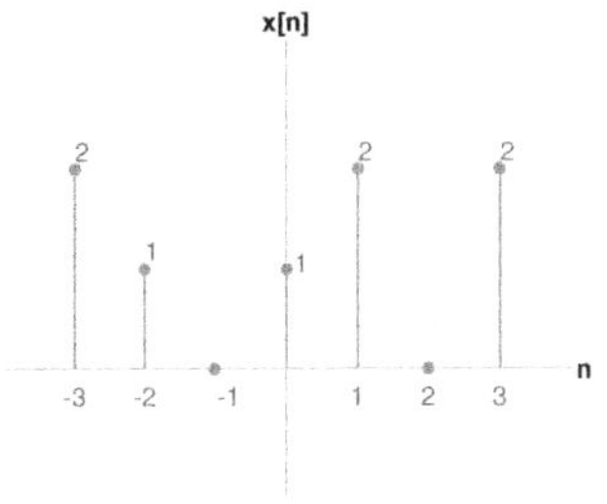

⇒

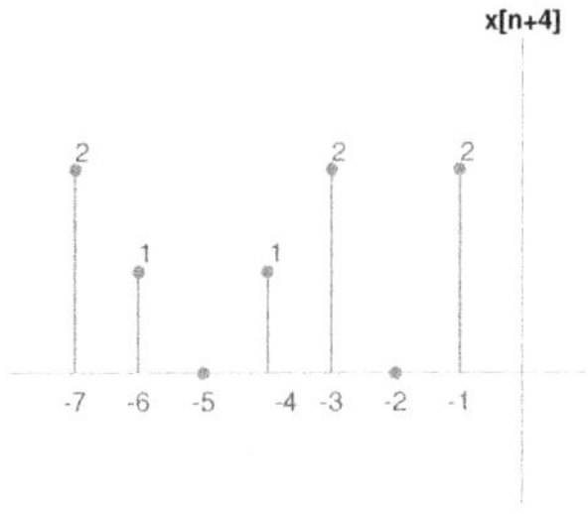

Signal $x[n]$ Signal $x[n+4]$

⇒

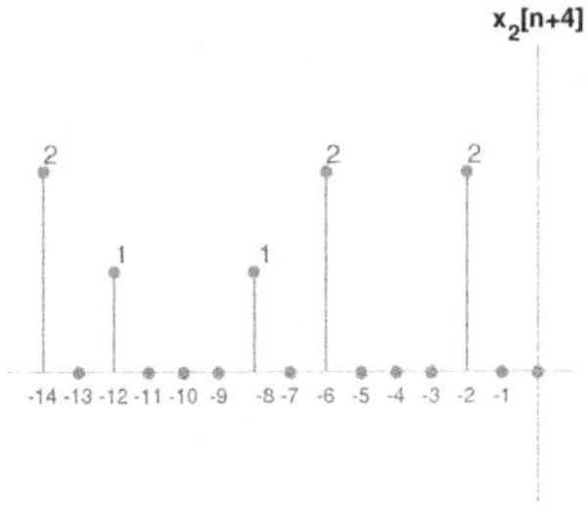

Signal $y[n]$

Figure 1.49: Problem 18 - Part (b).

Problem 19. Consider a discrete-time system with input $x[n]$ and output $y[n]$ related by

$$y[n] = x[n^2 + 1] - x[-n^2 + 2n - 1].$$

Determine the unit-step response of the system.
Solution. By substituting the unit-step signal into input-output relationship we get

$$y[n] = \mathrm{u}[n^2 + 1] - \mathrm{u}[-n^2 + 2n - 1] = \mathrm{u}[n^2 + 1] - \mathrm{u}[-(n-1)^2]$$

$$\rightarrow \quad \begin{cases} \forall n \quad \rightarrow \quad n^2 + 1 > 0 \quad \rightarrow \quad u[n^2 + 1] = 1 \\ \forall n \quad \rightarrow \quad (n-1)^2 \leq 0 \quad \rightarrow \quad u[-(n-1)^2] = 1 \quad \text{iff} \quad n = 1 \end{cases},$$

thus

$$y[n] = 1 - \delta[n-1].$$

Problem 20. Calculate energy and power for the following signals and determine whether the signals are energy signals, power signals, or neither.

(a) $x[n] = \begin{cases} (\frac{1}{3} + j\frac{\sqrt{3}}{3})^n, & n \geq 0 \\ 0, & n < 0 \end{cases}$

(b) $x[n] = (\frac{1}{3})^n \, \mathrm{u}[n]$

Solution.

(a) $x[n] = \begin{cases} (\frac{1}{3} + j\frac{\sqrt{3}}{3})^n, & n \geq 0 \\ 0, & n < 0 \end{cases}$.

$|x[n]|^2 = |(\frac{1}{3} + j\frac{\sqrt{3}}{3})^n|^2 = |(\frac{1}{3} + j\frac{\sqrt{3}}{3})|^{2n} = (\frac{1}{9} + \frac{1}{3})^n = (\frac{4}{9})^n \quad n \geq 0,$

$$E = \sum_{n=-\infty}^{\infty} |x[n]|^2 = \sum_{n=0}^{\infty} |x[n]|^2 = \sum_{n=0}^{\infty} (\frac{4}{9})^n$$

$$= \frac{(\frac{4}{9})^0 - (\frac{4}{9})^{\infty+1}}{1 - \frac{4}{9}} = \frac{1}{1 - \frac{4}{9}} = \frac{9}{5}.$$

$$P = \lim_{N\to\infty} \frac{1}{2N+1} \sum_{n=-N}^{N} |x[n]|^2 = \lim_{N\to\infty} \frac{\sum_{n=0}^{N} |x[n]|^2}{2N+1} = \lim_{N\to\infty} \frac{9/5}{2N+1} = 0.$$

Thus, $x[n]$ is an energy signal.

(b) $x[n] = (\frac{1}{3})^n \, \mathrm{u}[n]$.

$$E = \sum_{n=-\infty}^{\infty} |x[n]|^2 = \sum_{n=-\infty}^{\infty} |(\frac{1}{3})^n \, \mathrm{u}[n]|^2 .$$

$$= \sum_{n=0}^{\infty} |(\frac{1}{3})^n|^2 = \sum_{n=0}^{\infty} (\frac{1}{9})^n = \frac{9}{8}.$$

$$P = \lim_{N\to\infty} \frac{1}{2N+1} \sum_{n=-N}^{N} |x[n]|^2$$

$$= \lim_{N\to\infty} \frac{\sum_{n=-N}^{N} |x[n]|^2}{2N+1} = \lim_{N\to\infty} \frac{\sum_{n=0}^{N} |x[n]|^2}{2N+1} = \lim_{N\to\infty} \frac{\frac{9}{8}}{2N+1} = 0.$$

Thus, $x[n]$ is an energy signal.

Note: In both cases, once we come up with a finite value for energy, it is obvious that the power is 0, however, since the question asks us to obtain both power and energy, we acted accordingly.

Chapter 2

Properties of Continuous-time and Discrete-time Signals

2.1 Introduction

A system in general sense can be viewed as a mathematical model of a physical process in which input signals are transformed into the output signals.

A system can be an interconnection of several subsystems. Let x and y be the input and output signals, respectively. Then the relationship between input and output signal is represented by notation

$$y = Tx,$$

where T is defined as an operator by which x is transformed into y. Also systems can be multiple-input / multiple-output.

Figure 2.1: (a) System with single input and output signals; (b) System with multiple input and output signals.

2.2 Continuous-Time and Discrete-Time Systems

A continuous-time system is a system for which the input and output signals x and y are continuous-time signals. Such a system is displayed in Fig 2.2 (a). Similarly, a discrete-time system is a system for which the input and output signals are discrete-time signals which is depicted in Fig 2.2 (b).

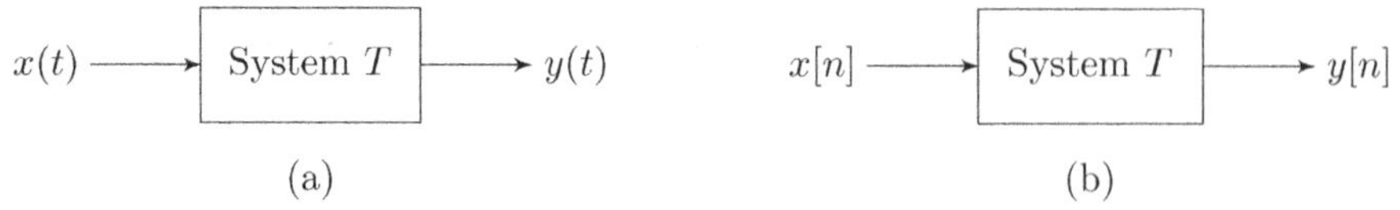

Figure 2.2: (a) Continuous-time system; (b) discrete-time system.

Note that some more complex systems might be considered with continuous-time input signals and discrete-time output signals or vice versa. We will see these kind of systems in Chapter 8 (Sampling).

2.3 Interconnections of Systems

We can see a lot of real systems that are made through interconnecting several subsystems. While there are several interconnections of systems, we focus on three basic ones: series or cascade, parallel, and feedback interconnections.

A series or cascade interconnection of two systems results from feeding the output of one system as an input ot the other one, i.e., as illustrated in Fig 2.3, the input signal x into System 1 results in an output which is used as the input for System 2. The output of System 2 is also the output y of the overall system.

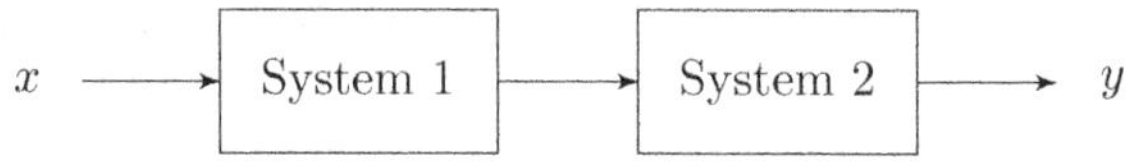

Figure 2.3: A series or cascade interconnection of two systems.

In a parallel interconnection of two systems, the same input signal x is fed into both Systems 1 and 2 simultaneously. The output from both systems are added together to produce the resulting output signal y as illustrated in Fig 2.4.

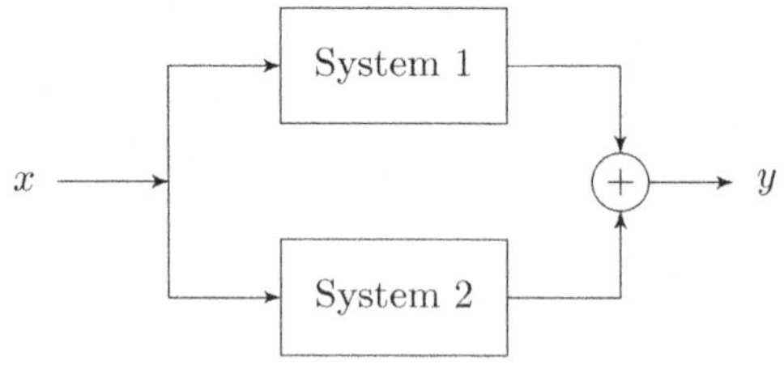

Figure 2.4: A parallel interconnection of two systems.

Another important type of system interconnection is a feedback interconnection. Here, the output of the System 1 is fed back to System 2 and the resulting output at System 2 is added (subtracted) to input $x[n]$ which will serve as the input to System 1 as illustrated in Fig 2.5.

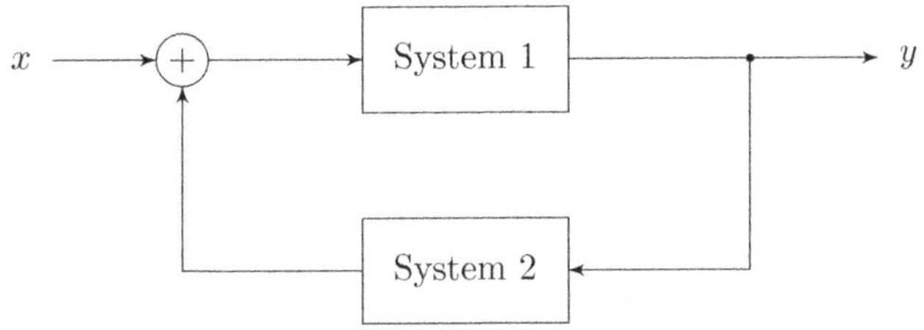

Figure 2.5: A feedback interconnection.

2.4 Basic System Properties

In this section we introduce and discuss a number of basic properties of continuous-time and discrete-time systems.

$$\begin{aligned}\text{A continuous-time system} &\Rightarrow\quad x(t) \xrightarrow{S} y(t)\\ \text{A discrete-time system} &\Rightarrow\quad x[n] \xrightarrow{S} y[n].\end{aligned}$$

2.4.1 Linear Systems

A linear system, in continuous time or discrete time, is a system that satisfies the following two conditions:

1. **Additivity:**

   If y_1 is the response of a system to an input x_1 and y_2 is the response of the system to an input x_2, then

$$x_1(t) + x_2(t) \xrightarrow{S} y_1(t) + y_2(t) \tag{2.1}$$

$$x_1[n] + x_2[n] \xrightarrow{S} y_1[n] + y_2[n]. \tag{2.2}$$

2. **Homogeneity (or Scaling):**

   If y is the response of a system to an input x, then

$$ax(t) \xrightarrow{S} ay(t) \tag{2.3}$$

$$ax[n] \xrightarrow{S} ay[n]. \tag{2.4}$$

   for any signal x and any scalar a.

The two properties of a linear system can be combined into a single condition:

$$a_1x_1(t) + a_2x_2(t) \xrightarrow{S} a_1y_1(t) + a_2y_2(t) \tag{2.5}$$

$$a_1x_1[n] + a_2x_2[n] \xrightarrow{S} a_1y_1[n] + a_2y_2[n]. \tag{2.6}$$

where a_1 and a_2 are any scaler. Equations (2.5) or (2.6) are also known as the superposition principle.

2.4.2 Time-Invariant and Time-Varying Systems

Let $y(t)$ be the response of a continuous-time system to an input $x_1(t)$. The system is called time-invariant if $y(t - t_0)$ is the output when $x(t - t_0)$ is applied. In other words, a time shift in the input signal results in the same time shift at the output signal.

$$\text{A continuous-time system} \quad \Rightarrow \quad x(t-t_0) \xrightarrow{S} y(t-t_0) \tag{2.7}$$

$$\text{A discrete-time system} \quad \Rightarrow \quad x[n-n_0] \xrightarrow{S} y[n-n_0], \tag{2.8}$$

for any real value of t_0 and any integer n_0. A system which does not satisfy (2.7) or (2.8) is called a time-varying system.

2.4.3 Systems with and without Memory

A system is called to be memoryless if the output at a given time is dependent on the input at the very same time only. In other words, the output should not depend on the past or future values of the input.

2.4.4 Causal Systems

A system is said to be causal if its output at the present time depends on only values of the input at the present time and/or in the past. Such a system is referred as non-anticipative, as it is not possible to determine an output before an input is applied to the system. A system is said noncausal if the output at the present time depends also on values of the input at future times. Note that all memoryless systems are causal, but not vice versa.

2.4.5 Stable Systems

A system is stable if any bounded-input results in a bounded-output (BIBO).

$$\text{If} \quad |x(t)| \le A < \infty \quad \Rightarrow \quad |y(t)| \le B < \infty \tag{2.9}$$

$$\text{If} \quad |x[n]| \le A < \infty \quad \Rightarrow \quad |y[n]| \le B < \infty. \tag{2.10}$$

where A and B are finite real constants. A system is unstable if the output becomes unbounded even if there is no unbounded input.

2.4.6 Invertibility and Inverse Systems

A system is said to be invertible if it produces distinct output signals according to distinct input signals.

$$x_1(t) \neq x_2(t) \quad \Rightarrow \quad y_1(t) \neq y_2(t)$$

$$x_1[n] \neq x_2[n] \quad \Rightarrow \quad y_1[n] \neq y_2[n]$$

or

$$\begin{aligned} y_1(t) = y_2(t) \quad &\Rightarrow \quad x_1(t) = x_2(t) \\ y_1[n] = y_2[n] \quad &\Rightarrow \quad x_1[n] = x_2[n] \end{aligned}$$

In addition to applying the above conditions, there are two ways to determine if a system is invertible or not.

1. If we can find two distinct inputs leading to the same outputs, then the system is not invertible.

2. If we can find the inverse of a system, it is indeed invertible.

2.5 Useful Tips

a) In a linear and memoryless system, the relationship between the input and the output is

$$\begin{aligned} y(t) &= A(t)x(t) \\ y[n] &= A[n]x[n], \end{aligned}$$

where A is a real constant. As such, in these systems, if the input is 0 at a specific time, the output will also be 0 at that time:

$$x(t_0) = 0 \quad \xrightarrow{S} \quad y(t_0) = 0$$

$$x[n_0] = 0 \quad \xrightarrow{S} \quad y[t_0] = 0.$$

b) Based on the above property, in a linear and memoryless system S for a specific time $t = t_0$ ($n = n_0$):

$$x_2(t_0) = ax_1(t_0) \quad \xrightarrow{S} \quad y_2(t_0) = ay_1(t_0).$$

$$x_2[n_0] = ax_1[n_0] \quad \xrightarrow{S} \quad y_2[n_0] = ay_1[n_0].$$

Note: You may wonder if the above conditions should hold for any linear system based on the homogeneity property of Section (2.4.1) and regardless of memory. Note, however, that the homogeneity holds when the input is scaled for all times, e.g., $\forall t$ or $\forall n$. In the above equation, the scaling can be applied to only one given time instant and still the output is scaled for that time instant.

2.6 Problems

2.6.1 Continuous-time Systems

Problem 21. Determine whether the following continuous-time systems are linear.

(a) $y(t) = 3x(2t) + 3$

(b) $y(t) = \begin{cases} t^2 x(t+5) & t < 0 \\ 4sin(\sqrt{t})x(t^2) & t \geq 0 \end{cases}$

(c) $y(t) = \begin{cases} x(t) & x(t) < 1 \\ x(t) + 2 & x(t) \geq 1 \end{cases}$

(d) $y(t) = x(t^2) + x(t+5)$

(e) $y(t) = \mathrm{Re}[x(t)]$

(f) $y(t) = \begin{cases} \frac{x^3(t-1)}{x^2(t-2)} & x(t-2) \neq 0 \\ 0 & x(t-2) = 0 \end{cases}$

(g) $y(t) = \int_{-\infty}^{t} x(\tau) d\tau$

(h) $y(t) = \sin(x(t))$

Solution.

(a) $y(t) = 3x(2t) + 3$

First we investigate the homogeneity property:

$$T\{x(t)\} = 3x(2t) + 3 \xrightarrow{x(t) \Rightarrow \alpha x(t)} T\{\alpha x(t)\} = 3\alpha x(2t) + 3,$$

$$\alpha T\{x(t)\} = \alpha[3x(2t) + 3] = 3\alpha x(2t) + 3\alpha,$$

$$\Rightarrow T\{\alpha x(t)\} \neq \alpha T\{x(t)\}.$$

The system does not have the homogeneity property, therefore it is a nonlinear system.

(b) $y(t) = \begin{cases} t^2 x(t+5) & t < 0 \\ 4\sin(\sqrt{t})x(t^2) & t \geq 0 \end{cases}$

First investigate homogeneity property:

$$T\{\alpha x(t)\} = \begin{cases} t^2 \alpha x(t+5) & t < 0 \\ 4\alpha\sin(\sqrt{t})x(t^2) & t \geq 0 \end{cases} = \alpha \begin{cases} t^2 x(t+5) & t < 0 \\ 4\sin(\sqrt{t})x(t^2) & t \geq 0 \end{cases} = \alpha T\{x(t)\}.$$

The system has the homogeneity property. To investigate the additivity property, we have

$$\text{input } x_1(t) \Rightarrow y_1(t) = \begin{cases} t^2 x_1(t+5) & t < 0 \\ 4\sin(\sqrt{t})x_1(t^2) & t \geq 0 \end{cases},$$

$$\text{input } x_2(t) \Rightarrow y_2(t) = \begin{cases} t^2 x_2(t+5) & t < 0 \\ 4\sin(\sqrt{t})x_2(t^2) & t \geq 0 \end{cases},$$

$$\text{input } x_3(t) = x_1(t) + x_2(t) \Rightarrow y_3(t) = \begin{cases} t^2 x_3(t+5) & t < 0 \\ 4\sin(\sqrt{t})x_3(t^2) & t \geq 0 \end{cases}$$

$$y_3(t) = \begin{cases} t^2(x_1(t+5) + x_2(t+5)) & t < 0 \\ 4\sin(\sqrt{t})(x_1(t^2) + x_2(t^2)) & t \geq 0 \end{cases}$$

$$= \begin{cases} t^2 x_1(t+5) & t < 0 \\ 4\sin(\sqrt{t}) x_1(t^2) & t \geq 0 \end{cases} + \begin{cases} t^2 x_2(t+5) & t < 0 \\ 4\sin(\sqrt{t}) x_2(t^2) & t \geq 0 \end{cases}$$

$$\Rightarrow y_3(t) = y_1(t) + y_2(t).$$

The system has the homogeneity property too. Therefore, the system is linear.

(c) $y(t) = \begin{cases} x(t) & x(t) < 1 \\ x(t) + 2 & x(t) \geq 1 \end{cases}$

First we investigate the homogeneity property:

$$T\{\alpha x(t)\} = \begin{cases} \alpha x(t) & \alpha x(t) < 1 \\ \alpha x(t) + 2 & \alpha x(t) \geq 1 \end{cases},$$

if $\alpha > 0$ and $\alpha \neq 1$, according to the conditions we have

$$T\{\alpha x(t)\} = \begin{cases} \alpha x(t) & x(t) < 1/\alpha \\ \alpha x(t-1) & x(t) \geq 1/\alpha \end{cases},$$

$$\alpha T\{x(t)\} = \alpha \begin{cases} x(t) & x(t) < 1 \\ x(t) + 2 & x(t) \geq 1 \end{cases}.$$

As the homogeneity property is not established (i.e $T\{\alpha x(t)\} \neq \alpha T\{x(t)\}$), the system is nonlinear.

(d) $y(t) = x(t^2) + x(t+5)$

Homogeneity property:

$$T\{\alpha x(t)\} = \alpha x(t^2) + \alpha x(t+5) = \alpha[x(t^2) + x(t+5)] = \alpha T\{x(t)\}.$$

Homogeneity property is established. To investigate additivity property, we have

$$x_1(t) \Rightarrow y_1(t) = x_1(t^2) + x_1(t+5),$$

$$x_2(t) \Rightarrow y_2(t) = x_2(t^2) + x_2(t+5),$$

$$x_3(t) = x_1(t) + x_2(t) \Rightarrow y_3(t) = x_3(t^2) + x_3(t+5),$$

$$\begin{aligned} y_3(t) &= (x_1(t^2) + x_2(t^2)) + (x_1(t+5) + x_2(t+5)) \\ &= (x_1(t^2) + x_1(t+5)) + (x_2(t^2) + x_2(t+5)) \\ &= y_1(t) + y_2(t). \end{aligned}$$

The system has additivity property too and as such, the system is linear.

(e) $y(t) = \text{Re}[x(t)]$

$$T\{\alpha x(t)\} = \text{Re}[\alpha x(t)] \neq \alpha T\{x(t)\}.$$

This system does not satisfy the homogeneity property, hence it is a nonlinear system. Note that α is a complex constant in general and can be factored out of the "Real" operator.

(f) $y(t) = \begin{cases} \frac{x^3(t-1)}{x^2(t-2)} & x(t-2) \neq 0 \\ 0 & x(t-2) = 0 \end{cases}$

$$T\{\alpha x(t)\} = \begin{cases} \frac{\alpha^3 x^3(t-1)}{\alpha^2 x^2(t-2)} & \alpha x(t-2) \neq 0 \\ 0 & \alpha x(t-2) = 0 \end{cases} = \alpha \begin{cases} \frac{x^3(t-1)}{x^2(t-2)} & x(t-2) \neq 0 \\ 0 & x(t-2) = 0 \end{cases} = \alpha T\{x(t)\}.$$

The system has the homogeneity property. To investigate additivity property, we have

$$x_1(t) \Rightarrow y_1(t) = \begin{cases} \frac{x_1^3(t-1)}{x_1^2(t-2)} & x_1(t-2) \neq 0 \\ 0 & x_1(t-2) = 0 \end{cases},$$

$$x_2(t) \Rightarrow y_2(t) = \begin{cases} \frac{x_2^3(t-1)}{x_2^2(t-2)} & x_2(t-2) \neq 0 \\ 0 & x_2(t-2) = 0 \end{cases},$$

$$x_3(t) = x_1(t) + x_2(t) \Rightarrow y_3(t) = \begin{cases} \frac{x_3^3(t-1)}{x_3^2(t-2)} & x_3(t-2) \neq 0 \\ 0 & x_3(t-2) = 0 \end{cases},$$

$$y_3(t) = \begin{cases} \frac{\left(x_1(t-1)+x_2(t-1)\right)^3}{\left(x_1(t-2)+x_2(t-2)\right)^2} & x_1(t-2) + x_2(t-2) \neq 0 \\ 0 & x_1(t-2) + x_2(t-2) = 0 \end{cases}$$

$$\Rightarrow y_3(t) \neq y_1(t) + y_2(t).$$

As the system does not satisfy the additivity property, it is a nonlinear system.

(g) $y(t) = \int_{-\infty}^{t} x(\tau)d\tau$

Homogeneity property:

$$T\{\alpha x(t)\} = \int_{-\infty}^{t} \alpha x(\tau)d\tau = \alpha \int_{-\infty}^{t} x(\tau)d\tau = \alpha T\{x(t)\}.$$

The homogeneity property is established. To investigate the additivity property, we have

$$x_1(t) \Rightarrow y_1(t) = \int_{-\infty}^{t} x_1(\tau)d\tau,$$

$$x_2(t) \Rightarrow y_2(t) = \int_{-\infty}^{t} x_2(\tau)d\tau,$$

$$x_3(t) = x_1(t) + x_2(t) \Rightarrow y_3(t) = \int_{-\infty}^{t} x_3(\tau)d\tau,$$

$$y_3(t) = \int_{-\infty}^{t} (x_1(\tau) + x_2(\tau))d\tau = \int_{-\infty}^{t} x_1(\tau)d\tau + \int_{-\infty}^{t} x_2(\tau)d\tau = y_1(t) + y_2(t).$$

The system satisfies the additivity property too and thus the system is linear.

(h) $y(t) = \sin(x(t))$

First let's investigate the homogeneity property:

$$T\{\alpha x(t)\} = \sin\left(\alpha x(t)\right) \neq \alpha \sin(t) = \alpha T\{x(t)\}.$$

The homogeneity property does not hold, therefore the system is not linear.

Problem 22. Consider the below continuous-time systems with the given input-output relationships. Determine whether these systems are time-invariant.

(a) $y(t) = 2x(t) - 1$

(b) $y(t) = x(t/4)$

(c) $y(t) = \int_{-\infty}^{t} x(\tau)d\tau$

(d) $y(t) = \int_{-\infty}^{2t} x(\tau)d\tau$

(e) $y(t) = x(1) + \int_{-\infty}^{t} x(\tau)d\tau$

Solution:

(a) $y(t) = 2x(t) - 1$

By definition

$$\begin{aligned} T\{x(t-t_0)\} &= 2x(t-t_0) - 1 \\ y(t-t_0) &= 2x(t-t_0) - 1 \end{aligned} \Rightarrow T\{x(t-t_0)\} = y(t-t_0).$$

Hence this system is time-invariant.

(b) $y(t) = x(t/4)$

By definition

$$\begin{aligned} T\{x(t-t_0)\} &= x(t/4 - t_0) \\ y(t-t_0) = x(\frac{t-t_0}{4}) &= x(t/4 - t_0/4) \end{aligned} \Rightarrow T\{x(t-t_0)\} \neq y(t-t_0).$$

Therefore this system is not time-invariant.

(c) $y(t) = \int_{-\infty}^{t} x(\tau)d\tau$

By definition

$$T\{x(t-t_0)\} = \int_{-\infty}^{t} x(\tau - t_0)d\tau \overset{\tau - t_0 = \alpha}{=\!=\!=\!=} \int_{-\infty}^{t-t_0} x(\alpha)d\alpha$$

$$y(t-t_0) = \int_{-\infty}^{t-t_0} x(\tau)d\tau = \int_{-\infty}^{t-t_0} x(\alpha)d\alpha$$

$$\Rightarrow T\{x(t-t_0)\} = y(t-t_0).$$

Therefore This system is time-invariant.

(d) $y(t) = \int_{-\infty}^{2t} x(\tau)d\tau$
By definition

$$T\{x(t-t_0)\} = \int_{-\infty}^{2t} x(\tau - t_0)d\tau \xlongequal{\tau - t_0 = \alpha} \int_{-\infty}^{2t-t_0} x(\alpha)d\alpha$$

$$y(t-t_0) = \int_{-\infty}^{2(t-t_0)} x(\tau)d\tau = \int_{-\infty}^{2t-2t_0} x(\tau)d\tau$$

$$\Rightarrow T\{x(t-t_0)\} \neq y(t-t_0).$$

This system is not time-invariant.

(e) $y(t) = x(1) + \int_{-\infty}^{t} x(\tau)d\tau$
By definition

$$T\{x(t-t_0)\} = x(1-t_0) + \int_{-\infty}^{t} x(\tau - t_0)d\tau \xlongequal{\tau - t_0 = \alpha} x(1-t_0) + \int_{-\infty}^{t-t_0} x(\alpha)d\alpha$$

$$y(t-t_0) = x(1) + \int_{-\infty}^{t-t_0} x(\tau)d\tau$$

$$\Rightarrow T\{x(t-t_0)\} \neq y(t-t_0).$$

This system is not time-invariant (due to the first term).

Problem 23. Determine whether the continuous-time systems described by the following input-output equations are causal or noncausal.

(a) $y(t) = x(t/3)$

(b) $y(t) = \int_{-\infty}^{t/3} x(\tau)d\tau$

(c) $y(t) = x(t)x(-2)$

(d) $y(t) = \frac{dx(t)}{dt}$

(e) $y(t) = x(-|t|)$

Solution.

(a) $y(t) = x(t/3)$
Let's set $t = -3$. We have

$$y(-3) = x(-3/3) = x(-1),$$

which means that the output at time $t = -3$ depends on the value of the input at time $t = -1$. Hence, the system is not causal.
Note: we could have tested some other values of t and come up with the same conclusion.

(b) $y(t) = \int_{-\infty}^{t/3} x(\tau)d\tau$
Let's set $t = -3$. We have

$$y(-3) = \int_{-\infty}^{-1} x(\tau)d\tau,$$

which means that the output at time $t = -3$ depends on the value of the input at time $t = -1$. Hence, the system is not causal.
Note: we could have tested some other values of t and come up with the same conclusion.

(c) $y(t) = x(t)x(-2)$

Let's set $t = -5$. We have

$$y(-5) = x(-5)x(-2),$$

which means the output at time $t = -5$ depends on the value of the input at time $t = -2$. Hence, the system is not causal.

(d) $y(t) = \frac{dx(t)}{dt}$ Based on the definition of differentiation we have:

$$y(t) = \frac{dx(t)}{dt} = \lim_{h \to 0} \frac{x(t+h) - x(t)}{h}, \tag{2.11}$$

In order for (2.11) to hold, h has to tend to 0 from both positive and negative sides, e.g., $h \to 0^+$ and $h \to 0^-$. Therefore, we see that the output at time t depends on the value of the input at time $t + 0^+$ or in fact t^+. Hence, the system is not causal in strict sense. However, since at the limit, it is, it has been assumed a causal system in some literature.

(e) $y(t) = x(-|t|)$

For any amount of t_0, we have

$$-|t_0| \leq t_0,$$

hence, the system is causal.

Problem 24. Determine whether the following continuous-time systems are memoryless.

(a) $y(t) = x(-t)$

(b) $y(t) = e^{t+3}x(t)$

(c) $y(t) = \cos(x(t) + 1)$

(d) $y(t) = x(t) + x(2)$

Solution.

(a) $y(t) = x(-t)$

It is clear that such a system is not memoryless. Since the output at time t_0 depends on the value of the input signal at time $-t_0$. So the output depends on the past and the future values of the input.

(b) $y(t) = e^{t+3}x(t)$

This system is memoryless, as the value of output, $y(t)$, at any particular time t_0 depends only on the value of input, $x(t)$ at that time. Note that e^{t+3} is only a coefficient and has no effect on system memory.

(c) $y(t) = \cos(x(t) + 1)$

It is seen that the output at time t_0 depends on the value of the output at this time. Hence, this system is a memoryless.

(d) $y(t) = x(t) + x(2)$

As can be seen, the output at time t_0 depends not only on the value of the input at time t_0 but also on time $t = 2$. Hence, the system is not memoryless.

Problem 25. Comment on the stability of the following continuous-time systems.

(a) $y(t) = 5x(t) + 2$

(b) $y(t) = \int_{-\infty}^{t} x(\tau)d\tau$

(c) $y(t) = \sin(t)x(t)\delta(t-1)$

(d) $y(t) = \begin{cases} \frac{1}{x(t)} & |x(t)| > 0 \\ x(t+1) & |x(t)| \leq 0 \end{cases}$

(e) $y(t) = \begin{cases} \frac{1}{x(t)} & |x(t)| > 1 \\ x(t+1) & |x(t)| \leq 1 \end{cases}$

Solution.

(a) $y(t) = 5x(t) + 2$

Let B be an arbitrary positive number, and let $x(t)$ be an arbitrary signal bounded by B; that is, we are making no assumption about $x(t)$, except that

$$|x(t)| \leq B,$$

or

$$-B \leq x(t) \leq B,$$

for all t. Then $y(t)$ must satisfy

$$-5B + 2 \leq y(t) \leq 5B + 2.$$

We see that if any input to the system is bounded by an arbitrary positive number B, the corresponding output is guaranteed to be bounded by $5B + 2$. Therefore, this system is stable.

(b) $y(t) = \int_{-\infty}^{t} x(\tau)d\tau$

Let $x(t) = u(t)$. Then

$$y(t) = \int_{-\infty}^{t} u(\tau)d\tau = \int_{0}^{t} d\tau = r(t),$$

where $r(t) = tu(t)$ is the unit ramp function. The input is clearly a bounded signal while the output $y(t)$ grows unboundedly as t grows. As such, the system is unstable.

(c) $y(t) = \sin(t)x(t)\delta(t-1)$

Due to the existence of the Dirac delta function, any bounded input that is not equal zero at $t = 1$, will result in an unbounded output at $t = 1$. Therefore, the system is not stable.

(d) $y(t) = \begin{cases} \frac{1}{x(t)} & |x(t)| > 0 \\ x(t+1) & |x(t)| \leq 0 \end{cases}$

The system is not stable. The reason is that for a bounded input $x(t) = \epsilon, \epsilon \to 0$ to the system, the output gets unbounded. We are not worried about the time when the input is exactly 0, as it will be forwarded to the second sub-function.

(e) $y(t) = \begin{cases} \frac{1}{x(t)} & |x(t)| > 1 \\ x(t+1) & |x(t)| \leq 1 \end{cases}$

The system is stable. The reason is that for all bounded inputs except when $x(t)$ is close to 0, both sub-functions are bounded. We are also not worried about the time when the input is exactly 0 as the input is forwarded to the second sub-function which is bounded when $x(t) = 0$.

Problem 26. Determine if each of the following continuous-time systems is invertible.

(a) $y(t) = \frac{dx(t)}{dt}$

(b) $y(t) = \text{even}\{x(t)\}$

(c) $y(t) = x(t) - \cos(y(t-1))$

(d) $y(t) = \sin\left(x(t)\right)$

Solution.

(a) $y(t) = \frac{dx(t)}{dt}$

This system is not invertible, because for example , two different inputs of $x_1(t) = 1$ and $x_2(t) = 2$, result in the same output of $y(t) = 0$.

(b) $y(t) = \text{even}\{x(t)\}$

Note: *even* part of signal $x(t)$ is defined as $(x(t) + x(-t))/2$. The system is not invertible as the output of this system equals to zero for any input signal which is an odd function.

(c) $y(t) = x(t) - \cos(y(t-1))$

Let inputs $x_1(t)$ and $x_2(t)$ result in outputs $y_1(t)$ and $y_2(t)$. If $y_1(t) = y_2(t)$, then

$$y_1(t) + \cos(y_1(t-1)) = y_2(t) + \cos(y_2(t-1)).$$

This means that $x_1(t) = x_2(t)$ and the system is invertible.

(d) $y(t) = \sin\left(x(t)\right)$

The system is not invertible. Take any $x_1(t)$ and set $x_2(t) = x_1(t) + 2\pi$. Then $x_1(t)$ and $x_2(t)$ are mapped to the same output.

Problem 27. The outputs of a continuous-time linear system S to the inputs e^{j3t} and e^{-j3t} are $y_1(t)$ and $y_2(t)$, respectively.

(a) Obtain the output of the system S based on $y_1(t)$ and $y_2(t)$ to the input $x(t) = \cos(3t)$.

(b) Obtain the output of the system S based on $y_1(t)$ and $y_2(t)$ to the input $x(t) = \cos(3t - 2)$.

Solution.

Here the input-output relationship is not given, so to obtain the output, one has to deploy the properties of linear systems. We have

$$x_1(t) = e^{j3t} \Rightarrow y_1(t)$$

$$x_2(t) = e^{-j3t} \Rightarrow y_2(t)$$

(a)

$$x(t) = \cos(3t) = \frac{e^{j3t} + e^{-j3t}}{2}$$
$$= \frac{e^{j3t}}{2} + \frac{e^{-j3t}}{2} = \frac{x_1(t)}{2} + \frac{x_2(t)}{2}$$

$$\Rightarrow y(t) = \frac{y_1(t)}{2} + \frac{y_2(t)}{2}.$$

(b)

$$x(t) = \cos(3t - 2) = \frac{e^{j(3t-2)} + e^{-j(3t-2)}}{2}$$
$$= \frac{e^{j3t}e^{-j2}}{2} + \frac{e^{-j3t}e^{+j2}}{2} = \frac{x_1(t)e^{-2j}}{2} + \frac{x_2(t)e^{+2j}}{2}$$

$$\Rightarrow y(t) = \frac{y_1(t)e^{-2j}}{2} + \frac{y_2(t)e^{+2j}}{2}.$$

Problem 28. Consider the continuous-time system shown in Fig. 2.6.

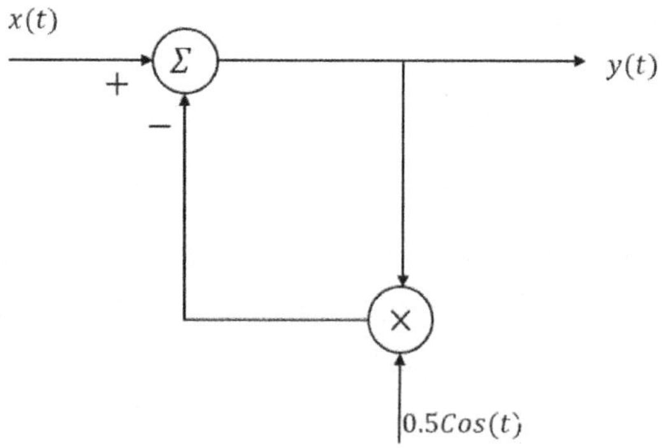

Figure 2.6: Problem 28.

(a) Find the relationship between the input and the output.

(b) Determine whether the system is (i) LTI, (ii) causal, (ii) memoryless, (iv) invertible, (v) stable.

Solution.

(a) From Fig. 2.6, we have

$$y(t) = x(t) - y(t)0.5\cos(t) \Rightarrow y(t) = \frac{x(t)}{1 + 0.5\cos(t)}.$$

(b) (i) As can be seen above, the system simply multiplies the input with a time-variant coefficient. So the system is linear but not time-invariant.

(ii) The system is casual. Since the output does not depend on the future inputs.

(iii) The system is memoryless. Since the output only depends on the current input.

(iv) This system is invertible as the coefficient can not be equal to zero at any time t.

(v) The system is stable. The coefficient is bounded for any value of t and so, any bounded input will result in a bounded output.

Problem 29. Determine whether the continuous-time system shown in Fig. 2.7 is (a) linear, (b) time-invariant, (c) stable, (d) memoryless.

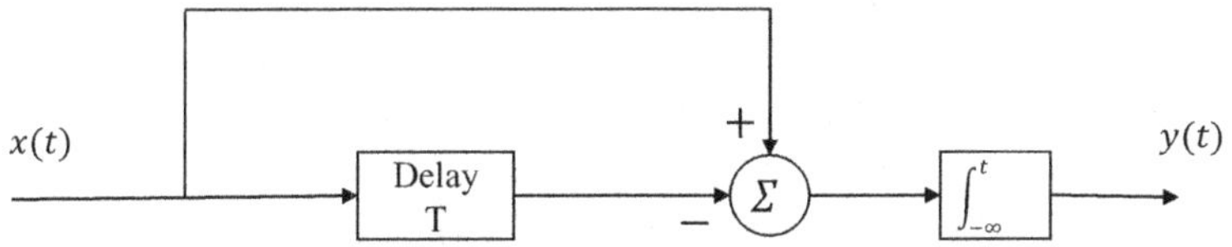

Figure 2.7: Problem 29.

Solution. From Fig. 2.8, we find input-output relationship as follows:

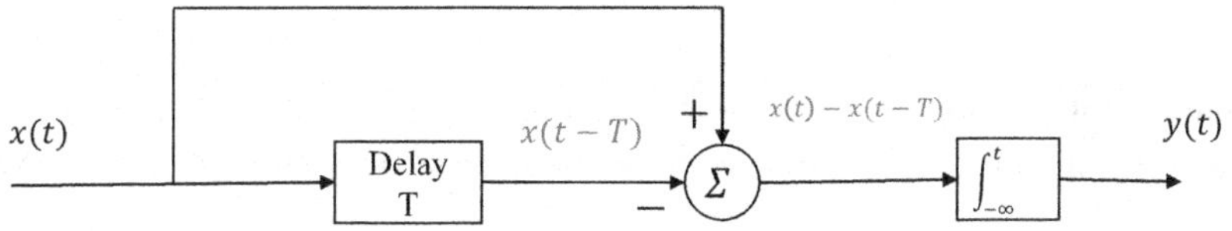

Figure 2.8: Problem 29.

$$\begin{aligned} y(t) &= \int_{-\infty}^{t} \big[x(\tau) - x(\tau - T)\big] d\tau \\ &= \int_{-\infty}^{t} x(\tau) d\tau - \int_{-\infty}^{t} x(\tau - T) d\tau. \end{aligned} \tag{2.12}$$

By substituting $\tau - T = \alpha$ in the second term of (2.12), we have

$$
\begin{aligned}
y(t) &= \int_{-\infty}^{t} x(\tau)d\tau - \int_{-\infty}^{t-T} x(\alpha)d\alpha \\
&= \int_{-\infty}^{t} x(\tau)d\tau - \int_{-\infty}^{t-T} x(\tau)d\tau \qquad (2.13) \\
&= \int_{t-T}^{t} x(\tau)d\tau.
\end{aligned}
$$

(a) From (2.13), we observe that the system is an integral operator which is a linear operator, so the system is linear. Note that we can also apply the 2 properties of linear systems directly and comp up with the same conclusion.

(b) This system is time-invariant:

By definition

$$T\{x(t-t_0)\} = \int_{t-T}^{t} x(\tau - t_0)d\tau \xlongequal{\tau - t_0 = \alpha} \int_{t-t_0-T}^{t-t_0} x(\alpha)d\alpha, \; (A)$$

$$y(t-t_0) = \int_{t-t_0-T}^{t-t_0} x(\tau)d\tau, \; (B)$$

$$(A) \text{ and } (B) \Rightarrow T\{x(t-t_0)\} = y(t-t_0).$$

(c) The system is stable. If $x(t) = B < \infty$, then we have

$$y(t) = \int_{t-T}^{t} x(\tau)d\tau = \int_{t-T}^{t} B\, d\tau = BT < \infty.$$

(d) The system is not memoryless. Since the output at time t_0 depends on the value of the input signal at time $t_0 - T$, the output depends on the past and the present values of the input.

Problem 30. Consider a continuous-time linear system with input $x(t)$ and output $y(t)$. Three inputs of this system and their corresponding outputs are shown in Fig. 2.9.

(a) Determine whether the system is (i) memoryless, (ii) causal, (iii) time-invariant.

(b) Determine and sketch carefully the response of the system to the input $x_4(t)$ depicted in Fig. 2.10.

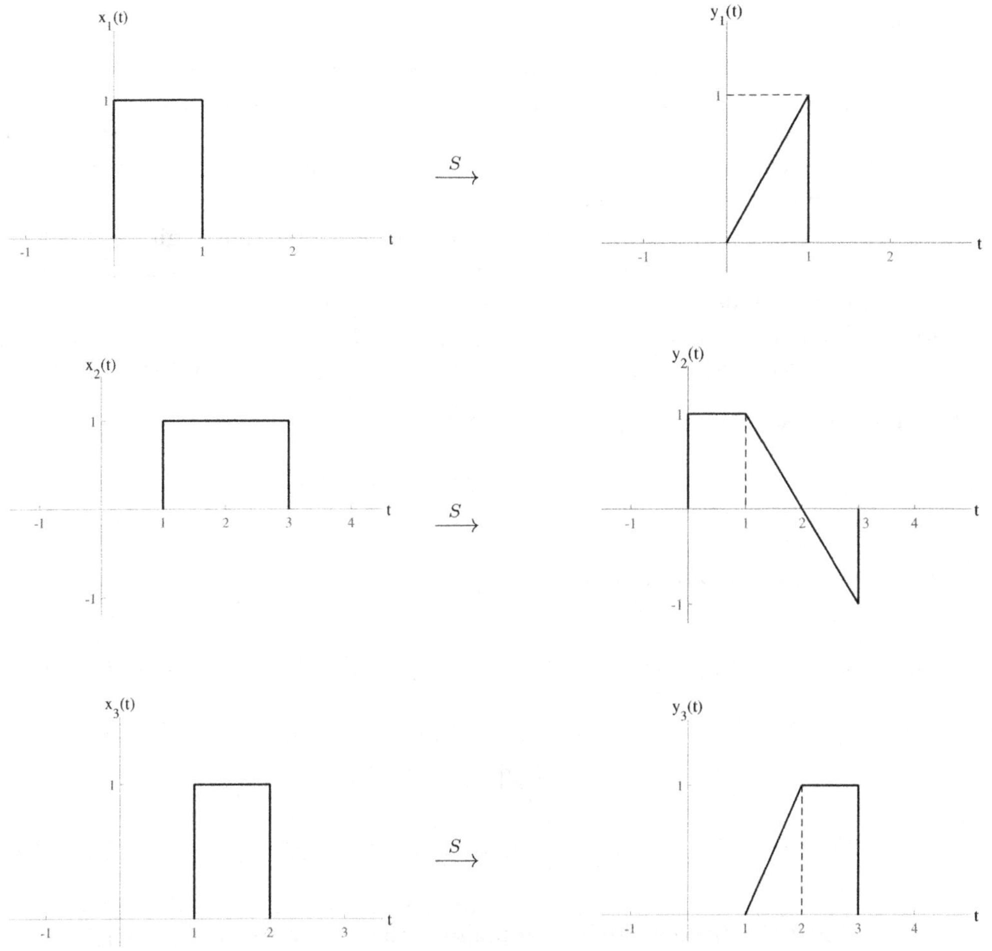

Figure 2.9: Problem 30 - Part (a).

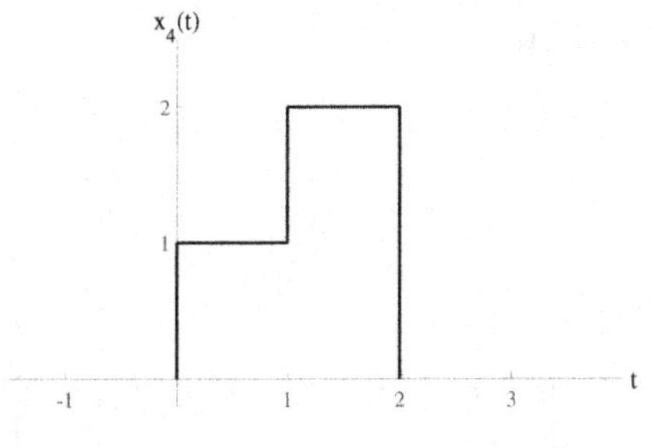

Figure 2.10: Problem 30 - Part (b), Signal $x_4(t)$.

Solution.

(a) (i) This system is not memoryless based on Fig. 2.9. For example, by comparing the second and third sub-figures at $t = 2$, $x_2(2) = 1$ as an input has resulted in output $y_2(2) = 0$. However, $x_3(2) = 1$ has resulted in $y_3(2) = 1$. This means that the output of the system at $t = 2$ can not be assumed to be depending only on the input at $t = 2$ and it also depends on other time instances as well.

(ii) This system is not causal. As can be seen in the figure, although for $0 < t < 1$, we have $x_2(t) = 0$, the system has prdocued a non-zero output on the same time interval, i.e., $y_2(t) \neq 0$ for $0 < t < 1$.

(iii) This system is not time-invariant, because we have $x_3(t) = x_1(t-1)$ but $y_3(t) \neq y_1(t-1)$, i.e., a time shift at the input has not resulted in a time shift at the output.

(b) Based on the linearity property, we can write:

$$x_4(t) = x_1(t) + 2x_3(t),$$
$$\Rightarrow y_4(t) = y_1(t) + 2y_3(t).$$

The output is plotted in Fig. 2.11.

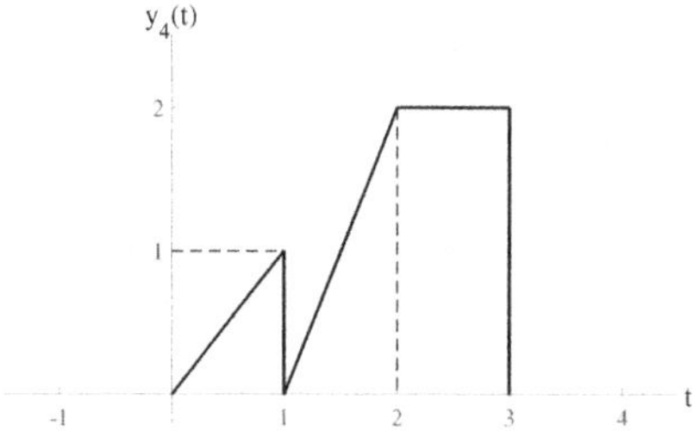

Figure 2.11: Problem 30, The response of the system to the input $x_4(t)$.

2.6.2 Discrete-time Signals

Problem 31. Determine whether the following discrete-time system is linear.

$$y[n] = \begin{cases} x[n] & n > 10 \\ 5 & -10 \leq n \leq 10 \\ -x[n] & n < 10 \end{cases}$$

Solution.

$$T\{\alpha x[n]\} = \begin{cases} \alpha x[n] & n > 10 \\ 5 & -10 \leq n \leq 10 \\ -\alpha x[n] & n < 10 \end{cases} \neq \alpha T\{x[n]\} = \begin{cases} \alpha x[n] & n > 10 \\ 5 & -10 \leq n \leq 10 \\ -\alpha x[n] & n < 10 \end{cases}.$$

The system does not have the homogeneity property, therefore it is nonlinear.

Problem 32. Consider the discrete-time systems with the following input-output relations. Determine whether these systems are time-invariant.

(a) $y[n] = x[-n]$

(b) $y[n] = x^2[n] + n^2$

(c) $y[n] = \begin{cases} x^2[n] & x[n] \geq 0 \\ x^3[n+1] & x[n] < 0 \end{cases}$

(d) $y[n] = \begin{cases} x^2[n] & n \geq 0 \\ x^3[n+1] & n < 0 \end{cases}$

(e) $y[n] = \sum_{k=n}^{n+2} x[k]$

Solution.

(a) $y[n] = x[-n]$

By definition

$$\begin{aligned} T\{x[n-n_0]\} &= x[-n-n_0] \\ y[n-n_0] &= x[-(n-n_0)] = x[-n+n_0] \end{aligned} \Rightarrow T\{x[n-n_0]\} \neq y[n-n_0].$$

This system is not time-invariant.

(b) $y[n] = x^2[n] + n^2$

By definition

$$\begin{aligned} T\{x[n-n_0]\} &= x^2[n-n_0] + n^2 \\ y[n-n_0] &= x^2[n-n_0] + (n-n_0)^2 \end{aligned} \Rightarrow T\{x[n-n_0]\} \neq y[n-n_0].$$

This system is not time-invariant.

(c) $y[n] = \begin{cases} x^2[n] & x[n] \geq 0 \\ x^3[n+1] & x[n] < 0 \end{cases}$

We have

$$T\{x[n-n_0]\} = \begin{cases} x^2[n-n_0] & x[n-n_0] \geq 0 \\ x^3[n+1-n_0] & x[n-n_0] < 0 \end{cases}$$

$$y[n-n_0] = \begin{cases} x^2[n-n_0] & x[n-n_0] \geq 0 \\ x^3[n-n_0+1] & x[n-n_0] < 0 \end{cases}$$

$$\Rightarrow T\{x[n-n_0]\} = y[n-n_0].$$

This system is time-invariant.

(d) $y[n] = \begin{cases} x^2[n] & n \geq 0 \\ x^3[n+1] & n < 0 \end{cases}$

We have

$$T\{x[n-n_0]\} = \begin{cases} x^2[n-n_0] & n \geq 0 \\ x^3[n+1-n_0] & n < 0 \end{cases}$$

$$y[n-n_0] = \begin{cases} x^2[n-n_0] & n-n_0 \geq 0 \\ x^3[n-n_0+1] & n-n_0 < 0 \end{cases}$$

$$\Rightarrow T\{x[n-n_0]\} \neq y[n-n_0].$$

This system is not time-invariant.

(e) $y[n] = \sum_{k=n}^{n+2} x[k]$

By definition

$$T\{x[n-n_0]\} = \sum_{k=n}^{n+2} x[k-n_0] \xlongequal{k-n_0=m} \sum_{m=n-n_0}^{n+2-n_0} x[m]$$

$$y[n-n_0] = \sum_{k=n-n_0}^{n-n_0+2} x[k]$$

$$\Rightarrow T\{x[n-n_0]\} = y[n-n_0].$$

This system is time-invariant.

Problem 33. Determine if the discrete-time systems described by the following input-output equations are causal or noncausal.

(a) $y[n] = \begin{cases} x[n-2]+n+7 & n \geq -1 \\ -x[n]-\cos(n) & n < -1 \end{cases}$

(b) $y[n] = \sum_{k=-\infty}^{n+4} x[k-1]2^{k+n+4}$

(c) $y[n] = \sum_{k=-\infty}^{n+5} x[k]\delta[k-n]k^2$

(d) $y[n] = \begin{cases} x^2[n] & x[-n+2] \geq 0 \\ x[n-1] & x[-n+2] < 0 \end{cases}$

Solution.

(a) $y[n] = \begin{cases} x[n-2]+n+7 & n \geq -1 \\ -x[n]-\cos(n) & n < -1 \end{cases}$

Output at time n_0 depends on the value of the input signal at time (n_0-2) from the first function and on the value of the input signal at time n_0 from the second function. So the output does not depend on future values of the input, hence, the system is causal.

(b) $y[n] = \sum_{k=-\infty}^{n+4} x[k-1]2^{k+n+4}$

Using change of variables we write:

$$y[n] = \sum_{k=-\infty}^{n+4} x[k-1]2^{k+n+4} \xlongequal{k-1=m} \sum_{m=-\infty}^{n+3} x[m]2^{m+n+5}$$

As can be seen, the upper bound of the summation is $n+3$. meaning that the value of output at n depends on the input at time $n+3$. Hence, the system is not causal.

Note: Instead of changing the variable, we could also directly replace $n+4$, the upper bound of the summation into the index of x inside the summation and come up with $x[n+3]$ to conclude the system is not casual.

(c) $y[n] = \sum_{k=-\infty}^{n+5} x[k]\delta[k-n]k^2$

Hint: At first look, the system does not seem to be causal as the output at time n seems to be depending on input at $n+5$. However, we should be careful in deriving such conclusions when there is a delta function around. In fact, the only nonzero term in the summation is when $k = n$ and we have:

$$y[n] = \sum_{k=-\infty}^{n+5} x[k]\delta[k-n]k^2 = x[n]n^2.$$

So the system is in fact causal.

(d) $y[n] = \begin{cases} x^2[n] & x[-n+2] \geq 0 \\ x[n-1] & x[-n+2] < 0 \end{cases}$

Let's set $n = -2$. For this value of n we have:

$$y[-2] = \begin{cases} x^2[-2] & x[4] \geq 0 \\ x[-3] & x[4] < 0 \end{cases}.$$

As can be seen, the output at time $n = -2$ can only be deiced if we know the input at time $n = 4$. Hence, the system is not causal.

Problem 34. Determine whether the following discrete-time systems are memoryless.

(a) $y[n] = y[n-3] + x[n]$

(b) $y[n] = x[n]\delta[n]$

(c) $y[n] = \sum_{k=-\infty}^{n+2} x[k-1]\delta[n-k]$

(d) $y[n] = \begin{cases} x[n] + \cos(n) & x[n-2] \geq 0 \\ x^3[n] + 3 & x[n-2] < 0 \end{cases}$

(e) $y[n] = \frac{1}{2N+1} \sum_{k=-N}^{N} x[n-k]$

Solution.

(a) $y[n] = y[n-3] + x[n]$

It is seen that the output at time n_0 depends on the value of the output at time $n_0 - 3$. Hence, this system is not memoryless.

(b) $y[n] = x[n]\delta[n]$

$$y[n] = x[n]\delta[n] = x[0]\delta[n] = x[0]. \tag{2.14}$$

From (2.14), the output at time $n = 0$ depends on the value of the input at that time. For other values of n, the output still depends on the value at $n = 0$. Hence, the system is not memoryless.

(c) $y[n] = \sum_{k=-\infty}^{n+2} x[k-1]\delta[n-k]$

$$y[n] = \sum_{k=-\infty}^{n+2} x[k-1]\delta[n-k] = x[n-1] \sum_{k=-\infty}^{n+2} \delta[n-k].$$

It is clear that the output at time n_0 depends on the value of the input at time $n_0 - 1$. Hence, the system is not memoryless.

(d) $y[n] = \begin{cases} x[n] + \cos(n) & x[n-2] \geq 0 \\ x^3[n] + 3 & x[n-2] < 0 \end{cases}$

To obtain the $y[n_0]$, although both sub-functions depend only on the value of $x[n_0]$, one has to know the value of $x[n_0 - 2]$ to determine the choice of sub-functions. Hence, the system is not memoryless.

(e) $y[n] = \frac{1}{2N+1} \sum_{k=-N}^{N} x[n-k]$

It is clear that the output depends on the past and future values of the input. For instance, output at time $n = 0$ is equal to

$$y[0] = \frac{1}{2N+1} \sum_{k=-N}^{N} x[-k] = \frac{1}{2N+1} \left[\cdots + x[-1] + x[0] + x[1] + \cdots \right].$$

Hence, the system is not memoryless.

Problem 35. Check the stability of following systems.

(a) $y[n] = x[2^{|n|}]$

(b) $y[n] = \sum_{k=n}^{n+5} x[5k+2]$

(c) $y[n] = \frac{\sqrt{|n|x[n]}}{5n-3}$

Solution.

(a) $y[n] = x[2^{|n|}]$

If $|x[n]| \leq B$, we have

$$-B' \leq x[2^{|n|}] \leq B' \Rightarrow -B' \leq y[n] \leq B'.$$

Therefore this system is stable. Note that the potentially unbounded term is inside the argument of $x[n]$ and has no effect on stability.

(b) $y[n] = \sum_{k=n}^{n+5} x[5k+2]$

As an input we select the bounded signal such that

$$|x[n]| < B, \forall n \Rightarrow x[5n+2] < B,$$

we have

$$y[n] = \sum_{k=n}^{n+5} x[5k+2] < \sum_{k=n}^{n+5} B = 6B < \infty.$$

It is seen that the output is bounded for any bounded input. As such, then the system is stable.

(c) $y[n] = \frac{\sqrt{|n|}x[n]}{5n-3}$

Since $5n-3$ does not approach zero for integer values, $\frac{\sqrt{|n|}}{5n-3}$ is bounded. Then, we have

$$|x[n]| < B \Rightarrow |y[n]| = |\frac{\sqrt{|n|}}{5n-3}||x[n]| < KB,$$

where K is the upper bound of $\frac{\sqrt{n}}{5n-3}$. Therefore the system is stable.

Problem 36. Determine if each of the following discrete-time systems is invertible.

(a) $y[n] = x[2n] + x[n]$

(b) $y[n] = \cos(\frac{\pi n}{7})x[n]$

(c) $y[n] = \begin{cases} x[n-1] & n \geq 1 \\ x^3[n] & n < 1 \end{cases}$

(d) $y[n] = \begin{cases} x[n/2] & n \text{ even} \\ 0 & n \text{ odd} \end{cases}$

(e) $y[n] = x[2n]$

(f) $y[n] = x[n]x[n-3]$

Solution.

(a) $y[n] = x[2n] + x[n]$

This system is not invertible. Let $x_1[n] = 0$ and $x_2[n]$ as shown in Fig. 2.12.

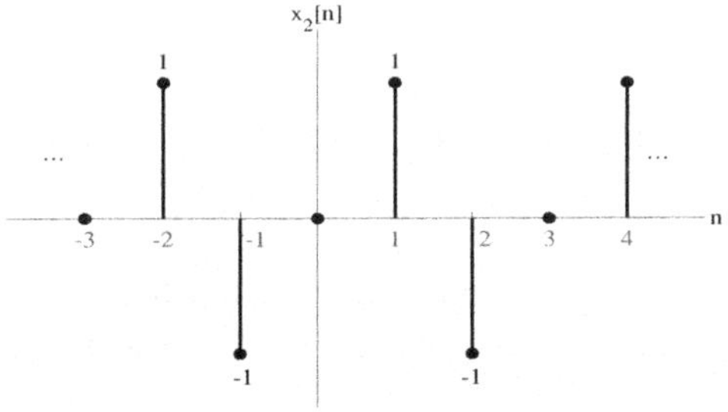

Figure 2.12: Problem 36 - Part (b).

Then $y_1[n] = y_2[n] = 0$ which means that x_1 and x_2 are mapped to the same output.

(b) $y[n] = \cos(\frac{\pi n}{7})x[n]$

This system is invertible, because the coefficient $\cos(\frac{\pi n}{7})$ is non-zero for any integer n. As such, there will be a unique input signal for any produced output.

(c) $y[n] = \begin{cases} x[n-1] & n \geq 1 \\ x^3[n] & n < 1 \end{cases}$

If $y_1[n] = y_2[n]$, then if $n \geq 1$, $x_1[n-1] = x_2[n-1]$ and so $x_1[n] = x_2[n]$. If $n < 1$, then

$x_1^3[n] = x_2^3[n]$, and so $x_1[n] = x_2[n]$. Therefore, in both cases, equal outputs, results in equal inputs and as such, the system is invertible.

(d) $y[n] = \begin{cases} x[n/2] & n, \text{ even} \\ 0 & n, \text{ odd} \end{cases}$
If $y_1[n] = y_2[n]$, then if n is even, $x_1[n/2] = x_2[n/2]$ and so $x_1[n] = x_2[n]$. If n is odd, $y_1[n] = y_2[n] = 0$ regardless of the input. Therefore, equal outputs, results in equal inputs and as such, the system is invertible.

(e) $y[n] = x[2n]$
In this case, in contrast to the above case, samples of $x[n]$ at odd times are eliminated. Therefore, there are infinitely many inputs which are equal at even time instants but have different values at odd instants of n. As such, the system is not invertible.

(f) $y[n] = x[n]x[n-3]$

Let the input be $x_1[n]$ where $x_1[n] = 1, \forall n$. Let $x_2[n]$ to be another input where $x_2[n] = -1, \forall n$. for all times. For both inputs, we will have $y[n] = 1, \forall n$, i.e., 2 inputs have resulted in 1 output. As such, the system is not invertible.

Problem 37. Consider a discrete-time linear memoryless system with input / output relationships as follows:

$$x_1[n] = \delta[n] + \delta[n-1] + 2\delta[n-2] + 3\delta[n-3] + 4\delta[n-4], \tag{2.15}$$

$$y_1[n] = 2\delta[n-1] + 2\delta[n-2]. \tag{2.16}$$

(a) Find the output of this system to the input $x_2[n] = \delta[n-1] + \delta[n-2]$.

(b) What can we say about the output of this system to the input $x_3[n] = \delta[n-5]$?

Solution.

For this system, we recall from Section 2.5 that in a linear memoryless system, the output at each time instant, can be written as the product of the input and a constant which may depend on time:

$$y[n_0] = A[n_0]x[n_0].$$

(a) $x_2[n] = \delta[n-1] + \delta[n-2]$
We can imagine the system has been fed 5 impulses at $n = 0, 1, 2, 3, 4$ as its input which has resulted in no impulse at $n = 0, 3, 4$, one impulse at $n = 1$ and one at $n = 2$. Based on the above relationship, we obtain $A[0] = A[3] = A[4] = 0, A[1] = 2, A[2] = 1$. As can be seen, $x_2[n]$ is zero for $n \neq 1, 2$. Using the above relationship, we conclude that $y_2[n]$ is also zero at $n \neq 1, 2$. Using the same values of $A[n]$ and the linearity property we have:

$$y_2[n] = 2\delta[n-1] + \delta[n-2].$$

(b) $x_3[n] = \delta[n-5]$

Using the memoryless assumption, we already know that $y_3[n]$ is zero for $n \neq 5$. For $y[5]$, we have to know the value of $A[5]$ but we can not comment on its value based on the given info. As such we can only state that

$$y_3[n] = A[5]\delta[n-5].$$

Problem 38. Consider a discrete-time linear system S. The output of the system equals $y[n] = \cos(mn)$ when the input is $x[n] = u[n-m]$ for any integer m. Determine the output of the system when the input is as follows:

$$x[n] = \delta[n+1] + \delta[n] + \delta[n-1].$$

Solution. In summary, we have

$$x_1[n] = u[n-m] \quad \xrightarrow{S} \quad y_1[n] = \cos(nm),$$

$$x_2[n] = \delta[n+1] + \delta[n] + \delta[n-1] \quad \xrightarrow{S} \quad y_2[n] = ?$$

We should write the input $x_2[n]$ based on the unit step function and its shifted versions. We know that $\delta[n]{=}u[n]-u[n-1]$. As such, $\delta[n-1]{=}u[n-1]-u[n-2]$ and $\delta[n+1]{=}u[n+1]-u[n]$. Thus,

$$x_2[n] = u[n+1] - u[n-2].$$

Now we have

$$y_2[n] = \cos(-n) + \cos(2n).$$

Problem 39. Consider a discrete-time linear system S. When the inputs to the system are $x_1[n]$, and $x_2[n]$, the outputs of the system are $y_1[n]$, and $y_2[n]$, as shown in Fig. 2.13, respectively. Determine whether the system is (a) memoryless, (b) time-invariant.

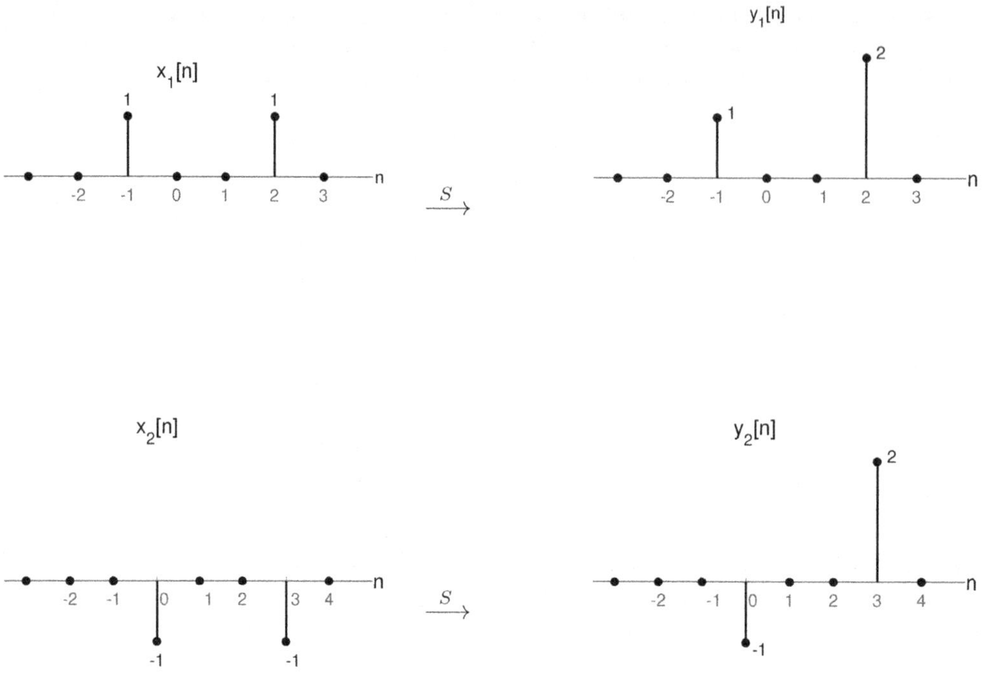

Figure 2.13: Problem 39.

Solution.

(a) If the system is linear and memoryless, using the tips in Section 2.5, we can write

$$Y[n] = X[n]A[n].$$

Now using the info for $x_1[n]$ and $y_1[n]$, we obtain $A[-1] = 1, A[2] = 2$. Based on these values, for $x_2[-1] = x_2[1] = -1$, we must have $y_2[-1] = -1$ and $y_2[2] = -2$. Based on the figure, this is not the case for $y_2[2]$. Thus this system can not be memoryless.

(b) To check the time-invariance note that we have

$$x_2[n] = -x_1[n-1].$$

Now if the system is time-invariant, as it is also assumed to be linear in the first place, we necessarily must have

$$y_2[n] = -y_1[n-1]. \tag{2.17}$$

Looking at the figures, (2.17) does not hold. As such, the system can not be time-invariant.

Important Note: Even if the values of $y_2[n]$ in above parts were such that we could not rule out the assumptions of the system being memoryless and/or time-invariant, it was still a necessary, and not a sufficient condition. In other words, we could not still explicitly say the system is memoryless and/or time-invariant with the given information.

Problem 40. Consider three systems with the following input-output relationships:

$$\begin{aligned} S_1: &\quad z[n] = \begin{cases} x[n/2] & n = 2k \\ 0 & \text{otherwise} \end{cases}, \\ S_2: &\quad w[n] = z[n] + z[n-1], \\ S_3: &\quad y[n] = w[2n]. \end{aligned} \tag{2.18}$$

Suppose that these systems are concatenated in series. Is the overall system linear? Is it time-invariant?

Solution.

These systems are connected in series as depicted in Fig. 2.14. We have:

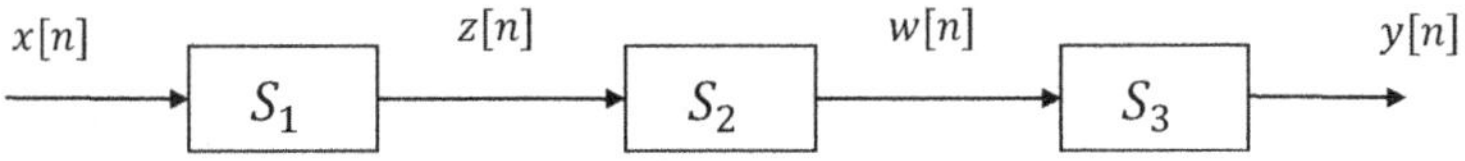

Figure 2.14: Problem 40.

$$z[n] = \begin{cases} x[n/2] & n = 2k \\ 0 & \text{otherwise} \end{cases}. \tag{2.19}$$

Applying S_2 and (2.19) we have:

$$\begin{aligned} w[n] &= z[n] + z[n-1] \\ &= x[n/2] + x[(n-1)/2] \\ &= \begin{cases} x[n/2] & n = 2k \\ x[(n-1)/2] & n = 2k+1 \\ 0 & \text{otherwise} \end{cases}. \end{aligned} \tag{2.20}$$

Given (2.20), by applying S_3, we sample at times $2k$. Then

$$y[n] = w[2n] = x[n].$$

As such, the overall system is linear and time-invariant. The above process is also depicted in Fig. 2.15.

Problem 41. Consider a linear system whose response to signal $x[n] = u[n-k]$ is the signal $y[n] = k^2\delta[n-k+1]$.

Is this system time-invariant?

Solution.

Let's find the response of the system to input $x[n] = \delta[n-k]$. We have

$$x_1[n] = \delta[n-k] = u[n-k] - u[n-k-1]. \tag{2.21}$$

Now using the system linearity we can write the output

$$y[n] = k^2\delta[n-k+1] - (k+1)^2\delta[n-k], \tag{2.22}$$

where we let $k \to k+1$ to obtain the second term. From (2.21) and (2.22):

$$k = 0 \longrightarrow x_0[n] = \delta[n] \quad \Rightarrow \quad y_0[n] = -\delta[n],$$
$$k = 1 \longrightarrow x_1[n] = \delta[n-1] \quad \Rightarrow \quad y_1[n] = \delta[n] - 4\delta[n-1].$$

So although for the inputs we have $x_0[n] = x_1[n-1]$, for the outputs we have

$$y_0[n] \neq y_1[n-1].$$

As such the system is not time-invariant.

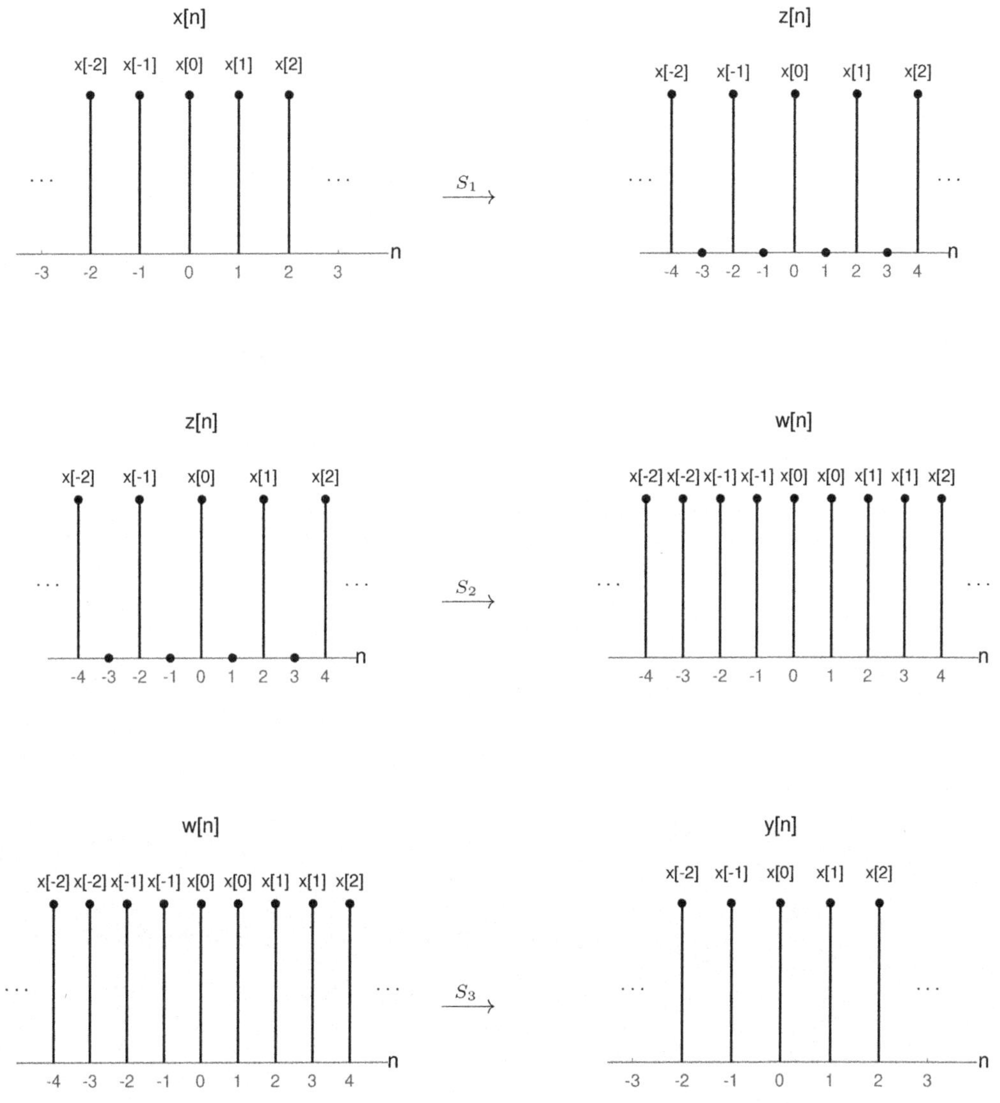

Figure 2.15: Problem 40, the process of obtaining $y[n]$ from a given $x[n]$.

Chapter 3

Liner Time-invariant Systems

3.1 Introduction

In the previous chapter we introduced a list of basic system properties. Two of these, linearity and time-invariance, are the most important attributes of systems and play a main role in signal and system analysis. In fact, a practical mathematical representation can be introduced for systems having these attributes. Many physical systems can be modeled as linear time-invariant (LTI) systems. One of the most important tools in signal and system analysis is the convolution operation. This operation is referred to as the convolution sum in the discrete-time case and the convolution integral in continuous time.

3.2 The Convolution Integral and the Convolution Sum

3.2.1 Convolution Integral

The impulse response $h(t)$ of a continuous-time LTI system is defined as the response of the system when the input is $\delta(t)$:

$$h(t) = T\{\delta(t)\}, \tag{3.1}$$

where T represents a continuous-time LTI system.

Also based on the sifting property of the impulse function, any input signal can be expressed as a linear *combination* of shifted unit impulses $\delta(t-\tau)$:

$$x(t) = \int_{-\infty}^{+\infty} x(\tau)\delta(t-\tau)d\tau. \tag{3.2}$$

Now according to the linearity property, the response $y(t)$ of the system to an arbitrary input $x(t)$ can be written as

$$y(t) = T\{x(t)\} = T\{\int_{-\infty}^{+\infty} x(\tau)\delta(t-\tau)d\tau\}$$

$$= \int_{-\infty}^{+\infty} x(\tau) T\{\delta(t-\tau)\} d\tau. \tag{3.3}$$

Now based on the time-invariance property, we have

$$h(t-\tau) = T\{\delta(t-\tau)\}. \tag{3.4}$$

Substituting (3.4) into (3.3), we obtain

$$y(t) = \int_{-\infty}^{+\infty} x(\tau) h(t-\tau) d\tau. \tag{3.5}$$

The convolution of two continuous-time signals $x(t)$ and $h(t)$ is defined and denoted as follows:

$$y(t) = x(t) * h(t) = \int_{-\infty}^{+\infty} x(\tau) h(t-\tau) d\tau. \tag{3.6}$$

Equation (3.6) is also referred to as the convolution integral. The main catch is that is the output of any continuous-time LTI system to the input $x(t)$ can be determined if we have the impulse response $h(t)$ at hand.

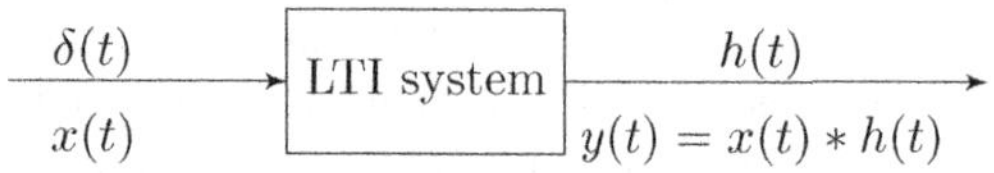

Figure 3.1: Continuous-time LTI system.

To obtain $y(t)$, we have to multiply signals $x(\tau)$ and $h(t-\tau)$ and then integrate over the resulting product from $\tau = -\infty$ to $\tau = +\infty$. From (3.6), the convolution integral evaluation contains the following steps:

Step 1. Plot the signals $x(\tau)$ and $h(\tau)$.

Step 2. Keep the signal $x(\tau)$ unchanged and obtain the signal $h(-\tau)$ from $h(\tau)$ by mirroring it with respect to the origin.

Step 3. Obtain the signal $h(t-\tau)$ from $h(\tau)$ by a shift to the right by t if $t > 0$ or a shift to the left by $|t|$ for $t < 0$.

Step 4. Now we change t from $-\infty$ to $+\infty$, multiply signals $x(\tau)$, and $h(t-\tau)$ and calculate the area under the resulting signal.

3.2.2 Convolution Sum

As in continuous time, the impulse response $h[n]$ of a discrete-time LTI system is denoted by

$$h[n] = T\{\delta[n]\}, \tag{3.7}$$

where T represent a discrete-time LTI system.

Also based on the sifting property of the impulse function, any input signal can be expressed as a linear combination of shifted unit impulses $\delta[n-k]$:

$$x[n] = \sum_{k=-\infty}^{+\infty} x[k]\delta[n-k]. \tag{3.8}$$

According to the linearity property, the response $y[n]$ of the system to an arbitrary input $x[n]$ can be written as

$$\begin{aligned} y[n] = T\{x[n]\} &= T\{\sum_{k=-\infty}^{+\infty} x[k]\delta[n-k]\} \\ &= \sum_{k=-\infty}^{+\infty} x[k]T\{\delta[n-k]\}. \end{aligned} \tag{3.9}$$

Based on the time-invariance property, we have

$$h[n-k] = T\{\delta[n-k]\}. \tag{3.10}$$

Substituting (3.10) into (3.9), we obtain

$$y[n] = \sum_{k=-\infty}^{+\infty} x[k]h[n-k]. \tag{3.11}$$

The convolution of two signals $x[n]$ and $h[n]$ is defined and denoted as follows:

$$y[n] = x[n] * h[n] = \sum_{k=-\infty}^{+\infty} x[k]h[n-k]. \tag{3.12}$$

Equation (3.12) is sometimes referred to as the convolution sum and the main catch is that the output of any discrete-time LTI system to the input $x[n]$ can be determined if impulse response $h[n]$ is at hand.

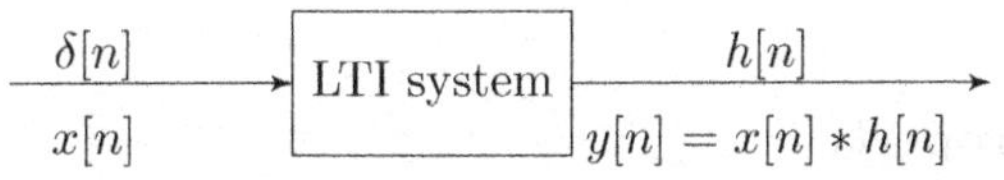

Figure 3.2: Discrete-time LTI system.

3.3 Properties of the Convolution Integral

The convolution integral has the following properties:

(a) The Commutative Property

$$x(t) * h(t) = h(t) * x(t). \tag{3.13}$$

According to this property, the output of an LTI system with the input $x(t)$ and unit impulse response $h(t)$ is the same as the output of an LTI system with input $h(t)$ and unit impulse response $x(t)$.

(b) The Associative Property

$$\{x(t) * h_1(t)\} * h_2(t) = x(t) * \{h_1(t) * h_2(t)\}. \tag{3.14}$$

According to this property, a cascade of two LTI systems with impulse response $h_1(t)$ and $h_2(t)$ can be replaced by a single LTI system with the impulse response $h_1(t) * h_2(t)$.

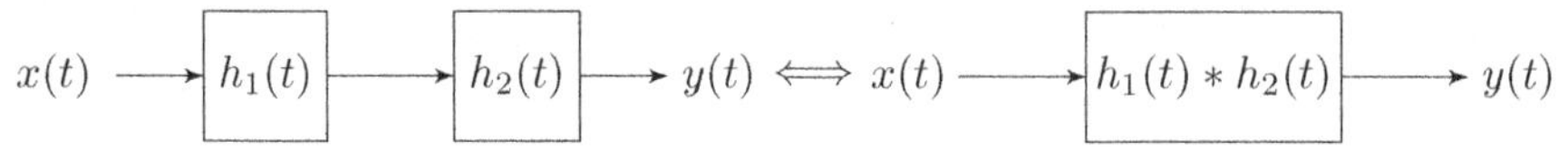

Figure 3.3: Associative property of convolution for the series interconnection of LTI systems.

(c) The Distributive Property

$$x(t) * \{h_1(t) + h_2(t)\} = x(t) * h_1(t) + x(t) * h_2(t). \tag{3.15}$$

According to this property, a parallel combination of two LTI systems with the impulse response $h_1(t)$ and $h_2(t)$ can be replaced by a single LTI system with the impulse response $h_1(t) + h_2(t)$.

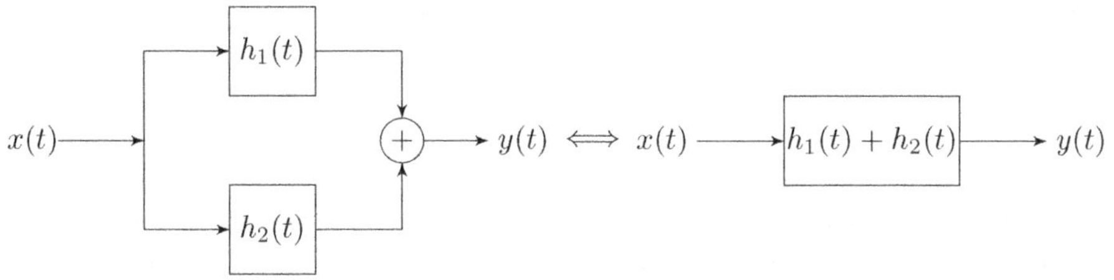

Figure 3.4: Distributive property of convolution for a parallel interconnection of LTI systems.

(d) The Derivative Property

If $y(t) = x(t) * h(t)$, we have

$$\frac{dy(t)}{dt} = x(t) * \frac{dh(t)}{dt} = \frac{dx(t)}{dt} * h(t). \tag{3.16}$$

According to this property, the output of an LTI system to the derivative of the input $x(t)$ and unit impulse response $h(t)$ is the derivative of output $y(t)$.

Not that this property is also true for other linear operators.

3.4 Properties of the Convolution Sum

Analogously, the convolution sum has the following properties:

(a) The Commutative Property

$$x[n] * h[n] = h[n] * x[n]. \tag{3.17}$$

(b) The Associative Property

$$x[n] * \{h_1[n] * h_2[n]\} = \{x[n] * h_1[n]\} * h_2[n]. \tag{3.18}$$

(c) The Distributive Property

$$x[n] * \{h_1[n] + h_2[n]\} = x[n] * h_1[n] + x[n] * h_2[n]. \tag{3.19}$$

3.5 Step Response of LTI Systems

We know that the behavior of an LTI system can be described using the impulse response of the system. There is another response signal that is also used quite often to describe the behavior of LTI systems: The unit step response which is shown by $s(t)$ or $s[n]$ (discrete case). Based on definition, the step response is the response of an LTI system to the unit step input, shown by $u(t)$ and $u[n]$ (discrete case).

Continuous-time LTI systems:

$$s(t) = h(t) * u(t) = \int_{-\infty}^{+\infty} h(\tau)u(t-\tau)d\tau = \int_{-\infty}^{t} h(\tau)d\tau. \tag{3.20}$$

Thus, the step response $s(t)$ can be determined by integrating the impulse response $h(t)$. Also the impulse response $h(t)$ can be determined by differentiating $s(t)$:

$$h(t) = s'(t) = \frac{ds(t)}{dt}. \tag{3.21}$$

Discrete-time LTI systems:

$$s[n] = h[n] * u[n] = \sum_{k=-\infty}^{+\infty} h[k]u[n-k] = \sum_{k=-\infty}^{n} h[k]. \tag{3.22}$$

From (3.22) we have:

$$h[n] = s[n] - s[n-1]. \tag{3.23}$$

Equations (3.22) and (3.23) are the discrete-time counterparts of (3.20) and (3.21), respectively.

3.6 Properties of Linear Time-Invariant Systems

In previous sections, we introduced properties of the convolution. In this section we want to discuss properties of LTI systems based on the convolution principle.

3.6.1 LTI Systems with and without Memory

We know that a system is memoryless if its output at any time depends only on the value of the present input. Since the system is linear and time-invariant, we have

$$\begin{aligned} y[n] &= \sum_{k=-\infty}^{+\infty} h[k]x[n-k] \\ &= h[0]x[n] + \sum_{k\neq 0} h[k]x[n-k], \end{aligned} \tag{3.24}$$

From (3.24), we see the first part depends only on the value of the present input, but the second part depends on the past and the future values of the input. Thus for the system to be memoryless, we must have

$$h[n] = 0, \quad n \neq 0 \quad \Rightarrow \quad h[n] = k\delta[n]. \tag{3.25}$$

where k is a constant.

So the system has memory, iff $h[n] \neq 0$ for $n \neq 0$.

The same conclusion holds for continuous-time systems as well, i.e.,

$$h(t) = 0, \quad t \neq 0 \quad \Rightarrow \quad h(t) = k\delta(t). \tag{3.26}$$

where k is a constant.

The system has memory if $h(t) \neq 0$ for $t \neq 0$.

3.6.2 Causality for LTI Systems

From Chapter 2, we know that a system is casual if the output depends only on the present and past values of the input to the system. From the definition, we have

$$\begin{aligned} y[n] &= \sum_{k=-\infty}^{+\infty} h[k]x[n-k] \\ &= \sum_{k=-\infty}^{-1} h[k]x[n-k] + \sum_{k=0}^{+\infty} h[k]x[n-k]. \end{aligned} \tag{3.27}$$

From (3.27), we see the first term depends on the future values of the input, but the second part depends only the past and present values of the input. As such, for a discrete-time causal LTI system we must have

$$h[n] = 0 \quad \text{for } n < 0. \tag{3.28}$$

The same argument holds for continuous-time systems as well. Therefore, the causality condition for a continuous-time LTI system is

$$h(t) = 0 \quad \text{for } t < 0. \tag{3.29}$$

3.6.3 Stability for LTI Systems

From Chapter 2 we know that a system is stable if bounded-input leads to bounded-output (BIBO), i.e.,

$$\text{if} \quad |x[n]| \leq A < \infty,$$

where A is a finite real constant, we must have

$$\begin{aligned} |y[n]| = \left| \sum_{k=-\infty}^{+\infty} h[k]x[n-k] \right| &\leq \sum_{k=-\infty}^{+\infty} |h[k]| \, |x[n-k]| \\ &\leq A \sum_{k=-\infty}^{+\infty} |h[k]| \end{aligned} \tag{3.30}$$

From (3.30), a discrete-time LTI system is stable if the impulse response is absolutely summable, i.e., if

$$\sum_{k=-\infty}^{+\infty} |h[k]| < \infty. \tag{3.31}$$

The same argument holds for continuous-time systems as well. Therefore, a continuous-time LTI system is stable if the impulse response is absolutely integrable, i.e., if

$$\int_{-\infty}^{+\infty} |h(\tau)| \, d\tau < \infty. \tag{3.32}$$

3.6.4 Invertibility of LTI Systems

An LTI system is invertible only if an inverse system exists such that when connected in series with the original system, the overall system is identical to the identity system, i.e., if

$$h[n] * h_I[n] = \delta[n]. \tag{3.33}$$

Similarly, in continuous case, the impulse response $h_I(t)$ of the inverse system of an LTI system with impulse response $h(t)$ must satisfy

$$h(t) * h_I(t) = \delta(t). \tag{3.34}$$

3.7 Casual LTI Systems Described by Differential and Difference Equations

3.7.1 Linear Constant-Coefficient Differential Equations

A general Nth-order linear constant-coefficient differential equation is given by

$$\sum_{k=0}^{N} a_k \frac{d^k y(t)}{dt^k} = \sum_{k=0}^{M} b_k \frac{d^k x(t)}{dt^k}, \tag{3.35}$$

where coefficients a_k and b_k are real number constants. The order N refers to the highest degree of derivation of $y(t)$.

In general, to solve this differential equation, we must specify N auxiliary conditions. In order for the system described by (3.35) to be linear and causal, we must assume the condition of initial rest. That is, if $x(t) = 0$ for $t \le t_0$, then $y(t) = 0$ for $t \le t_0$. The solution of (3.35) for a given input $x(t)$ consists of the sum of two terms:

$$y(t) = y_p(t) + y_h(t), \tag{3.36}$$

where $y_p(t)$ is the *particular solution* satisfying (3.35) and $y_h(t)$ is the *homogeneous solution* which is determined by N auxiliary conditions.

3.7.2 linear Constant-Coefficient Difference Equations

The discrete-time counterpart of (3.35) is the Nth-order linear constant-coefficient difference equation given by

$$\sum_{k=0}^{N} a_k y[n-k] = \sum_{k=0}^{M} b_k x[n-k]. \tag{3.37}$$

Equation (3.37) can be solved in the same way as in differential equations. Specifically, the solution of $y[n]$ for a particular input $x[n]$ consists of the sum of a particular solution, $y_p[n]$, and a homogeneous solution, $y_h[n]$.

3.8 Important Points

(a) Discrete-time systems with linear constant-coefficient difference equations are often divided into two types:

1. *Finite impulse response (FIR) systems*: the impulse response of such systems has finite duration and the corresponding difference equation is *non-recursive.*

2. *infinite impulse response (IIR) system*: the impulse response of such systems has infinite duration and the corresponding difference equation is *recursive.*

(b) Periodic Convolution of Continuous-Time Signals

Consider two continuous-time periodic signals $x(t)$ and $y(t)$ with periods T_x and T_y, respectively. The periodic convolution of two signals can be determined if a value T exists such that

$$\frac{T}{T_x} \in \mathbb{Z} \text{ and } \frac{T}{T_y} \in \mathbb{Z}.$$

Then, the periodic convolution of continuous-time signals is given by

$$z(t) = x(t) \circledast y(t) = x_T(t) * y(t), \tag{3.38}$$

where

$$x_T(t) = \begin{cases} x(t) & |t| < \frac{T}{2} \\ 0 & \text{other} \end{cases}. \tag{3.39}$$

(c) Periodic Convolution of Discrete-Time Signals

Consider two discrete-time periodic signals $x[n]$ and $y[n]$ with periods N_x and N_y, respectively. The periodic convolution of two signals can always be defined since both N_x and N_y are integer numbers as well as their least common multiple, N. The periodic convolution of discrete-time signals is given by

$$z[n] = x[n] \circledast y[n] = x_N[n] * y[n], \tag{3.40}$$

where

$$x_N[n] = \begin{cases} x[n] & |n| \leq \frac{N}{2} \\ 0 & \text{other} \end{cases}. \tag{3.41}$$

(d) Time Scaling Property

If

$$y(t) = x(t) * h(t),$$

then

$$x(at) * h(at) = \frac{1}{|a|} y(at); \tag{3.42}$$

where a is a constant.

(e) In LTI systems:

If a relationship between a given input and output exists that only contains linear operates and time shifts, the relationship golfs for any input and output. The rational behind this statement can be explained using system transfer functions which will be visited in upcoming chapters.

3.9 Problems

3.9.1 Continuous-time LTI systems

Problem 42. Obtain and plot the convolution $y(t) = x(t) * h(t)$ for a continuous-time LTI system whose impulse response $h(t)$ and inputs $x_1(t)$ and $x_2(t)$ are given by

$$h(t) = \begin{cases} 1 & |t| \leq 1 \\ 0 & |t| > 0 \end{cases}.$$

(a) $x_1(t) = \begin{cases} 1 & 1 \leq t \leq 3 \\ 0 & \text{otherwise} \end{cases}$,

(b) $x_2(t) = e^{-t}u(t)$,

Solution.

(a) $x_1(t) = \begin{cases} 1 & 1 \leq t \leq 3 \\ 0 & \text{otherwise} \end{cases}$

In order to obtain the convolution of $x_1(t)$ and $h(t)$, it is convenient to consider four separate intervals for t. First we depict signals $x_1(\tau)$ and $h(-\tau)$ in Fig. 3.5.

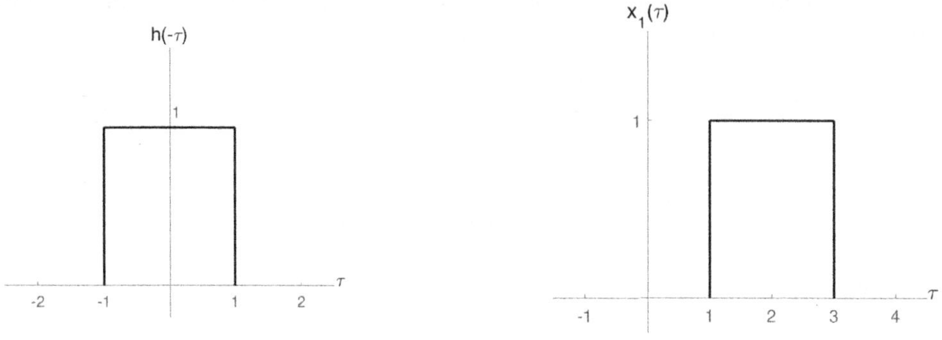

Figure 3.5: Problem 42: Signals $h(-\tau)$ and $x_1(\tau)$.

Interval 1. For $t < 0$, the signals $x_1(\tau)$ and $h(t-\tau)$ do not overlap, and consequently, $y(t) = 0$ for $t < 0$ (see Fig. 3.6).

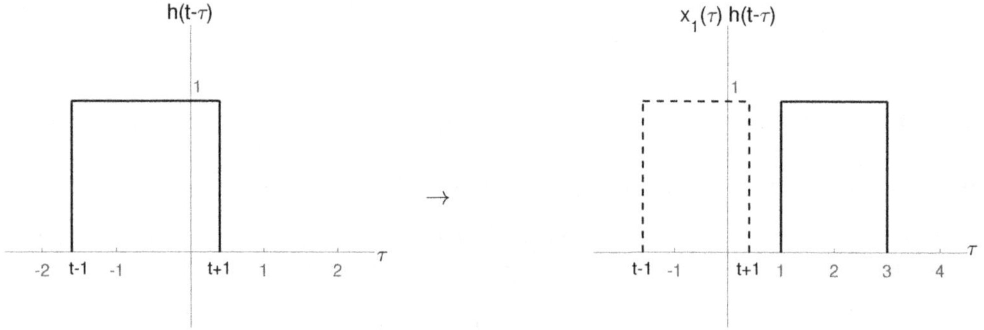

Figure 3.6: Problem 42: Signal $h(t-\tau)$ and the product $x_1(\tau)h(t-\tau)$ for $t < 0$.

Interval 2. For $0 \leq t \leq 2$, signals $x_1(\tau)$ and $h(t-\tau)$ overlap, and thus, we compute the area under the rectangular pulses for these interval (see Fig. 3.7). We obtain

$$y(t) = x_1(t) * h(t) = \int_{-\infty}^{+\infty} x_1(\tau)h(t-\tau)d\tau = \int_{+1}^{t+1} 1 d\tau = t.$$

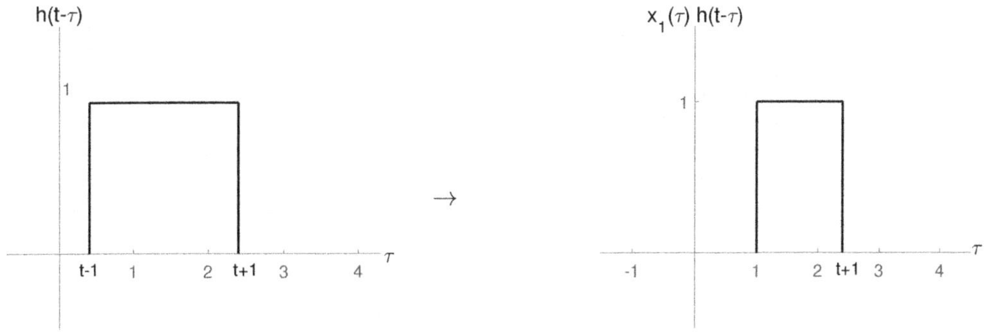

Figure 3.7: Problem 42 - Signal $h(t-\tau)$ and the product $x_1(\tau)h(t-\tau)$ for $0 \leq t \leq 2$.

Interval 3. For $2 \leq t \leq 4$, signals $x_1(\tau)$ and $h(t-\tau)$ overlap, and thus, we compute the area under the rectangular pulses for these interval (see Fig. 3.8). We obtain

$$y(t) = x_1(t) * h(t) = \int_{-\infty}^{+\infty} x_1(\tau)h(t-\tau)d\tau = \int_{t-1}^{3} 1d\tau = 4-t$$

.

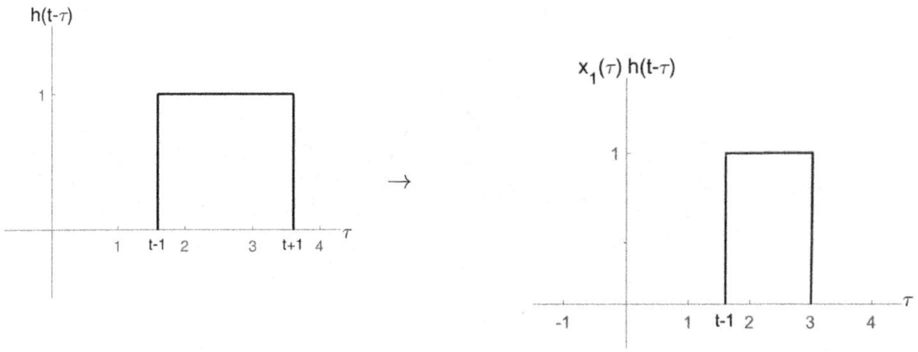

Figure 3.8: Problem 42: Signal $h(t-\tau)$ and the product $x_1(\tau)h(t-\tau)$ for $2 \leq t \leq 4$.

Interval 4. For $t > 4$, there is no overlap between two signals $x_1(\tau)$ and $h(t-\tau)$, and thus, $y(t) = 0$ for $t > 4$ (see Fig. 3.9).

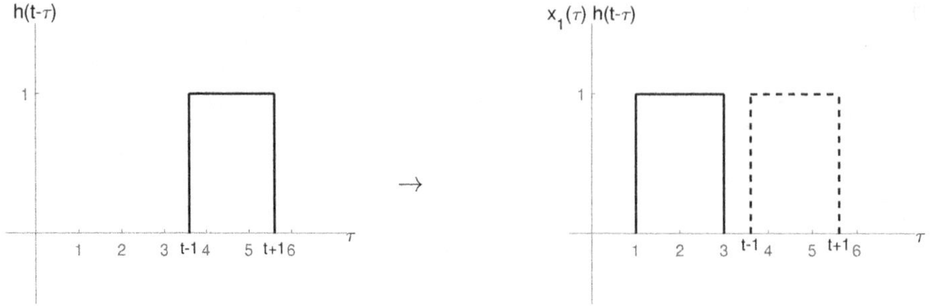

Figure 3.9: Problem 42: Signal $h(t-\tau)$ and the product $x_1(\tau)h(t-\tau)$ for $t > 4$.

Finally, we obtain

$$y(t) = \begin{cases} 0 & t < 0 \\ t & 0 \le t \le 2 \\ 4-t & 2 \le t \le 4 \\ 0 & t > 4 \end{cases},$$

which is depcited in Fig. 3.10.

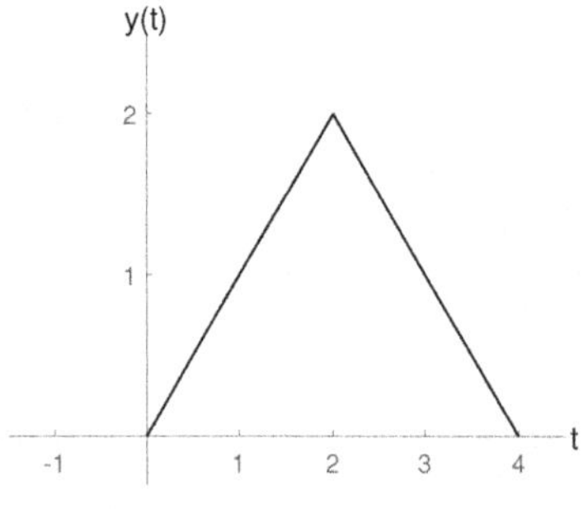

Figure 3.10: Problem 42: Signal $y(t) = x_1(t) * h(t)$.

(b) $x_2(t) = e^{-t}u(t)$

To calculate the convolution of $x_2(t)$ and $h(t)$, it is convenient to consider three separate intervals for t. First we depict signals $x_2(\tau)$ and $h(-\tau)$ in Fig. 3.11.

Figure 3.11: Problem 42: Signals $h(-\tau)$ and $x_2(\tau)$.

Interval 1. For $t < -1$, from Fig. 3.12, we see that signals $x_2(\tau)$ and $h(t-\tau)$ do not overlap, and consequently, $y(t) = 0$.

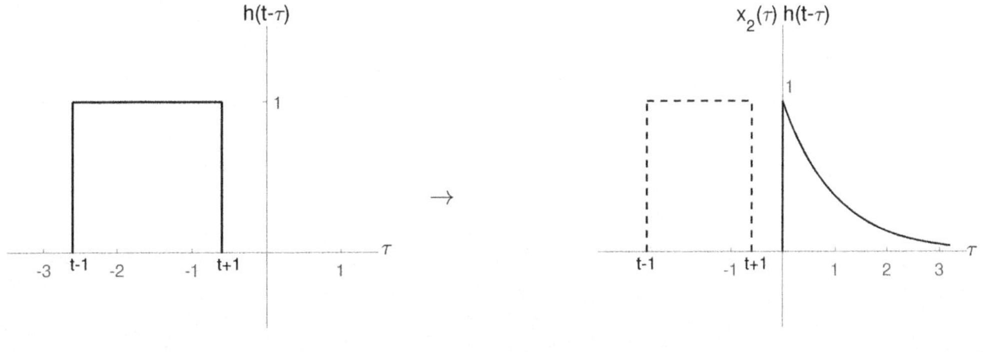

Figure 3.12: Problem 42: Signal $h(t-\tau)$ and the product $x_2(\tau)h(t-\tau)$ for $t < -1$.

Interval 2. For $-1 \le t \le 1$, signals $x_1(\tau)$, $h(t-\tau)$ and the product $x_1(\tau)h(t-\tau)$ are shown in Fig. 3.13. We obtain

$$y(t) = x_2(t) * h(t) = \int_{-\infty}^{+\infty} x_2(\tau)h(t-\tau)d\tau = \int_0^{t+1} e^{-\tau} d\tau = 1 - e^{-(t+1)}.$$

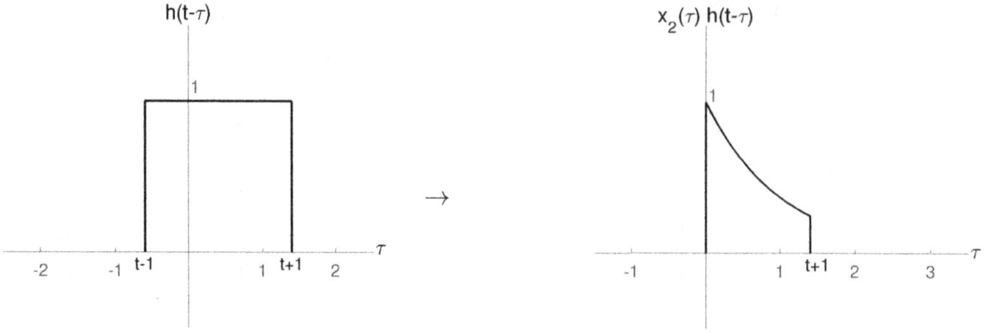

Figure 3.13: Problem 42: Signal $h(t-\tau)$ and the product $x_2(\tau)h(t-\tau)$ for $-1 \leq t \leq 1$.

Interval 3. For $t > 1$, we have (see Fig. 3.14)

$$y(t) = x_2(t) * h(t) = \int_{-\infty}^{+\infty} x_2(\tau)h(t-\tau)d\tau = \int_{-1+t}^{1+t} e^{-\tau}d\tau = e^{-(-1+t)} - e^{-(1+t)}.$$

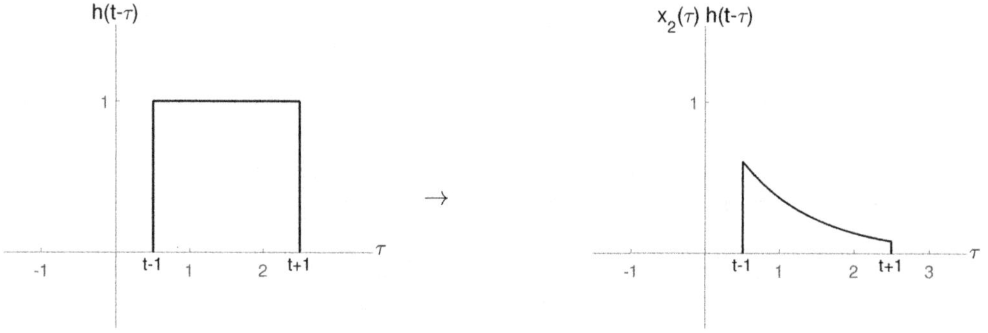

Figure 3.14: Problem 42: Signal $h(t-\tau)$ and the product $x_2(\tau)h(t-\tau)$ for $t > 1$.

As a result, we have

$$y(t) = \begin{cases} 0 & t < -1 \\ 1 - e^{-(t+1)} & -1 \leq t \leq 1 \\ e^{-(-1+t)} - e^{-(1+t)} & t > 1 \end{cases},$$

which is depicted in Fig. 3.15.

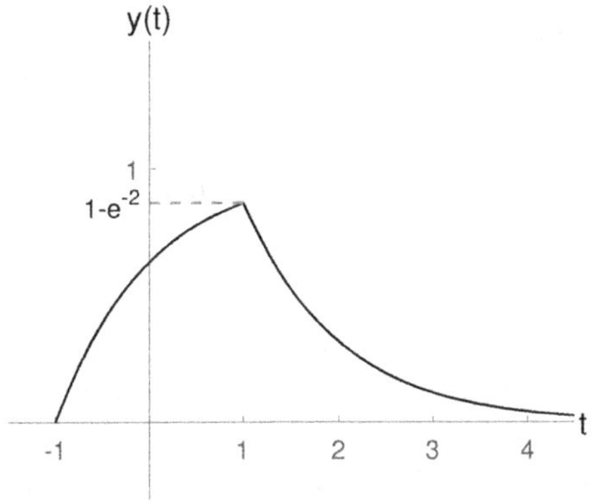

Figure 3.15: Problem 42: Signal $y(t) = x_2(t) * h(t)$.

Problem 43. Compute the convolution of the following pairs of continuous-time signals.

(a) $x(t) = 2\sin(t)u(t), \quad h(t) = 2\cos(t)u(t)$

(b) $x(t) = tu(t), \quad h(t) = \sin(2t)u(t)$

(c) $x(t) = u(t-2), \quad h(t) = e^{-t}$

(d) $x(t) = \cos(t), \quad h(t) = e^{-|t|}$

Solution.

(a) $x(t) = 2\sin(t)u(t), \quad h(t) = 2\cos(t)u(t)$

According to the convolution integral

$$y(t) = x(t) * h(t) = \int_{-\infty}^{+\infty} x(\tau)h(t-\tau)d\tau = \int_{-\infty}^{+\infty} x(t-\tau)h(\tau)d\tau,$$

since the impulse response is nonzero for $t \geq 0$, we write

$$\begin{aligned} y(t) &= \int_{-\infty}^{+\infty} x(t-\tau)h(\tau)d\tau \\ &= 4\int_0^t \sin(t-\tau)\cos(\tau)d\tau \\ &= 2\int_0^t (\sin(t-2\tau) + \sin(t))\, d\tau \\ &= 2\left(\frac{1}{2}\cos(t-2\tau) + \tau\sin(t)\right)\Big|_0^t \\ &= (\cos(-t) - \cos(t)) + 2t\sin(t) = 2t\sin(t). \end{aligned}$$

(b) $x(t) = tu(t), \quad h(t) = \sin(2t)u(t)$

$$\begin{aligned}
y(t) = x(t) * h(t) &= \int_{-\infty}^{+\infty} x(t-\tau)h(\tau)d\tau \\
&= \int_0^t (t-\tau)\sin(2\tau)d\tau \\
&= \frac{1}{2}\left(-t\cos(2\tau) - \frac{1}{2}\sin(2\tau) + \tau\cos(2\tau)\right)\Big|_0^t \\
&= \frac{1}{2}(-t\cos(2t) - \frac{1}{2}\sin(2t) + t\cos(2t) + t) = \frac{1}{2}t - \frac{1}{4}\sin(2t).
\end{aligned}$$

(c) $x(t) = u(t-2), \quad h(t) = e^{-t}u(t)$

Instead of directly obtaining the convolution, we can first we find the response of this system to the step input as discussed in Section 3.5:

$$s(t) = \int_{-\infty}^{t} e^{-\lambda}u(\lambda)d\lambda = \int_0^t e^{-\lambda}d\lambda = (1-e^{-t})u(t). \tag{3.43}$$

Now since $x(t) = u(t-2)$ and the system is LTI, from (3.43), we obtain

$$y(t) = s(t-2) = (1-e^{-(t-2)})u(t-2).$$

(d) $x(t) = \cos(t), \quad h(t) = e^{-|t|}$

We write $h(t-\tau)$ as follows:

$$h(t-\tau) = e^{-|t-\tau|} = \begin{cases} e^{t-\tau} & \tau > t \\ e^{\tau-t} & \tau \le t \end{cases}, \tag{3.44}$$

thus,

$$\begin{aligned}
y(t) = x(t) * h(t) &= \int_{-\infty}^{+\infty} x(\tau)h(t-\tau)d\tau \\
&= \int_{-\infty}^{t} \cos(\tau)e^{\tau-t}d\tau + \int_t^{+\infty} \cos(\tau)e^{t-\tau}d\tau \\
&= e^{-t}\int_{-\infty}^{t} \cos(\tau)e^{\tau}d\tau + e^{t}\int_t^{+\infty} \cos(\tau)e^{-\tau}d\tau \\
&= \frac{1}{2}e^{-t}\Big[e^{\tau}\left(\cos(\tau)+\sin(\tau)\right)\Big]_{-\infty}^{t} + \frac{1}{2}e^{t}\Big[e^{-\tau}\left(-\cos(\tau)+\sin(\tau)\right)\Big]_{t}^{+\infty} \\
&= \cos(t).
\end{aligned}$$

Problem 44. Consider the following linear time-invariant system.

(a) The response of this system to signal $x_1(t)$ in Fig. 3.19 (a) is signal $y_1(t)$ illustrated in Fig. 3.19 (b). Determine and sketch carefully the response of the system to the input $x_2(t)$ depicted in Fig. 3.19 (c).

(b) Find the impulse response $h(t)$ of this system and then work out Part (a) again.

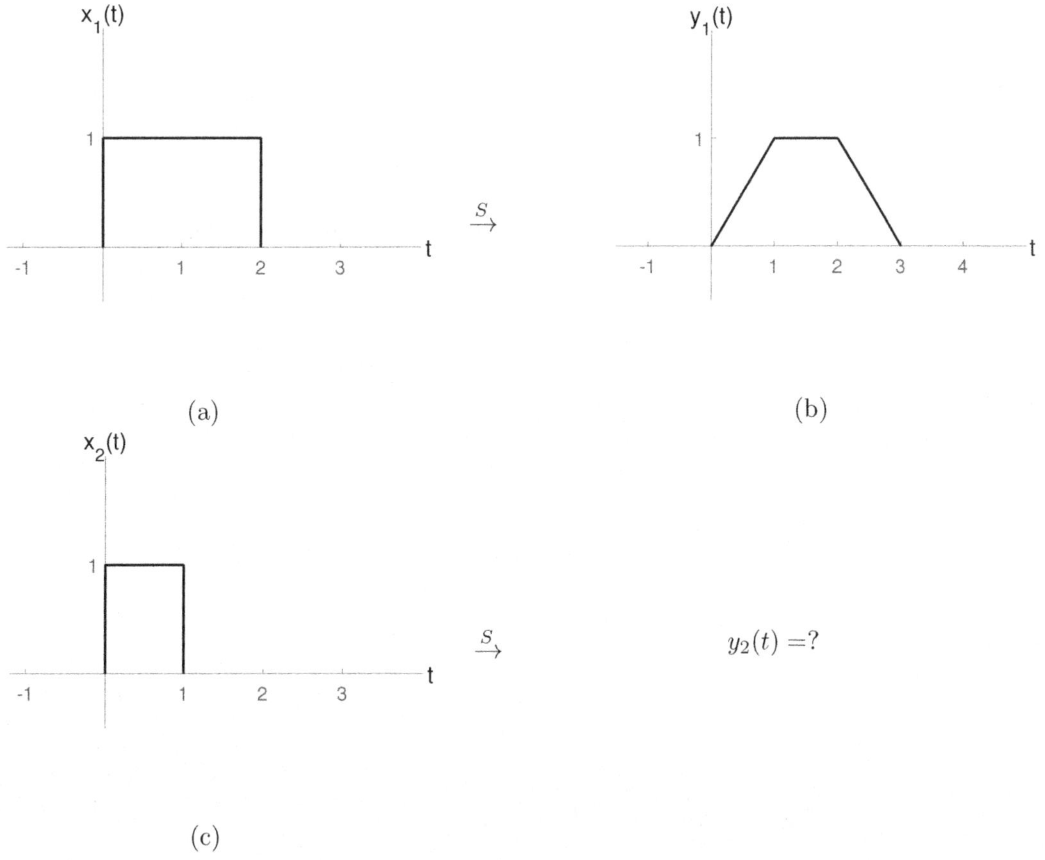

Figure 3.19: Problem 44.

Solution.

(a) To solve this part, we take the following two steps:

Step 1. First we should write $x_2(t)$ based on $x_1(t)$ as follows (using linear and time-invariant operators):

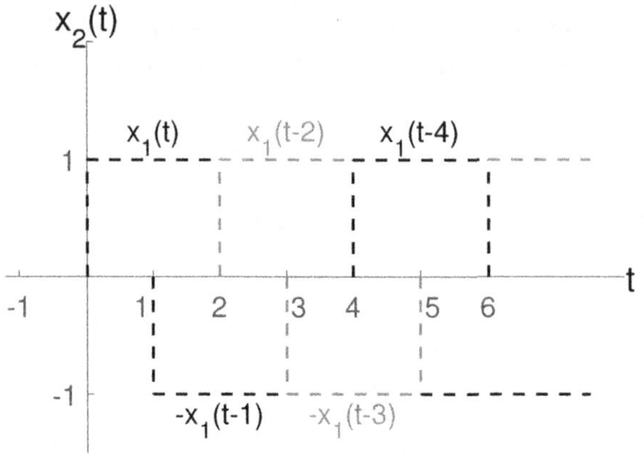

Figure 3.20: Problem 44: Finding $x_2(t)$ based on $x_1(t)$.

$$x_2(t) = x_1(t) - x_1(t-1) + x_1(t-2) - x_1(t-3) + x_1(t-4) - \cdots \tag{3.45}$$

Step 2. Since the system is LTI, a relationship that is established between the inputs, also holds between the outputs. From (3.45), we obtain output $y_2(t)$ as follows (see Fig. 3.23):

$$y_2(t) = y_1(t) - y_1(t-1) + y_1(t-2) - y_1(t-3) + y_1(t-4) - \cdots$$

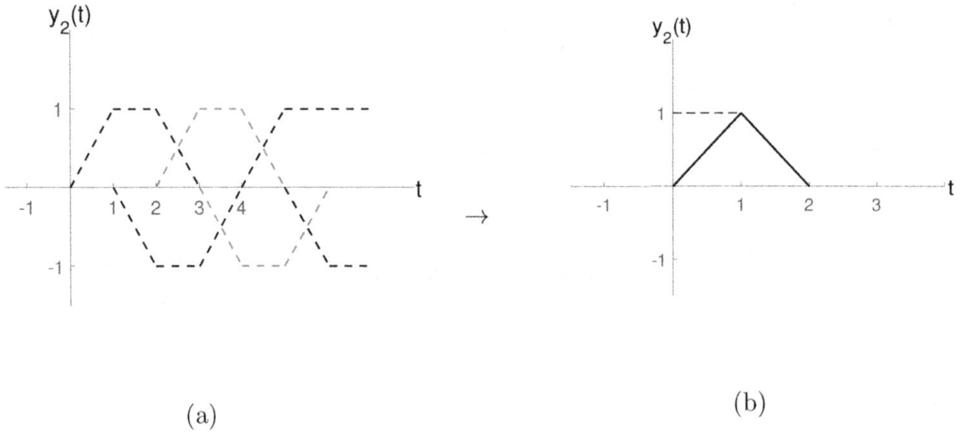

Figure 3.23: Problem 44: (a) Signal $y_2(t)$ based on $y_1(t)$; (b) Output $y_2(t)$.

(b) First we should find the impulse response $h(t)$ of this system using $x_1(t)$ and $y_1(t)$. By plotting the derivative of $y_1(t)$ in Fig. 3.24, we can see that

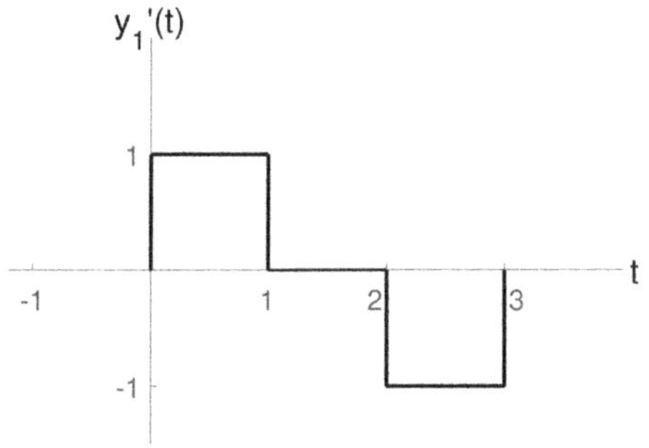

Figure 3.24: Problem 44: The derivative of $y_1(t)$.

$$y_1'(t) = x_1(t) - x_1(t-1). \tag{3.46}$$

Now based on the discussion in part e of Section 3.8, we claim that (3.46) holds for any arbitrary input / output, i.e.,

$$y'(t) = x(t) - x(t-1). \tag{3.47}$$

By substituting $x(t) = \delta(t)$ into (3.47), we obtain the derivative of the impulse response $h(t)$ of this system follows:

$$h'(t) = \delta(t) - \delta(t-1),$$

so

$$\begin{aligned} h(t) &= \int_{-\infty}^{t} h'(\tau)d\tau = \int_{-\infty}^{t} \big(\delta(\tau) - \delta(\tau-1)\big)d\tau \\ &= u(t) - u(t-1). \end{aligned}$$

Using the convolution of signals $x_2(t)$ and $h(t)$ and convolution properties, we evaluate and plot $y_2(t)$ as follows:

$$\begin{aligned} y_2(t) = h(t) * x_2(t) &= h'(t) * \int_{-\infty}^{t} x_2(\tau)d\tau \\ &= \big(\delta(t) - \delta(t-1)\big) * \big(r(t) - r(t-1)\big) \\ &= r(t) - r(t-1) - r(t-1) + r(t-2) \\ &= r(t) - 2r(t-1) + r(t-2), \end{aligned}$$

where $r(t)$ is the unit ramp function. As expected, $y_2(t)$ is the same as Part (a).

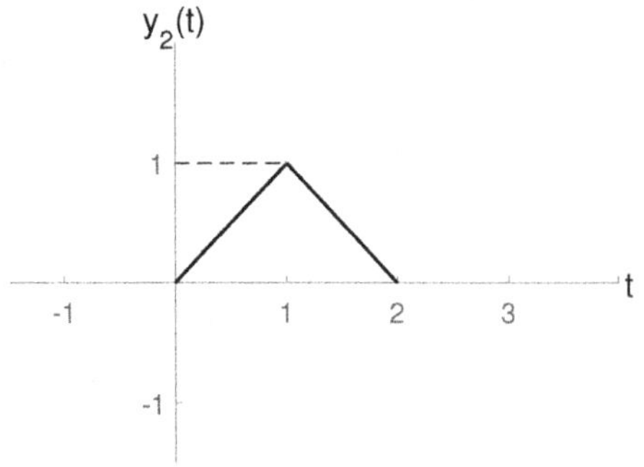

Figure 3.25: Problem 44: The output $y_2(t)$.

Problem 45. Consider the following impulse responses of continuous-time LTI systems. Determine whether each system is (i) causal, (ii) memoryless, (iii) stable.

(a) $h(t) = t^2 e^{-t}\ u(t+2)$

(b) $h(t) = e^t\ u(-t-1)$

(c) $h(t) = e^{jat}\ u(2t+1)\quad (a>0)$

Solution.

(a) $h(t) = t^2 e^{-t} u(t+2)$

(i) It is clear that such a system is not causal, as $h(t) \neq 0$ for $t \geq -2$.

(ii) Since the impulse response $h(t)$ is non-zero for $t \neq 0$, the system is not memoryless.

(iii) From Section 3.6.3, we write:

$$\int_{-\infty}^{+\infty} |h(t)|dt = \int_{-\infty}^{+\infty} |t^2 e^{-t} u(t+2)|dt = \int_{-2}^{+\infty} t^2 e^{-t} dt = \left[e^{-t}(-t^2-2t-2)\right]_{-2}^{+\infty},$$

which is a bounded value as when t gets large, the exponential term gets extremely small compared to polynomial terms. Therefore the system is stable.

(b) $h(t) = e^t u(-t-1)$

(i) $h(t)$ is non-zero for $t < -1$. As such, the system is not causal.

(ii) Since the impulse response $h(t)$ is non-zero for $t \neq 0$, the system is not memoryless.

(iii) From Section 3.6.3, we write:

$$\begin{aligned}\int_{-\infty}^{+\infty} |h(t)|dt &= \int_{-\infty}^{+\infty} |e^t u(-t-1)|dt \\ &= \int_{-\infty}^{-1} |e^t|dt = \int_{-\infty}^{-1} e^t dt < \infty.\end{aligned}$$

Since the impulse response is absolutely integrable, the system is stable.

(c) $h(t) = e^{jat}\, u(2t+1) \;\; (a > 0)$

(i) From the expression $u(2t+1)$, we have

$$u(2t+1) \;\;\Rightarrow\;\; 2t+1 > 0 \;\;\Rightarrow\;\; h(t) \neq 0 \quad \text{for} \quad t > -\frac{1}{2},$$

thus this LTI system is not causal.

(ii) Since the impulse response $h(t)$ is non-zero for $t \neq 0$, the system is not memoryless.

(iii) From Section 3.6.3, we write:

$$\begin{aligned}\int_{-\infty}^{+\infty} |h(t)|dt &= \int_{-\infty}^{+\infty} |e^{jat}\, u(3t+1)|dt \\ &= \int_{-\frac{1}{3}}^{+\infty} |e^{jat}|dt = \int_{-\frac{1}{3}}^{+\infty} (1)dt = \infty.\end{aligned}$$

Therefore the system is not stable.

Problem 46. Consider the following integrals which represent systems with input $x(t)$ and output $y(t)$. Try to write each of the integrals as a convolution integral and then determine whether each system is LTI. Comment on system memory, causality, and stability.

(a) $y(t) = \int_{-\infty}^{t} e^{-(t-\tau)} x(\tau) d\tau$

(b) $y(t) = \int_{t+a}^{t+b} x(\tau) d\tau, \qquad b > a$

(c) $y(t) = \int_{0}^{+\infty} e^{-2\tau} x(\tau - t) d\tau$

(d) $y(t) = \int_{-\infty}^{-t} x(-\tau) d\tau$

Solution.

(a) $y(t) = \int_{-\infty}^{t} e^{-(t-\tau)} x(\tau) d\tau$

First, we rewrite the integral as follows:

$$y(t) = \int_{-\infty}^{t} e^{-(t-\tau)} x(\tau) d\tau = \int_{-\infty}^{+\infty} e^{-(t-\tau)} x(\tau) u(t-\tau) d\tau.$$

Note that by adding the expression $u(t-\tau)$, we changed the upper limit of the integral to $+\infty$. Then we obtain

$$y(t) = \int_{-\infty}^{+\infty} e^{-(t-\tau)} x(\tau) u(t-\tau) d\tau = x(t) * e^{-t} u(t). \tag{3.48}$$

From (3.48), $h(t)$ is given as follows

$$h(t) = e^{-t}u(t). \tag{3.49}$$

From (3.48), since the input-output relationship is in the form $y(t) = x(t) * h(t)$, this system is LTI.

From (3.49) we have $h(t) = 0$ for $t < 0$, thus the system is causal and has memory. To verify the stability condition we write:

$$\int_{-\infty}^{+\infty} |h(t)|dt = \int_{-\infty}^{+\infty} |e^{-t}u(t)|dt = \int_{0}^{+\infty} e^{-t}dt < \infty.$$

Thus the system is stable.

(b) $y(t) = \int_{t+a}^{t+b} x(\tau)d\tau, \qquad b > a$

First, we rewrite the integral as follows

$$\begin{aligned} y(t) &= \int_{t+a}^{t+b} x(\tau)d\tau \\ &= \int_{-\infty}^{t+b} x(\tau)d\tau - \int_{-\infty}^{t+a} x(\tau)d\tau \\ &= \int_{-\infty}^{+\infty} x(\tau)u(t+b-\tau)d\tau - \int_{-\infty}^{+\infty} x(\tau)u(t+a-\tau)d\tau. \end{aligned} \tag{3.50}$$

Using the distributive property of convolution, (3.50) becomes:

$$\begin{aligned} y(t) &= \big[x(t) * u(t+b)\big] - \big[x(t) * u(t+a)\big] \\ &= x(t) * \big[u(t+b) - u(t+a)\big] \\ &= x(t) * h(t), \end{aligned} \tag{3.51}$$

where the impulse response $h(t)$ is given as

$$h(t) = u(t+b) - u(t+a). \tag{3.52}$$

From (3.51), it is clear that this system is LTI. To check the causality, if a and b are negative, $h(t) = 0$ for $t < 0$, and the system is causal. Otherwise, $h(t) \neq 0$ for $t < 0$ and the system is not casual.

For any value of a and b, the system can not be memoryless.

Using (3.52), we have

$$\int_{-\infty}^{+\infty} |h(t)|dt = \int_{a}^{b} (1)dt < \infty.$$

Therefore, the system is stable.

(c) $y(t) = \int_{0}^{+\infty} e^{-2\tau}x(\tau - t)d\tau$

We rewrite the integral as follows:

$$\begin{aligned} y(t) &= \int_{0}^{+\infty} e^{-2\tau} x(\tau - t) d\tau \\ &= \int_{-\infty}^{+\infty} e^{-2\tau} x(\tau - t) u(\tau) d\tau \\ &= e^{-2t} u(t) * x(-t). \end{aligned} \tag{3.53}$$

We are very close to the convolution format except that we should have $x(t)$ in place not $x(-t)$. To deal with this, we can assume that $x(t)$ is first given to a system to generate $x(-t)$ and then it is fed to an LTI system with the impulse response $h(t) = e^{-t}u(t)$ (see Fig. 3.26).

Figure 3.26: The overall system for Problem 46 - Part (c).

Both systems are linear, however, the first system is not time-invariant. As such, the overall system is not LTI. Moreover, the first system is not causal. As such, the overall system is not casual as well. Also, since both parts of this system are stable, the overall system is stable. The system is obviously not memoryless.

(d) $y(t) = \int_{-\infty}^{-t} x(-\tau) d\tau$
By substituting $\lambda = -\tau$ we obtain

$$\begin{aligned} y(t) &= \int_{t}^{+\infty} x(\lambda) d\lambda \\ &= \int_{-\infty}^{+\infty} x(\lambda) u(\lambda - t) d\lambda \\ &= x(t) * u(-t). \end{aligned} \tag{3.54}$$

Then the impulse response $h(t)$ is given as follows:

$$h(t) = u(-t). \tag{3.55}$$

From (3.54), we see that this system is LTI and since the impulse response $h(t) \neq 0$ for $t < 0$, the system is not causal and has memory.

From (3.55), we have

$$\int_{-\infty}^{+\infty} |h(t)| dt = \int_{-\infty}^{+\infty} |u(-t)| dt = \infty,$$

which shows the system is not stable.

Problem 47. Consider the system initially at rest condition, shown in Fig. 3.27.

(a) Find the input-output relationship for this system.

(b) Find the impulse response $h(t)$ of this system.

(c) Is this system memoryless?

(d) Is this system causal?

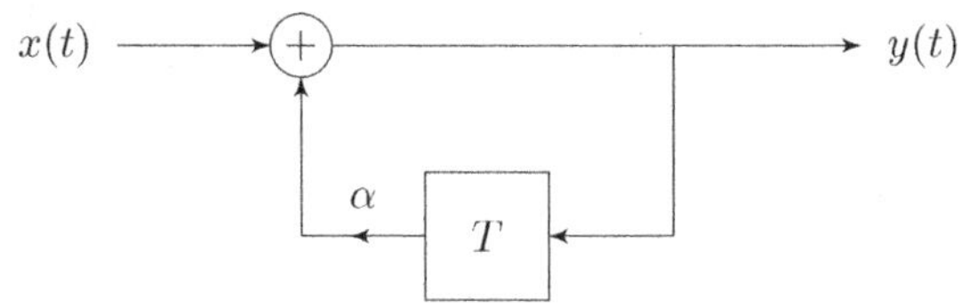

Figure 3.27: Problem 47.

Solution.

(a)

$$x(t) + \alpha y(t - T) = y(t). \tag{3.56}$$

(b) Since the system satisfies the condition of initial rest, then $y(t) = 0$ for $t < 0$. We apply the input

$$x(t) = \delta(t).$$

Starting from this initial condition and this input, we can solve for successive values of $y(t)$ for $t \geq 0$ as follows:

$$\begin{aligned}
t = 0 \quad &\Rightarrow \quad \delta(0) + \alpha y(-T) = y(0) \quad \Rightarrow \quad y(0) = \delta(0) \\
t = T \quad &\Rightarrow \quad \delta(T) + \alpha y(0) = y(T) \quad \Rightarrow \quad y(T) = \alpha y(0) = \alpha\delta(0) \\
t = 2T \quad &\Rightarrow \quad \delta(2T) + \alpha y(T) = y(2T) \quad \Rightarrow \quad y(2T) = \alpha y(T) = \alpha^2\delta(0) \\
&\;\;\vdots \qquad\qquad\qquad\qquad\qquad\qquad \vdots \\
t = nT \quad &\Rightarrow \quad \delta(nT) + \alpha y((n-1)T) = y(nT) \quad \Rightarrow \quad y(nT) = \alpha^n y(0) = \alpha^n\delta(0).
\end{aligned}$$

We see that the impulse response for this system is

$$h(t) = \alpha^n \delta(t - nT) \quad n = 0, 1, 2, \cdots \tag{3.57}$$

which is depicted in Fig. 3.28.

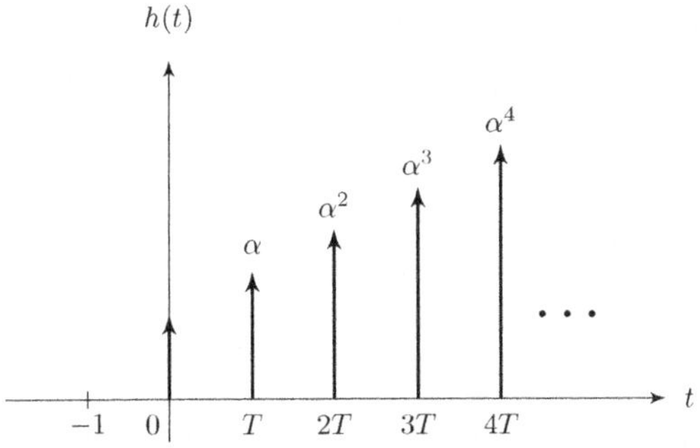

Figure 3.28: Problem 47, The impulse response $h(t)$ of the system.

(c) This system is not memoryless.

(d) From Fig. 3.28 we see that $h(t) = 0$ for $t < 0$, thus this system is causal.

Problem 48. For the following continuous-time LTI systems:
Plot the block diagram for the system described by the following differential equation:

$$\frac{dy}{dt} + ay(t) = \frac{dx}{dt} + bx(t). \tag{3.58}$$

Solution.
Let's divide both sides of (3.58) by a and rearrange the terms as

$$y(t) = \frac{1}{a}\left[\frac{dx(t)}{dt} + bx(t) - \frac{dy(t)}{dt}\right]. \tag{3.59}$$

From (3.59), we can assume that the differential equation is the combination of the following two equations:

(i) $y_1(t) = \frac{dx(t)}{dt} + bx(t)$,

(ii) $y_2(t) = -\frac{dy(t)}{dt}$.

Based on this assumption, the block diagram is obtained as Fig. 3.29.

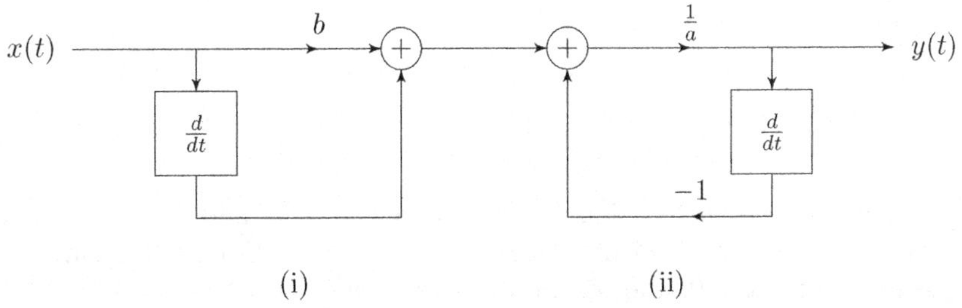

Figure 3.29: Problem 48, Block diagram of the system.

The obtained block digram can be compressed into the one in Fig. 3.30.

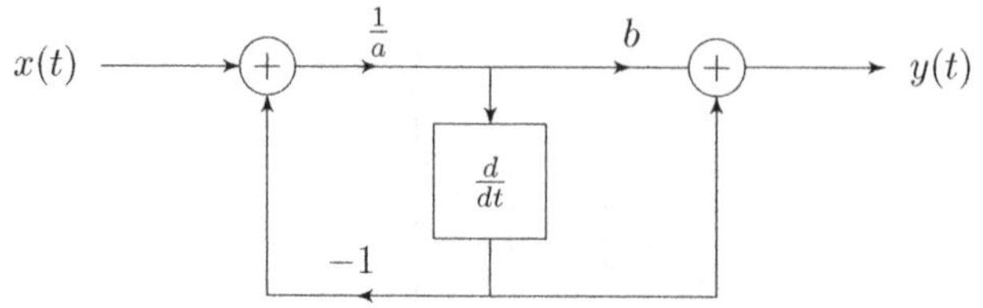

Figure 3.30: Problem 48, Modified bock diagram of the system.

3.9.2 Discrete-time LTI systems

Problem 49. Compute the convolution of the following pairs of discrete-time signals.

(a) $x[n] = (\frac{1}{6})^{n-5}u[n], \quad h[n] = (\frac{1}{5})^n u[n-2]$

(b) $x[n] = n(\frac{1}{2})^n \sin(n\pi - \frac{\pi}{2})u[n], \quad h[n] = u[n]$

Solution.

Hint:

$$\sum_{k=0}^{m} kr^k = \frac{r-(m+1)r^{(m+1)}+mr^{(m+2)}}{(1-r)^2}. \tag{3.60}$$

$$\sum_{k=0}^{m} r^k = \frac{1-r^{m+1}}{1-r}. \tag{3.61}$$

(a) $x[n] = (\frac{1}{6})^{n-5}u[n], \quad h[n] = (\frac{1}{5})^n u[n-2]$

In Fig. 3.31 we have depicted signal $x[k]$ followed by $h[k]$ and $h[n-k]$ for an arbitrary positive value of n and an arbitrary negative value of n. From this figure, we note that signal $x[k]$ is zero for $k < 0$, and signal $h[n-k]$ is zero for $k > n-2$. Therefore, $x[k]$ and $h[n-k]$ do not overlap for $n < 2$ and $x[k]h[n-k] = 0$ for all values of k. For $n \geq 2$ we have

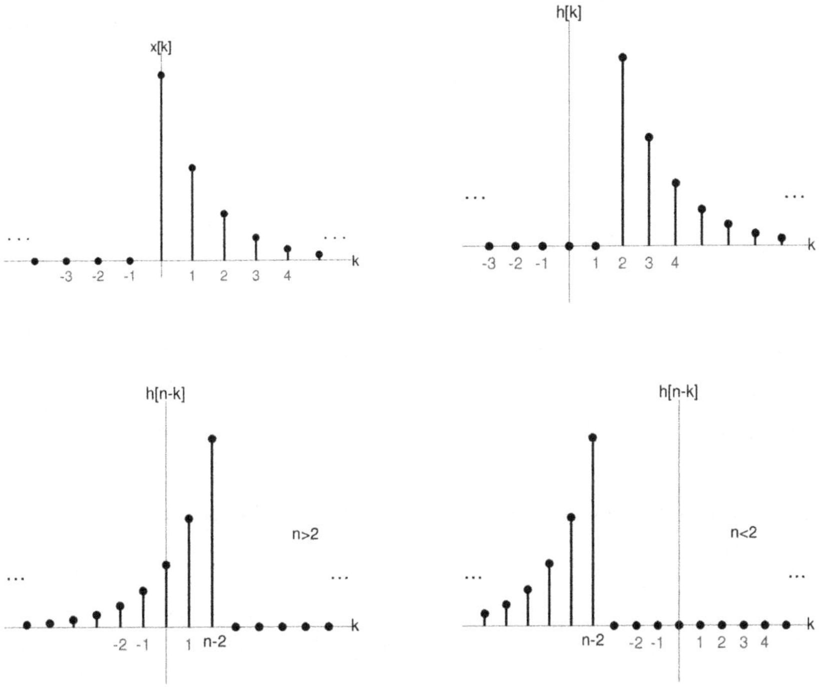

Figure 3.31: Problem 49 - Part (a), Graphical representation of $x[n]$ and $h[n-k]$.

$$
\begin{aligned}
y[n] = x[n] * h[n] &= \sum_{k=-\infty}^{+\infty} x[k]h[n-k] \\
&= \sum_{k=0}^{n-2} (\frac{1}{5})^{n-k} (\frac{1}{6})^{k-5} \\
&= 6^5 (\frac{1}{5})^n \sum_{k=0}^{n-2} (\frac{1}{5})^{-k} (\frac{1}{6})^k \\
&= 6^5 (\frac{1}{5})^n \sum_{k=0}^{n-2} (\frac{5}{6})^k.
\end{aligned}
$$

From (3.61) we have

$$
y[n] = 6^5 (\frac{1}{5})^n \frac{1-(\frac{5}{6})^{n-1}}{1-\frac{5}{6}} \qquad \text{for } n \geq 2. \tag{3.62}
$$

Finally, for all values of n,

$$y[n] = 6^5(\frac{1}{5})^n \left(\frac{1-(\frac{5}{6})^{n-1}}{1-\frac{5}{6}}\right) u[n-2].$$

(b) $x[n] = n(\frac{1}{2})^n \sin(n\pi - \frac{\pi}{2})u[n], \quad h[n] = u[n]$

We know that

$$\sin(n\pi - \pi/2) = -\cos(n\pi) = -(-1)^n.$$

So

$$x[n] = -n(\frac{1}{2})^n(-1)^n u[n] = -n(-\frac{1}{2})^n u[n].$$

Since the impulse response is nonzero for $n \geq 0$, then the output is nonzero only for $n \geq 0$ and we obtain

$$y[n] = x[n] * h[n] = \sum_{k=-\infty}^{+\infty} x[k]h[n-k]$$
$$= \sum_{k=0}^{n} x[k] = \sum_{k=0}^{n} -k(-\frac{1}{2})^k.$$

Problem 50. A system is formed by connecting 3 systems as shown in Fig. 3.32. The impulse responses of the systems are given by

$$h_1[n] = \delta[n] + 3\delta[n-1], \quad h_2[n] = u[n] \quad \text{and} \quad h_3[n] = \delta[n-3].$$

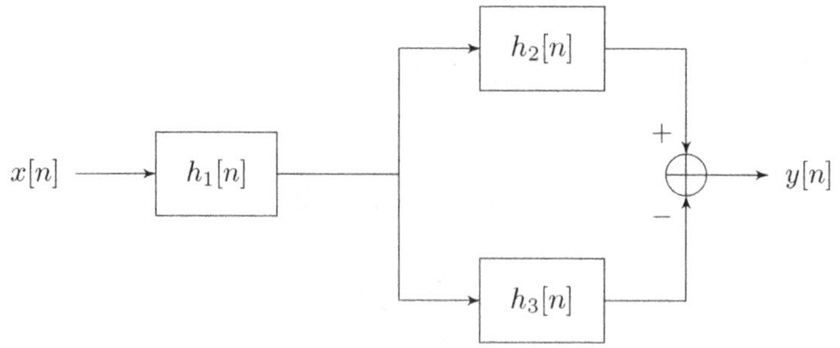

Figure 3.32: Problem 50.

Find the impulse response $h[n]$ of the overall system.

Solution.

First, we have simplified the overall system as in Fig. 3.33):

Figure 3.33: Problem 50, The simplified system diagram.

We know that the impulse response of the cascade of two LTI systems is the convolution of their individual impulse responses. So we have

$$\begin{aligned} h[n] &= h_1[n] * (h_2[n] - h_3[n]) \\ &= (\delta[n] + 3\delta[n-1]) * (u[n] - \delta[n-3]) \\ &= (\delta[n] + 3\delta[n-1]) * u[n] - (\delta[n] + 3\delta[n-1]) * \delta[n-3] \\ &= u[n] + 3u[n-1] - \delta[n-3] - 3\delta[n-4]. \end{aligned}$$

Problem 51. Consider the cascade of the following two discrete-time LTI systems S_1 and S_2, as depicted in Fig. 3.34. Let the step response of both systems be $s[n] = 3\delta[n-1] + \delta[n-2]$.

Figure 3.34: Problem 51.

(a) Find the impulse response of the overall system.

(b) Find the step response of the overall system.

Solution.

(a) We know that the first difference of the unit step function is the unit impulse function. As such, the first difference of the step response is in fact the impulse response. We write

$$\begin{aligned} h[n] &= s[n] - s[n-1] \\ &= 3\delta[n-1] + \delta[n-2] - (3\delta[n-2] + \delta[n-3]) \\ &= 3\delta[n-1] - 2\delta[n-2] - \delta[n-3]. \end{aligned} \tag{3.63}$$

We know that if two LTI system are cascaded, the impulse response of the overall system is the convolution of their individual impulse responses. From (3.63) we can obtain the impulse response of the overall system as follows:

$$\begin{aligned} h_T[n] &= h[n] * h[n] \\ &= \big(s[n] - s[n-1]\big) * \big(s[n] - s[n-1]\big) \\ &= \big(3\delta[n-1] - 2\delta[n-2] - \delta[n-3]\big) * \big(3\delta[n-1] - 2\delta[n-2] - \delta[n-3]\big) \\ &= 9\delta[n-2] - 12\delta[n-3] - 2\delta[n-4] + 4\delta[n-5] + \delta[n-6]. \end{aligned}$$

(b) To find the step response of the overall system, we write:

$$\begin{aligned} s_T[n] &= h_T[n] * u[n] \\ &= \big(s[n] - s[n-1]\big) * \big(s[n] - s[n-1]\big) * u[n], \end{aligned}$$

where

$$s[n] - s[n-1] = s[n] * \big(\delta[n] - \delta[n-1]\big).$$

Thus,

$$\begin{aligned} s_T[n] &= \big(s[n] - s[n-1]\big) * s[n] * \big(\delta[n] - \delta[n-1]\big) * u[n] \\ &= \big(s[n] - s[n-1]\big) * s[n] * \big(u[n] - u[n-1]\big) \\ &= \big(s[n] - s[n-1]\big) * s[n] * \delta[n] \\ &= \big(s[n] - s[n-1]\big) * s[n] \\ &= \big(3\delta[n-1] - 2\delta[n-2] - \delta[n-3]\big) * \big(3\delta[n-1] + \delta[n-2]\big) \\ &= 9\delta[n-2] - 3\delta[n-3] - 5\delta[n-4] - \delta[n-5]. \end{aligned}$$

Problem 52. Consider the following impulse responses of discrete-time LTI systems. Determine whether each system is (i) causal, (ii) memoryless and/or (iii) stable.

(a) $h[n] = (\frac{1}{3})^n u[n]$

(b) $h[n] = (1.001)^n u[n-3] - (\frac{1}{4})^n u[n]$

(c) $h[n] = n(\frac{1}{3})^n u[n-1]$

Solution.

(a) $h[n] = (\frac{1}{3})^n u[n]$

(i) Since $h[n] = 0$ for $n < 0$, the system is causal.

(ii) Since a discrete-time LTI system has an impulse response $h[n]$ that is non-zero for $n \neq 0$, the system has memory.

(iii) From Section (3.6.3), we have

$$\sum_{n=-\infty}^{+\infty} |h[n]| = \sum_{n=-\infty}^{+\infty} |(\frac{1}{3})^n u[n]| = \sum_{n=0}^{+\infty} |(\frac{1}{3})^n| = \sum_{n=0}^{+\infty} (\frac{1}{3})^n = \frac{3}{2} < \infty.$$

Therefore the system is stable.

(b) $h[n] = (1.001)^n u[n-3] - (\frac{1}{4})^n u[n]$

(i) Since $h[n] = 0$ for $n < 0$, the system is causal.

(ii) Since $h[n] \neq 0$ for $n \neq 0$, the system has memory.

(iii) To verify the stability, from Section (3.6.3) we should investigate if the impulse response is absolutely summable:

$$h[+\infty] = (1.001)^{+\infty} u[+\infty - 3] - (\frac{1}{4})^{+\infty} u[+\infty] = +\infty - 0 = +\infty,$$

and

$$\sum_{n=-\infty}^{+\infty} |h[n]| > h[+\infty] = +\infty.$$

Since the impulse response is not absolutely summable, the system is not stable.

(c) $h[n] = n(\frac{1}{3})^n u[n-1]$

(i) Since $h[n] = 0$ for $n < 1$, the system is causal.

(ii) Since $h[n] \neq 0$ for $n \neq 1$, the system has memory.

(iii) To verify the stability condition, we have

$$\sum_{n=-\infty}^{+\infty} |h[n]| = \sum_{n=-\infty}^{+\infty} |n(\frac{1}{3})^n u[n-1]| = \sum_{n=1}^{+\infty} n(\frac{1}{3})^n < \infty.$$

Therefore the system is stable. The validity of the last inequality can be justified by remembering that as n tends to infinity, the exponential term becomes smaller at a much faster rate than the growth rate of the polynomial term.
This can also be rigorously proved using (3.60):

$$\sum_{n=1}^{m} n(1/3)^n = \frac{1/3 - (n+1)(1/3)^{(n+1)} + n(1/3)^{(n+2)}}{(2/3)^2}.$$

Now as m gets large, the 2nd and 3rd terms tend to zero and the sum tends to $3/4$ which is a finite number.

Problem 53. Consider a discrete-time LTI system S with the following impulse response $h[n]$:

$$h[n] = (\frac{1}{4})^n u[n], \tag{3.64}$$

Find the impulse response of the inverse system.
Hint: The system corresponds to the following difference equation:

$$h[n] - \frac{1}{4} h[n-1] = \delta[n]. \tag{3.65}$$

Solution.
As we know, the impulse response $h[n]$ and the impulse response of the inverse system $h_r[n]$ should satisfy the following equation:

$$h[n] * h_r[n] = \delta[n]. \tag{3.66}$$

Using the impulse signal property as well as the distributive property of convolution, we can write (3.65) as

$$\begin{aligned} h[n] - \frac{1}{4}h[n-1] &= h[n] * \delta[n] - \frac{1}{4}h[n] * \delta[n-1] \\ &= h[n] * \left(\delta[n] - \frac{1}{4}\delta[n-1]\right) = \delta[n]. \end{aligned}$$

Comparing the above equation and (3.66), the impulse response of the inverse system $h_r[n]$ is given by

$$h_r[n] = \delta[n] - \frac{1}{4}\delta[n-1].$$

Problem 54. Consider a discrete-time system shown in Fig. 3.35.

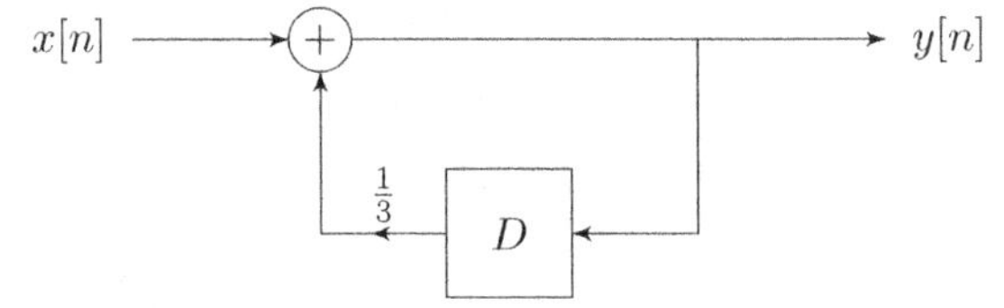

Figure 3.35: Problem 54, Block diagram of the system.

(a) Find the input-output relationship for this system.

(b) Find $y[n]$ with the initial condition $y[-1] = y_0$ and $x[n] = K\delta[n]$.

(c) For what conditions this system is LTI and causal.

Solution.

(a) The output of the unit delay element is $y[n-1]$. So, from Fig. 3.35 we obtain

$$y[n] = x[n] + \frac{1}{3}y[n-1]. \tag{3.67}$$

(b) To find the response of the discrete-time systems with feedback, we must solve the difference equations recursively:

$$\begin{aligned} n = -1 \;&\Rightarrow\; y[-1] = y_0 \\ n = 0 \;&\Rightarrow\; y[0] = K\delta[0] + \frac{1}{3}y_0 = K + \frac{1}{3}y_0 \\ n = 1 \;&\Rightarrow\; y[1] = K\delta[1] + \frac{1}{3}y[0] = \frac{1}{3}y[0] = \frac{1}{3}(K + \frac{1}{3}y_0) \\ n = 2 \;&\Rightarrow\; y[2] = K\delta[2] + \frac{1}{3}y[1] = \frac{1}{3}y[1] = \frac{1}{9}(K + \frac{1}{3}y_0) \\ &\;\;\vdots \end{aligned}$$

$$\Rightarrow \quad y[n] = (\frac{1}{3})^n (K + \frac{1}{3} y_0) u[n] \tag{3.68}$$

(c) If the initial condition is zero (i.e., $y[-1] = y_0 = 0$ or $x[n] = 0$ for $n < n_0$, then $y[n] = 0$ for $n < n_0$), then this system is LTI and causal and we have

$$y[n] = K(\frac{1}{3})^n u[n].$$

If the initial condition is not zero (i.e., $y[-1] = y_0 \neq 0$), the system is not linear. The reason is that based on (3.68), if we scale the input by α (which means K will be replaced by αK), the output will also be scaled by α only if $y[-1] = y_0 = 0$.

Problem 55. Consider the following discrete-time LTI system. Plot the block diagram for this system.

$$y[n] - \frac{1}{3} y[n-1] = x[n] + x[n-1] \tag{3.69}$$

Solution.
Let's rearrange (3.69) as

$$y[n] = x[n] + x[n-1] + \frac{1}{3} y[n-1]. \tag{3.70}$$

From (3.70), we see that the difference equation can be decomposed into the following 2 equations:

(i) $y_1[n] = x[n] + x[n-1]$,

(ii) $y_2[n] = \frac{1}{3} y[n-1]$.

The block diagram of the system is shown in Fig. 3.36.

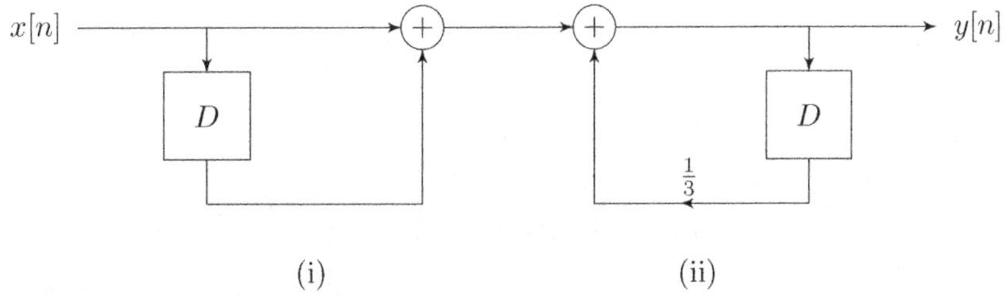

Figure 3.36: Problem 55 .

Fig. 3.36 can be further compressed into the block digram in Fig. 3.37.

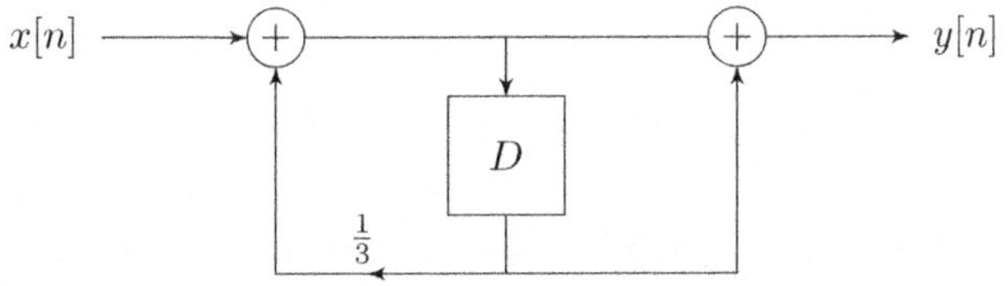

Figure 3.37: Problem 55, the modified block diagram.

Problem 56. Write the difference equation for the block diagram shown in Fig. 3.38.

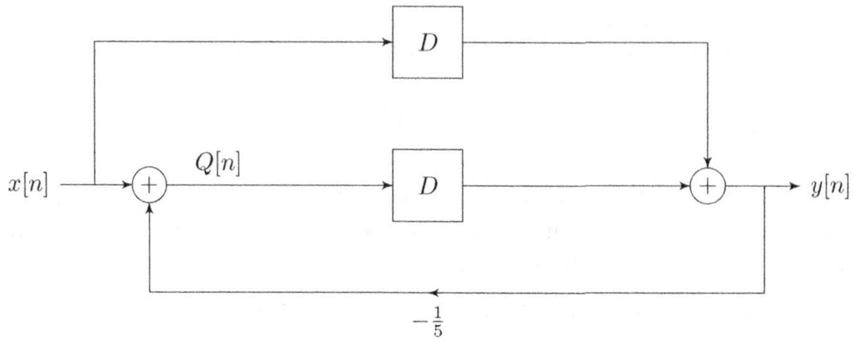

Figure 3.38: Problem 56, Block diagram of the system.

Solution. Let the output of the first summation be $Q[n]$. Then from Fig. 3.38 we see that

$$Q[n] = x[n] - \frac{1}{5}y[n], \tag{3.71}$$

$$y[n] = Q[n-1] + x[n-1]. \tag{3.72}$$

Changing n to $(n-1)$ in (3.71), we have

$$Q[n-1] = x[n-1] - \frac{1}{5}y[n-1]. \tag{3.73}$$

Substituting (3.73) into (3.72), we have

$$y[n] = x[n-1] - \frac{1}{5}y[n-1] + x[n-1].$$

Rearranging terms we obtain

$$y[n] + \frac{1}{5}y[n-1] = 2x[n-1], \tag{3.74}$$

which is the requested difference equation.

Problem 57. Consider the following discrete-time LTI systems.

(a) The step response $s_1[n]$ of the first system is given by

$$s_1[n] = \frac{3^n}{n+2}\cos(\frac{n\pi}{2})u[n]. \tag{3.75}$$

Determine the impulse response $h_1[n]$ of the system and the output of this system to the input $x_1[n]$ given by

$$x_1[n] = \delta[n] - \delta[n-1]. \tag{3.76}$$

(b) The step response $s_2[n]$ of the second system is given by

$$s_2[n] = (1 - (\frac{1}{2})^{n+1})u[n]. \tag{3.77}$$

Determine the impulse response $h_2[n]$ of the system and the output of this system to the input $x_2[n]$ given by

$$x_2[n] = (-1)^n u[n]. \tag{3.78}$$

Solution.

(a) As we know the impulse response of a discrete-time LTI system is the first difference of its step response, i.e.,

$$h[n] = s[n] - s[n-1]. \tag{3.79}$$

From (3.75) and (3.79) we obtain

$$h_1[n] = \frac{3^n}{n+2}\cos(\frac{n\pi}{2})u[n] - \frac{3^{n-1}}{n+1}\cos(\frac{(n-1)\pi}{2})u[n-1].$$

Thus $h_1[0] = \frac{1}{2}$ and for $n \geq 0$, we have $\left(\cos(\frac{(n-1)\pi}{2}) = \sin\frac{n\pi}{2}\right)$

$$h_1[n] = \Big(\frac{3^n}{n+2}\cos(\frac{n\pi}{2}) - \frac{3^{n-1}}{n+1}\sin\frac{n\pi}{2}\Big)u[n-1]. \tag{3.80}$$

Using the properties of LTI systems, we can determine the output $y_1[n]$ of the system as

$$\begin{aligned}
y_1[n] &= h_1[n] - h_1[n-1] \\
&= \Big(\frac{3^n}{n+2}\cos(\frac{n\pi}{2}) - \frac{3^{n-1}}{n+1}\sin\frac{n\pi}{2}\Big)u[n-1] \\
&\quad - \Big(\frac{3^{n-1}}{n+1}\cos(\frac{(n-1)\pi}{2}) - \frac{3^{n-2}}{n}\sin\frac{(n-1)\pi}{2}\Big)u[n-2].
\end{aligned}$$

(b) From (3.77) and (3.79), the impulse response $h_2[n]$ of the system is given by

$$\begin{aligned}
h_2[n] &= s_2[n] - s_2[n-1] \\
&= (1 - (\frac{1}{2})^{n+1})u[n] - (1 - (\frac{1}{2})^n)u[n-1].
\end{aligned}$$

For $n = 0$ we have

$$h_2[0] = (1 - \frac{1}{2}) = \frac{1}{2},$$

and for $n \geq 1$

$$\begin{aligned}
h_2[n] &= (1 - (\frac{1}{2})^{n+1} - 1 + (\frac{1}{2})^n)u[n-1] \\
&= (\frac{1}{2})^{n+1}.
\end{aligned}$$

Thus,

$$h_2[n] = (\frac{1}{2})^{n+1} u[n]. \tag{3.81}$$

Using the convolution of $x_2[n]$ and $h_2[n]$, we can determine the output $y_2[n]$ of the system as

$$\begin{aligned} y_2[n] &= h_2[n] * x_2[n] \\ &= \sum_{k=-\infty}^{+\infty} x_2[n-k] h_2[k] \\ &= \sum_{k=0}^{n} (-1)^{n-k} (\frac{1}{2})^{k+1} \\ &= \frac{1}{2}(-1)^n \sum_{k=0}^{n} (-\frac{1}{2})^k \\ &= \frac{1}{2}(-1)^n \frac{1-(-\frac{1}{2})^{n+1}}{1+(\frac{1}{2})}. \end{aligned}$$

Problem 58. The input $x[n]$ and the output $y[n]$ of a discrete-time LTI system are shown in Fig. 3.39 and Fig. 3.40, respectively.

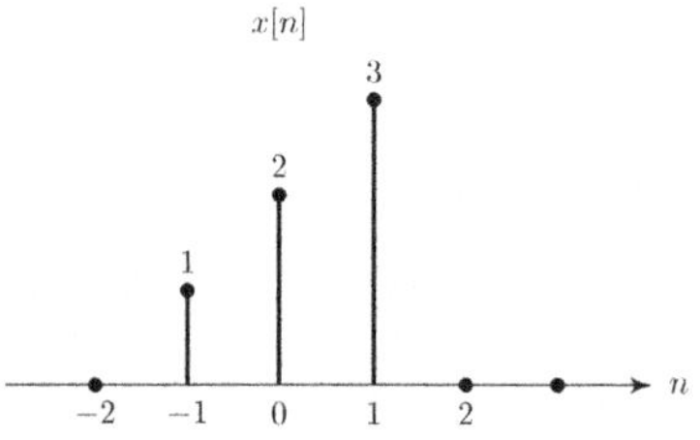

Figure 3.39: Problem 58, The input of the system.

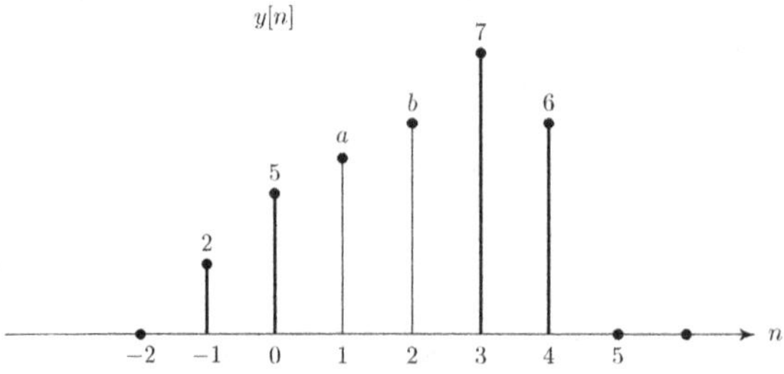

Figure 3.40: Problem 58, the output of the system.

(a) Find the impulse response for this system using the assumption that the length of the impulse response is 4.

(b) Find the values of the output $y[n]$ at $n = 1, 2$ (i.e. $y[1] = a$ and $y[2] = b$) using the impulse response $h[n]$ obtained in Part (a).

Solution.

(a) Since both input and output signals start from time $n = -1$, then the impulse response $h[n]$ of the system should start at time $n = 0$ (see Fig. 3.41).

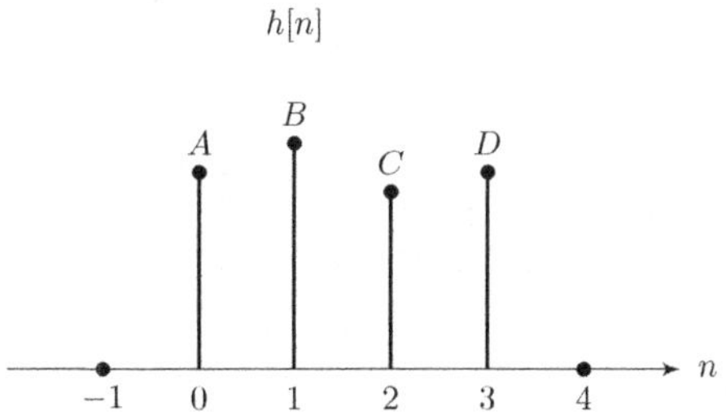

Figure 3.41: Problem 58, The impulse response $h[n]$ of the system.

From Fig. 3.39, we have

$$\begin{aligned} y[n] &= x[n] * h[n] \\ &= \Big(\delta[n+1] + 2\delta[n] + 3\delta[n-1]\Big) * h[n] \\ &= h[n+1] + 2h[n] + 3h[n-1]. \end{aligned} \tag{3.82}$$

From (3.82), Fig. 3.40 and Fig. 3.41, we obtain

$$\begin{aligned} n = -1 \ &\Rightarrow \ h[0] + 2h[-1] + 3h[-2] = 2 \ &\Rightarrow \ A = 2 \\ n = 0 \ &\Rightarrow \ B + 2A + (3 \times 0) = 5 \ &\Rightarrow \ B = 1 \\ n = 4 \ &\Rightarrow \ 0 + 0 + 3D = 6 \ &\Rightarrow \ D = 2 \\ n = 3 \ &\Rightarrow \ 0 + 2D + 3C = 7 \ &\Rightarrow \ C = 1. \end{aligned}$$

The impulse response $h[n]$ of the system is plotted in Fig. 3.42.

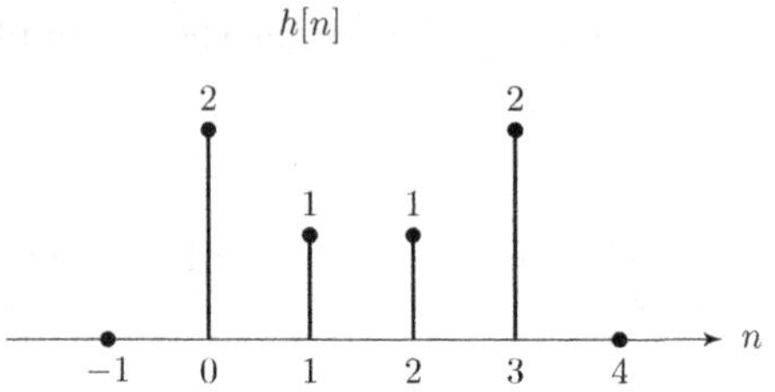

Figure 3.42: Problem 58, The impulse response $h[n]$ of the system.

(b) To obtain the values of $y[1]$ and $y[2]$ from Fig. 3.42 and (3.82), we write

$$a = y[1] = h[2] + 2h[1] + 3h[0] = 9,$$

$$b = y[2] = h[3] + 2h[2] + 3h[1] = 7.$$

Problem 59. The system shown in Fig. 3.43 is formed by cascading two causal LTI systems. The input-output relationships for S_1 and S_2 are given by

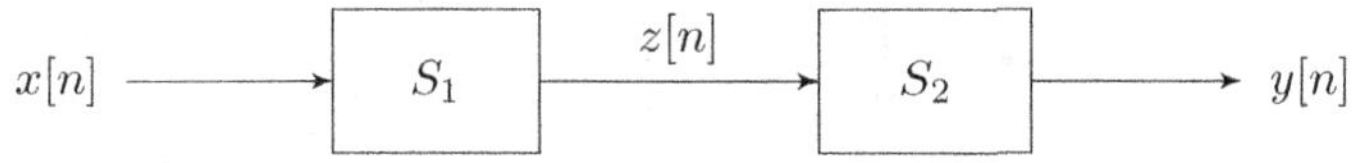

Figure 3.43: Problem 59, The cascade interconnection of two causal LTI systems.

$$S_1 : z[n] = \frac{1}{3}z[n-1] + x[n]$$
$$S_2 : y[n] = \alpha y[n-1] + \beta z[n]$$

and the input-output relationship of the overall system is

$$y[n] = -\frac{1}{6}y[n-2] + \frac{5}{6}y[n-1] + x[n]. \tag{3.83}$$

Find the values of α and β.

Solution.

From the first and second system relationships we have

$$z[n] = \frac{1}{3}z[n-1] + x[n], \tag{3.84}$$
$$y[n] = \alpha y[n-1] + \beta z[n]. \tag{3.85}$$

Dividing both sides of (3.85) by β and rearranging the terms we obtain

$$z[n] = \frac{1}{\beta}y[n] - \frac{\alpha}{\beta}y[n-1]. \tag{3.86}$$

Changing n to $(n-1)$ in (3.86), we have

$$z[n-1] = \frac{1}{\beta}y[n-1] - \frac{\alpha}{\beta}y[n-2]. \tag{3.87}$$

Substituting (3.86) and (3.87) into (3.84) we obtain

$$\frac{1}{\beta}y[n] - \frac{\alpha}{\beta}y[n-1] = \frac{1}{3\beta}y[n-1] - \frac{\alpha}{3\beta}y[n-2] + x[n].$$

Rearranging the terms, we get

$$y[n] = -\frac{\alpha}{3}y[n-2] + (\frac{1}{3} + \alpha)y[n-1] + \beta x[n]. \tag{3.88}$$

Comparing (3.88) and (3.83) reveals that

$$-\frac{\alpha}{3} = -\frac{1}{6} \quad \Rightarrow \quad \alpha = \frac{1}{2}, \qquad \beta = 1.$$

Problem 60. Consider a discrete-time LTI system where the input and output are related through the equation

$$y[n] = x[n] + \frac{1}{2}x[n-1]. \tag{3.89}$$

(a) Find the impulse response $h[n]$ of the system.

(b) Is the system causal?

(c) Find the output of this system to the input $x[n] = u[n+2]$.

(d) Consider the system below depicted in Fig. 3.44 in which the above system is a part of. Find the impulse response $h_{eq}[n]$ of this system assuming that

$$h_1[n] = h[n], \quad h_2[n] = \delta[n+1], \quad h_3[n] = -h[n-3].$$

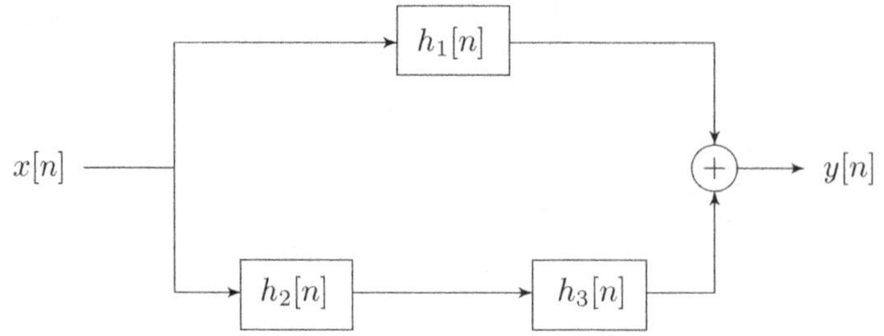

Figure 3.44: The LTI system for Problem 60 - Part (d).

(e) Consider the system of Part (d). Obtain the output of this system for the input $x[n] = u[n]$.

Solution.

(a) If we set $x[n] = \delta[n]$ in (3.89), then the impulse response $h[n]$ is obtained as follows:

$$h[n] = \delta[n] + \frac{1}{2}\delta[n-1]. \tag{3.90}$$

(b) Since $h[n] = 0$ for $n < 0$, the system is causal.

(c) Since the system is LTI and by the definition of the impulse response, we obtain

$$\begin{aligned} y[n] &= x[n] * h[n] \\ &= x[n] + \frac{1}{2}x[n-1] \\ &= u[n+2] + \frac{1}{2}u[n+1]. \end{aligned}$$

(d) Using the associative and the distributive properties of convolution and from (3.90), we have

$$\begin{aligned} h_{eq}[n] &= h_1[n] + (h_2[n] * h_3[n]) \\ &= h[n] - (\delta[n+1] * h[n-3]) \\ &= \delta[n] + \frac{1}{2}\delta[n-1] - \left(\delta[n-2] + \frac{1}{2}\delta[n-3]\right) \\ &= \delta[n] + \frac{1}{2}\delta[n-1] - \delta[n-2] - \frac{1}{2}\delta[n-3]. \end{aligned} \tag{3.91}$$

(e) From (3.91) and setting $x[n] = u[n]$, we see that the output $y[n]$ is given by

$$
\begin{aligned}
y[n] &= x[n] * h_{eq}[n] \\
&= x[n] * \left(\delta[n] + \frac{1}{2}\delta[n-1] - \delta[n-2] - \frac{1}{2}\delta[n-3]\right) \\
&= x[n] + \frac{1}{2}x[n-1] - x[n-2] - \frac{1}{2}x[n-3] \\
&= u[n] + \frac{1}{2}u[n-1] - u[n-2] - \frac{1}{2}u[n-3].
\end{aligned}
$$

Problem 61. Consider a discrete-time LTI system whose input $x[n]$ and output $y[n]$ satisfy the second-order difference equation

$$y[n] = \frac{5}{6}y[n-1] - \frac{1}{6}y[n-2] + x[n] - x[n-1]. \tag{3.92}$$

The system also satisfies the condition of initial rest.

(a) Find the impulse response $h[n]$ of the system.

(b) Determine the step response of the system.

Solution.

(a) The **Characteristic** equation of this system can be written as

$$r^2 - \frac{5}{6}r + \frac{1}{6} = (r - \frac{1}{2})(r - \frac{1}{3}) = 0 \quad \Rightarrow \quad \begin{cases} r_1 = \frac{1}{2} \\ r_2 = \frac{1}{3} \end{cases}.$$

Then the impulse response $h[n]$ of the system for $n \geq 0$ is given by

$$h[n] = Ar_1^n + Br_2^n = A(\frac{1}{2})^n + B(\frac{1}{3})^n. \tag{3.93}$$

Using the initial conditions $y[-2] = y[-1] = 0$ and substituting $x[n] = \delta[n]$ into (3.92), we obtain

$$
\begin{aligned}
y[0] &= \frac{5}{6}y[-1] - \frac{1}{6}y[-2] + \delta[0] - \delta[-1] = 1 \\
y[1] &= \frac{5}{6}y[0] - \frac{1}{6}y[-1] + \delta[1] - \delta[0] = -\frac{1}{6}.
\end{aligned}
$$

From (3.93) and the above conditions, we have

$$
\begin{aligned}
h[0] &= A + B = 1 \\
h[1] &= \frac{A}{2} + \frac{B}{3} = -\frac{1}{6}
\end{aligned}
\quad \Rightarrow \quad \begin{cases} A = -3 \\ B = 4 \end{cases}.
$$

Finally the impulse response of the system is obtained as

$$h[n] = \Big(-3(\frac{1}{2})^n + 4(\frac{1}{3})^n\Big)u[n]. \tag{3.94}$$

(b) As we know the step response of a discrete-time LTI system is the running sum of its impulse response. From (3.94) we obtain

$$\begin{aligned} s[n] &= \sum_{k=-\infty}^{n} h[k] = \sum_{k=0}^{n} h[k] \\ &= -3\sum_{k=0}^{n}(\frac{1}{2})^k + 4\sum_{k=0}^{n}(\frac{1}{3})^k \\ &= \Big(-3\frac{1-(\frac{1}{2})^{n+1}}{1-\frac{1}{2}} + 4\frac{1-(\frac{1}{3})^{n+1}}{1-\frac{1}{3}}\Big)u[n] \\ &= \big(3(\frac{1}{2})^n - 2(\frac{1}{3})^n\big)u[n]. \end{aligned} \tag{3.95}$$

Problem 62. Consider a discrete-time LTI system whose the impulse response is given by

$$h[n] = (\frac{1}{2})^n u[n]. \tag{3.96}$$

Determine the response of this system to the input

$$x[n] = \cos(n\pi)u[n].$$

Solution.

$$x[n] = \cos(n\pi)u[n] = (-1)^n u[n].$$

Since the system is linear and time-invariant, the response is given by

$$\begin{aligned} y[n] &= h[n] * x[n] \\ &= \sum_{k=-\infty}^{+\infty} h[k]x[n-k] \\ &= \sum_{k=0}^{n}(\frac{1}{2})^k(-1)^{n-k} \\ &= (-1)^n\sum_{k=0}^{n}(\frac{1}{2})^k(-1)^{-k} \\ &= (-1)^n\sum_{k=0}^{n}(-\frac{1}{2})^k. \end{aligned}$$

From (3.61), we obtain

$$\begin{aligned} y[n] &= (-1)^n \frac{1-(-\frac{1}{2})^{n+1}}{1-(-\frac{1}{2})} \\ &= \frac{(-1)^n - (-1)^n(-\frac{1}{2})(-\frac{1}{2})^n}{1+\frac{1}{2}} \\ &= (-1)^n \frac{2}{3} + \frac{1}{3}(\frac{1}{2})^n. \end{aligned}$$

Problem 63. Find the output $y[n]$ for the following difference equation.

$$y[n] + \frac{1}{3}y[n-1] = 4, \tag{3.97}$$

with the auxiliary condition $y[-1] = -3$.

Solution.

Let

$$y[n] = y_h[n] + y_p[n].$$

where $y_p[n]$ is the particular solution satisfying (3.97) and $y_h[n]$ is the homogeneous solution which satisfies

$$y_h[n] + \frac{1}{3}y_h[n-1] = 0, \tag{3.98}$$

A common method for finding the particular solution for a constant input signal is to look for a so-called forced response. Then the particular solution is a signal of the same form as the input. We assume that

$$y_p[n] = A. \tag{3.99}$$

Substituting (3.99) into (3.97) yields

$$A + \frac{1}{3}A = 4 \ \Rightarrow \ A = 3.$$

So that

$$y_p[n] = 3.$$

In order to determine $y_h[n]$, we form the **Characteristic** equation corresponding to (3.97) as follows:

$$r + \frac{1}{3} = 0 \ \Rightarrow \ r = -\frac{1}{3}.$$

So that

$$y_h[n] = B(-\frac{1}{3})^n.$$

Combining $y_p[n]$ and $y_h[n]$, we have

$$y[n] = B(-\frac{1}{3})^n + 3. \tag{3.100}$$

To obtain B in (3.100) we need the value of $y[0]$. Using auxiliary condition $y[-1] = -3$ and setting $n = 0$ in (3.97), we get

$$y[0] + \frac{1}{3}y[-1] = 4 \quad \Rightarrow \quad y[0] = 5,$$

Setting $n = 0$ in (3.100) and equating with $y[0] = 5$, we obtain

$$y[0] = B + 3 = 5 \quad \Rightarrow \quad B = 2.$$

Thus:

$$y[n] = 2(-\frac{1}{3})^n + 3. \tag{3.101}$$

Chapter 4

Fourier Series Representation of Continuous-time Signals

4.1 Introduction

In previous chapters, we introduced the response of an LTI system which is given by convolution of the input and the impulse response of the system. In this chapter, we will introduce one of the most important tools on the representation of continuous-time periodic signals known as the Fourier series. It is also demonstrated that one can obtain the response of an LTI system to a sinusoidal signal with a specific frequency using its so-called frequency response. As such, having the Fourier series representation on a periodic input to an LTI system will open an alternative way to convolution, to find the output.

4.2 The Response of LTI Systems to Complex Exponentials

Consider a continuous-time LTI system with impulse response $h(t)$ and the input $x(t)$. The response of the LTI system is given by the convolution integral, so with $x(t) = e^{st}$ we get:

$$\begin{aligned} y(t) = x(t) * h(t) &= \int_{-\infty}^{\infty} h(\tau)x(t-\tau)d\tau \\ &= \int_{-\infty}^{\infty} h(\tau)e^{s(t-\tau)}d\tau \\ &= e^{st}\int_{-\infty}^{\infty} h(\tau)e^{-s\tau}d\tau = e^{st}H(s), \end{aligned} \tag{4.1}$$

where s is in general a complex variable and $H(s)$ is called the system transfer function.

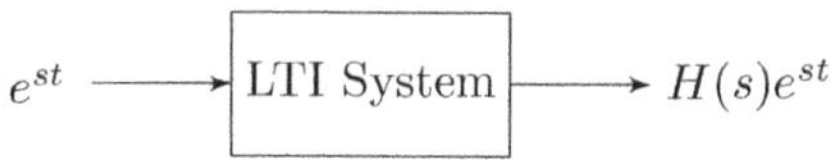

Figure 4.1: The response of LTI systems to complex exponentials.

The input signal $x(t) = e^{st}$ has a nice advantage: the very same version of the signal is observed at the output multiplied by $H(s)$. Using the linearity, if the input to an LTI system can be represented as a linear combination of complex exponentials, the output can also be represented as a linear combination of the same complex exponential signals, that is, if

$$x(t) = a_1 e^{s_1 t} + a_2 e^{s_2 t} + \cdots \tag{4.2}$$

then

$$y(t) = a_1 H(s_1) e^{s_1 t} + a_2 H(s_2) e^{s_2 t} + \cdots \tag{4.3}$$

If we set $s = jw$, then $H(jw)$ is called the frequency response of the system.

As will be seen, it will be very useful to expend a signal based on linear combination of complex expeoeitals e^{st} or e^{jwt}.

In general, we have 2 types of signal representation based on complex exponentials:

1. **The Continuous-Time Fourier Series**

   If the signal can be expanded solely based on linear sum of e^{st} with $s = jkw_0$ and coefficients a_k where
   $$k = 0,\ \pm 1,\ \pm 2, \cdots,$$
   then we call the resulting equivalence the Fourier series expansion of the signal.

2. **The Continuous-Time Fourier Transform**

   If the signal can not be expanded solely based on linear sum of $s = e^{jkw_0 t}$, the signal can still be presented based on $s = jwt$ using the continuous-time Fourier transform which will be discussed in the next chapters.

4.3 Fourier Series Representation of Continuous-Time Periodic Signals

In this section, we introduce the continuous-time Fourier series and in the next sections we will introduce its applications in the analysis of continuous-time signals and systems.

From Chapter 1, we know that a periodic signal is defined as

$$x(t) = x(t+T) \quad \text{for all } t. \tag{4.4}$$

The fundamental period of $x(t)$ is the smallest positive, nonzero value of T for which (4.4) holds. The value f_0 is referred to as the *fundamental frequency* for which we have

$$f_0 = \frac{1}{T}, \qquad w_0 = 2\pi f_0 = \frac{2\pi}{T}.$$

4.3.1 How to Obtain the Fourier Series Representation

The pair of the Fourier series of a periodic continuous-time signal is defined as

$$x(t) = \sum_{k=-\infty}^{+\infty} a_k e^{jkw_0 t} = \sum_{k=-\infty}^{+\infty} a_k e^{jk(2\pi/T)t}, \tag{4.5}$$

$$a_k = \frac{1}{T}\int_T x(t)e^{-jkw_0 t}dt = \frac{1}{T}\int_T x(t)e^{-jk(2\pi/T)t}dt. \tag{4.6}$$

Note that the interval of integration is over any interval of length T where 0 to T or $-T/2$ to $T/2$ are commonly used for the integration. The coefficient a_0 is the DC or the average value of $x(t)$ which is obtained by setting $k = 0$ in (4.6), i.e.,

$$a_0 = \frac{1}{T}\int_T x(t)dt. \tag{4.7}$$

According to the definition of a_k, the coefficients a_k are complex in general. So the coefficients are shown by the *amplitude spectrum*, $|a_k|$, and the *phase spectrum*, $\angle a_k$, which are plotted versus the angular frequency w.

4.3.2 Convergence of the Fourier Series

According to the definition of the Fourier series, a periodic signal $x(t)$ has a Fourier series representation if it satisfies the following Dirichlet conditions:

1. $x(t)$ must be *absolutely integrable* over any period; that is,

$$\int_T |x(t)|dt < \infty. \tag{4.8}$$

2. In any finite interval of t, $x(t)$ has no more than a finite number of maxima and minima.

3. In any finite interval of t, $x(t)$ has only a finite number of discontinuities, and each of them is finite.

4.3.3 Trigonometric Form of the Fourier Series

If periodic signal $x(t)$ is real, it can be written in terms of trigonometric functions; that is,

$$\begin{aligned} x(t) &= a_0 + 2\sum_{k=1}^{+\infty} A_k \cos(kw_0 t + \theta_k), \\ &= a_0 + 2\sum_{k=1}^{+\infty} \left[B_k \cos(kw_0 t) - C_k \sin(kw_0 t)\right]. \end{aligned} \tag{4.9}$$

The relationships between the coefficients A_k, B_k, C_k and θ_k with a_k are

$$a_k = A_k e^{j\theta_k} = B_k + jC_k, \tag{4.10}$$

where B_k and C_k are both real values.

Note that if a periodic signal $x(t)$ is even, then $C_k = 0$ and we have

$$x(t) = a_0 + 2\sum_{k=1}^{+\infty} B_k \cos(kw_0 t). \tag{4.11}$$

Also if a periodic signal $x(t)$ is odd, then $B_k = 0$ and $a_0 = 0$. So we have

$$x(t) = -2\sum_{k=1}^{+\infty} C_k \sin(kw_0 t). \tag{4.12}$$

4.3.4 The Fourier Series of Important Signals

Some common signals and their Fourier Series are:

(a)

$$x(t) = e^{jw_0 t} \quad \Rightarrow \quad a_k = \begin{cases} 1 & k = 1 \\ 0 & k \neq 1 \end{cases}. \tag{4.13}$$

(b)

$$x(t) = \cos w_0 t \quad \Rightarrow \quad a_k = \begin{cases} \frac{1}{2} & k = \pm 1 \\ 0 & k \neq \pm 1 \end{cases}. \tag{4.14}$$

(c)

$$x(t) = \sin w_0 t \quad \Rightarrow \quad a_k = \begin{cases} \frac{1}{2j} & k = 1 \\ -\frac{1}{2j} & k = -1 \\ 0 & k \neq \pm 1 \end{cases}. \tag{4.15}$$

(d)

$$x(t) = 1 \quad \Rightarrow \quad a_k = \begin{cases} 1 & k = 0 \\ 0 & k \neq 0 \end{cases}. \tag{4.16}$$

(e) Impulse Train

$$x(t) = \sum_{n=-\infty}^{+\infty} \delta(t - nT) \quad \Rightarrow \quad a_k = \frac{1}{T}. \tag{4.17}$$

(f) Pulse Train

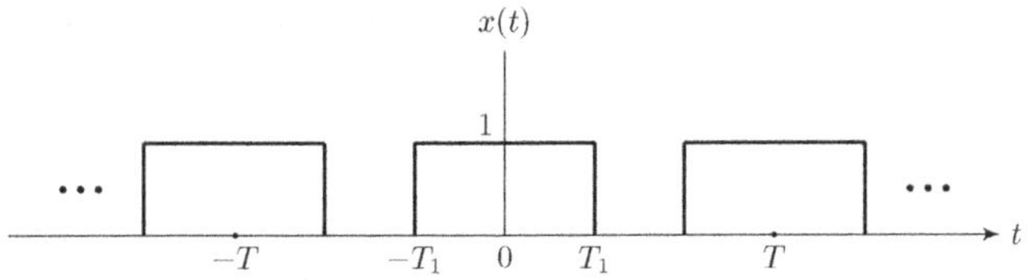

$$a_k = \frac{w_0 T_1}{\pi} \,\text{sinc}(\frac{k w_0 T_1}{\pi}) = \frac{\sin(k w_0 T_1)}{k\pi} = \frac{\sin(2k\pi \frac{T_1}{T})}{k\pi}. \tag{4.18}$$

4.4 The Fundamental Frequency of the Continuous-Time Periodic Signals

The first step associated to the Fourier analysis is to find the fundamental frequency of the periodic signal. Let $x(t)$ be a periodic signal with period T. The fundamental frequency is given by $f_0 = 1/T$ or $w_0 = 2\pi/T$. Now if a signal $y(t)$ is as

$$y(t) = x_1(t) + x_2(t),$$

where $x_1(t)$ and $x_2(t)$ are both periodic with period T_1 and T_2, respectively, the fundamental period of this signal is the Least Common Multiple (LCM) of the two fundamental periods. In other words, the fundamental period of signal $y(t)$ is

$$T_0 = LCM\ (T_1, T_2).$$

Therefore, the fundamental frequency of $y(t)$ is given by

$$w_0 = \frac{2\pi}{T_0}.$$

4.5 Properties of the Continuous-Time Fourier Series

(a) Linearity

Consider two given signals $x(t)$ and $y(t)$ with period T where Fourier series coefficients are

$$x(t) \overset{\mathcal{F.S.}}{\longleftrightarrow} a_k,$$
$$y(t) \overset{\mathcal{F.S.}}{\longleftrightarrow} b_k.$$

Then we have

$$z(t) = Ax(t) + By(t) \overset{\mathcal{F.S.}}{\longleftrightarrow} c_k = Aa_k + Bb_k. \tag{4.19}$$

(b) Time Shifting

Consider a periodic signal $x(t)$ with period T and the Fourier series coefficients a_k. The

Fourier series coefficients b_k of the signal $y(t) = x(t - t_0)$ is given by

$$y(t) = x(t - t_0) \quad \stackrel{\mathcal{F.S.}}{\longleftrightarrow} \quad b_k = e^{-jkw_0 t_0} a_k. \tag{4.20}$$

Eq. (4.20) indicates that when a periodic signal is shifted in time domain, the magnitudes of its Fourier series coefficients remain unchanged but the phase is rotated by $kw_0 t_0$. That is,

$$|b_k| = |a_k|, \qquad \angle b_k = -kw_0 t_0 \; + \angle a_k.$$

(c) Frequency Shifting

Consider a periodic signal $x(t)$ with period T and the Fourier series coefficients a_k. The frequency shifting property indicates that the multiplication of $x(t)$ by a complex exponential signal $e^{jMw_0 t}$ results in an index shift of M for a_k:

$$e^{jMw_0 t}\, x(t) \quad \stackrel{\mathcal{F.S.}}{\longleftrightarrow} \quad a_{k-M}. \tag{4.21}$$

(d) Time Differentiation

Consider a periodic signal $x(t)$ with period T and Fourier series coefficients a_k. Under certain differentiability conditions, the Fourier series coefficients of $x'(t)$ are equal to $jkw_0 a_k$, i.e.,

$$x(t) \quad \stackrel{\mathcal{F.S.}}{\longleftrightarrow} \quad a_k,$$
$$x'(t) \quad \stackrel{\mathcal{F.S.}}{\longleftrightarrow} \quad jkw_0 a_k.$$

(e) Time Reversal

If a periodic signal undergoes time reversal, its Fourier series coefficients will see a sign change in index:

$$x(t) \quad \stackrel{\mathcal{F.S.}}{\longleftrightarrow} \quad a_k,$$
$$x(-t) \quad \stackrel{\mathcal{F.S.}}{\longleftrightarrow} \quad a_{-k}.$$

From this property, if signal $x(t)$ is even, then we have

$$x(-t) = x(t) \quad \Rightarrow \quad a_{-k} = a_k.$$

If signal $x(t)$ is odd, then we have

$$x(-t) = -x(t) \quad \Rightarrow \quad a_{-k} = -a_k.$$

(f) Conjugation

The complex conjugation of a periodic signal $x(t)$ causes the complex conjugation and time reversal on the corresponding Fourier series coefficients. That is, if

$$x(t) \quad \stackrel{\mathcal{F.S.}}{\longleftrightarrow} \quad a_k,$$

$$x^*(t) \stackrel{\mathcal{F.S.}}{\longleftrightarrow} a^*_{-k}.$$

(g) Time Scaling

Generally, time scaling operation causes a change in the signal period. The coefficients a_k remain unchanged but now a_k corresponds to the frequency $k\alpha w_0$ rather than kw_0:

$$x(\alpha t) = \sum_{k=-\infty}^{+\infty} a_k e^{jk\alpha w_0 t}, \tag{4.22}$$

where α is a real constant.

(h) Multiplication

Let $x(t)$ and $y(t)$ be both periodic with period T and the Fourier coefficients a_k and b_k, respectively. Then the product $x(t).y(t)$ is also periodic with period T and its Fourier coefficients is given by the convolution sum as below:

$$z(t) = x(t).y(t) \stackrel{\mathcal{F.S.}}{\longleftrightarrow} c_k = \sum_{l=-\infty}^{+\infty} a_k b_{k-l}. \tag{4.23}$$

(i) Parseval's Relation for Continuous-Time Periodic Signals

Parseval's relation for continuous-time periodic signals is

$$\frac{1}{T}\int_T |x(t)|^2 dt = \sum_{k=-\infty}^{+\infty} |a_k|^2. \tag{4.24}$$

From Parseval's relation, we can obtain the average power of a periodic signal by summing over the average power of its harmonic components.

(j) Conjugate Symmetry for Real Signals

If a signal $x(t)$ is real, then its Fourier series coefficients are *conjugate symmetric*, i.e.,

$$x(t) = x^*(t) \quad \Rightarrow \quad a_k = a^*_{-k} \quad \Rightarrow \quad a_{-k} = a^*_k.$$

As such we have

$$\begin{array}{lcl} |a_k| \text{ is even} & & |a_k| = |a_{-k}| \\ \angle a_k \text{ is odd} & : & \angle a_k = -\angle a_{-k} \\ \Re\{a_k\} \text{ is even} & : & \Re\{a_k\} = \Re\{a_{-k}\} \\ \Im\{a_k\} \text{ is odd} & : & \Im\{a_k\} = -\Im\{a_{-k}\} \\ x_e(t) \stackrel{\mathcal{F.S.}}{\longleftrightarrow} \Re\{a_k\} & & \\ x_o(t) \stackrel{\mathcal{F.S.}}{\longleftrightarrow} j\Im\{a_k\} & & \end{array}$$

Note that if $x(t)$ is real and even, so are its Fourier series coefficients, and if $x(t)$ is real and odd, then its Fourier series coefficients are imaginary and odd. Also if $x(t)$ is even or odd, then its Fourier series coefficients are even or odd, respectively.

4.6 Fourier Series and LTI Systems

Consider an LTI system with the impulse response $h(t)$. Let the input be a periodic signal $x(t)$ with period T. Based on the properties of the LTI systems, the output signal $y(t)$ is a periodic signal with the same period of $x(t)$. The Fourier series coefficients of $y(t)$, and eventually $y(t)$ itself, can be obtained based on the Fourier coefficients of $x(t)$ and the frequency response of the system. That is, if

$$x(t) = \sum_{k=-\infty}^{+\infty} a_k e^{jkw_0 t}, \tag{4.25}$$

and

$$y(t) = \sum_{k=-\infty}^{+\infty} b_k e^{jkw_0 t}, \tag{4.26}$$

then

$$b_k = a_k.H(jkw_0), \tag{4.27}$$

where $H(jkw_0)$ is the frequency response of the system which is given by

$$H(jkw_0) = \int_{-\infty}^{+\infty} h(t)e^{-jkw_0 t}dt. \tag{4.28}$$

So, we obtain

$$\begin{aligned} |b_k| &= |a_k|.|H(jkw_0)|, \\ \angle b_k &= \angle a_k + \angle H(jkw_0). \end{aligned} \tag{4.29}$$

If the input of an LTI system is

$$x(t) = A_x \cos(w_0 t + \phi_x),$$

then the output of the system with the frequency response $H(jkw_0)$ is given by

$$y(t) = A_y \cos(w_0 t + \phi_y) \;\Rightarrow\; \begin{cases} A_y = A_x.|H(jkw_0)|, \\ \phi_y = \phi_x + \angle H(jkw_0) \end{cases}. \tag{4.30}$$

4.7 Problems

Problem 64. Determine and sketch the complex exponential Fourier series representation for each of the following signals

(a) $x(t) = \cos(t + \theta)$

(b) $x(t) = \sin(t) + \cos(t)$

(c) $x(t) = \sin^2(4t)$

(d) $x(t) = \cos(t).\sin(3t)$

(e) $x(t) = \cos^2(4t)$

Solution.

(a) $x(t) = \cos(t + \theta)$

The fundamental period T_0 of $x(t)$ is 2π and $w_0 = \frac{2\pi}{T_0} = 1$. Thus,

$$x(t) = \cos(t + \theta) = \sum_{k=-\infty}^{\infty} a_k e^{jkw_0 t} = \sum_{k=-\infty}^{\infty} a_k e^{jkt}.$$

Now, using Euler's formula, we get

$$x(t) = \cos(t + \theta) = \frac{1}{2} e^{jt} e^{j\theta} + \frac{1}{2} e^{-jt} e^{-j\theta} = \sum_{k=-\infty}^{\infty} a_k e^{jkt}.$$

Thus, the complex Fourier coefficients for $x(t)$ are

$$a_{-1} = \frac{1}{2} e^{-j\theta} \quad , \quad a_1 = \frac{1}{2} e^{j\theta}.$$

For other K's, $a_k = 0$.

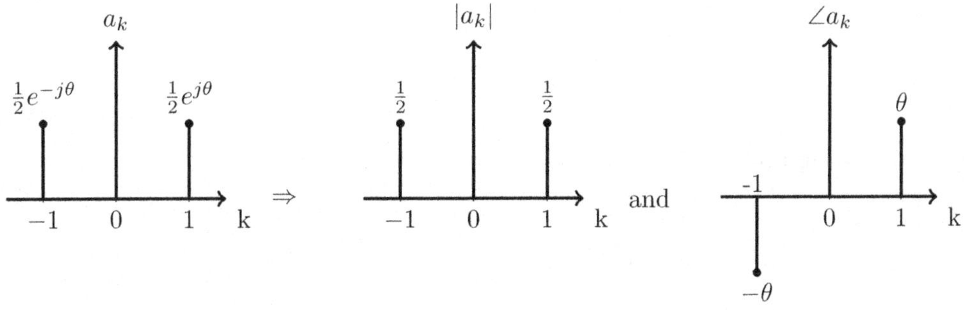

Figure 4.2: Magnitude and Phase of the Fourier series coefficients of the signal considered in Problem 64 - Part (a).

(b) $x(t) = \sin(t) + \cos(t)$

The fundamental period T_0 of $x(t)$ is 2π and $w_0 = \frac{2\pi}{T_0} = 1$. Thus,

$$x(t) = \sin(t) + \cos(t) = \sum_{k=-\infty}^{\infty} a_k e^{jkw_0 t} = \sum_{k=-\infty}^{\infty} a_k e^{jkt}.$$

Now, using Euler's formula, we get

$$x(t) = \sin(t) + \cos(t) = \frac{1}{2j}e^{jt} - \frac{1}{2j}e^{-jt} + \frac{1}{2}e^{jt} + \frac{1}{2}e^{-jt} = \sum_{k=-\infty}^{\infty} a_k e^{jkt}.$$

Thus, the complex Fourier coefficients for $x(t)$ are

$$a_{-1} = \frac{1}{2} - \frac{1}{2j} \quad , \quad a_1 = \frac{1}{2} + \frac{1}{2j}.$$

For other K's, $a_k = 0$.

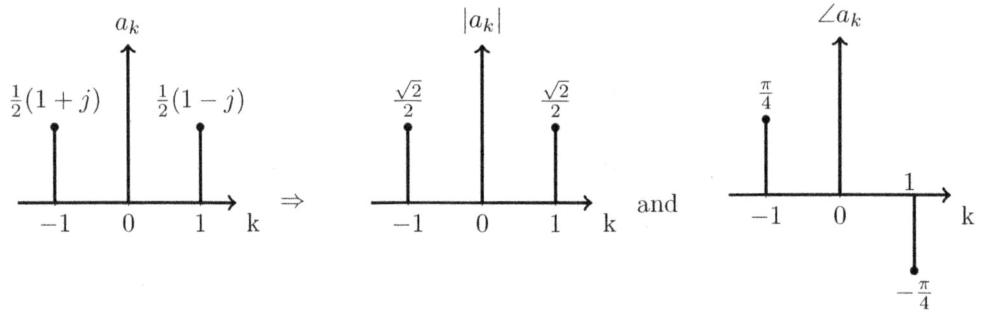

Figure 4.3: Magnitude and Phase of the Fourier series coefficients of the signal considered in Problem 64 - Part (b).

(c) $x(t) = \sin^2(4t)$

$$x(t) = \sin^2(4t) = \frac{1}{2} - \frac{1}{2}\cos(8t).$$

The fundamental period T_0 of $x(t)$ is $\frac{\pi}{4}$ and $w_0 = \frac{2\pi}{T_0} = 8$. Thus,

$$x(t) = \frac{1}{2} - \frac{1}{2}\cos(8t) = \sum_{k=-\infty}^{\infty} a_k e^{jkw_0 t} = \sum_{k=-\infty}^{\infty} a_k e^{j8kt}.$$

Now, using Euler's formula, we get

$$x(t) = \frac{1}{2} - \frac{1}{2}\cos(8t) = \frac{1}{2} - \frac{1}{4}e^{j8t} - \frac{1}{4}e^{-j8t} = \sum_{k=-\infty}^{\infty} a_k e^{j8kt}.$$

Thus, the complex Fourier coefficients for $x(t)$ are

$$a_{-1} = -\frac{1}{4} \quad , \quad a_0 = \frac{1}{2} \quad , \quad a_1 = -\frac{1}{4}.$$

For other K's, $a_k = 0$.

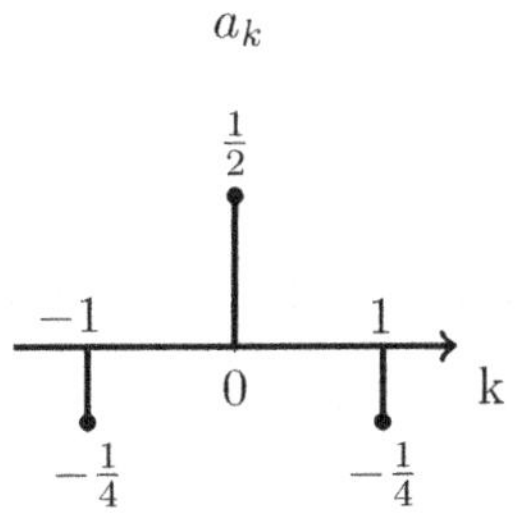

Figure 4.4: The Fourier series coefficients of the signal considered in Problem 64 - Part (c).

(d) $x(t) = \cos(t).\sin(3t)$

Hint: $\sin a.\cos b = \frac{1}{2}\sin(a+b) + \frac{1}{2}\sin(a-b)$

$$x(t) = \cos(t).\sin(3t) = \frac{1}{2}\sin(4t) + \frac{1}{2}\sin(2t).$$

The fundamental period T_0 of $x(t)$ is π and $w_0 = \frac{2\pi}{T_0} = 2$. Thus,

$$x(t) = \frac{1}{2}\sin(4t) + \frac{1}{2}\sin(2t) = \sum_{k=-\infty}^{\infty} a_k e^{jkw_0 t} = \sum_{k=-\infty}^{\infty} a_k e^{j2kt}.$$

Now, using Euler's formula, we get

$$x(t) = \frac{1}{2}\sin(4t) + \frac{1}{2}\sin(2t) = \frac{1}{4j}e^{j4t} - \frac{1}{4j}e^{-j4t} + \frac{1}{4j}e^{j2t} - \frac{1}{4j}e^{-j2t} = \sum_{k=-\infty}^{\infty} a_k e^{j2kt},$$

Thus, the complex Fourier coefficients for $x(t)$ are

$$a_{-2} = \frac{j}{4} \quad , \quad a_{-1} = \frac{j}{4} \quad , \quad a_1 = -\frac{j}{4} \quad , \quad a_2 = -\frac{j}{4}.$$

For other K's, $a_k = 0$.

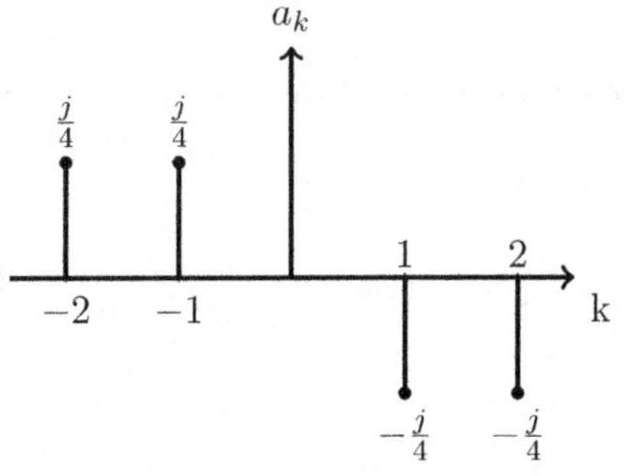

Figure 4.5: The Fourier series coefficients of the signal considered in Problem 64 - Part (d).

(e) $x(t) = \cos^2(4t)$

$$x(t) = \cos^2(4t) = \frac{1}{2} + \frac{1}{2}\cos(8t)$$

The fundamental period T_0 of $x(t)$ is $\frac{\pi}{4}$ and $w_0 = \frac{2\pi}{T_0} = 8$. Thus,

$$x(t) = \frac{1}{2} + \frac{1}{2}\cos(8t) = \sum_{k=-\infty}^{\infty} a_k e^{jkw_0 t} = \sum_{k=-\infty}^{\infty} a_k e^{j8kt}.$$

Now, using Euler's formula, we get

$$x(t) = \frac{1}{2} + \frac{1}{2}\cos(8t) = \frac{1}{2} + \frac{1}{4}e^{j8t} + \frac{1}{4}e^{-j8t} = \sum_{k=-\infty}^{\infty} a_k e^{j8kt}.$$

Thus, the complex Fourier coefficients for $x(t)$ are

$$a_{-1} = \frac{1}{4} \quad , \quad a_0 = \frac{1}{2} \quad , \quad a_1 = \frac{1}{4}.$$

For other K's, $a_k = 0$.

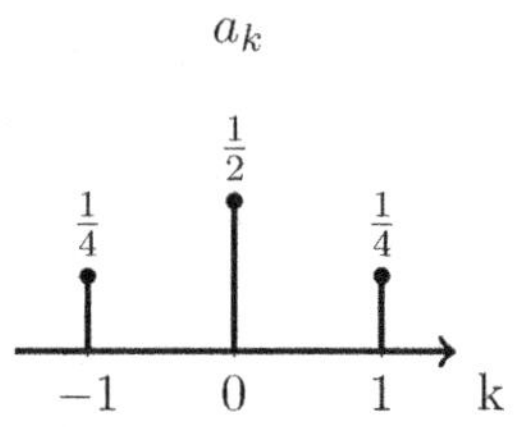

Figure 4.6: The Fourier series coefficients of the signal considered in Problem 64 - Part (e).

Problem 65. Determine the complex exponential Fourier series representation for the following signal:

$$x(t) = \sum_{m=-\infty}^{\infty} (-1)^m [\delta(t - \frac{1}{3}m) + \delta(t + \frac{2}{3}m)]$$

Solution.

First we sketch signal $x(t)$:

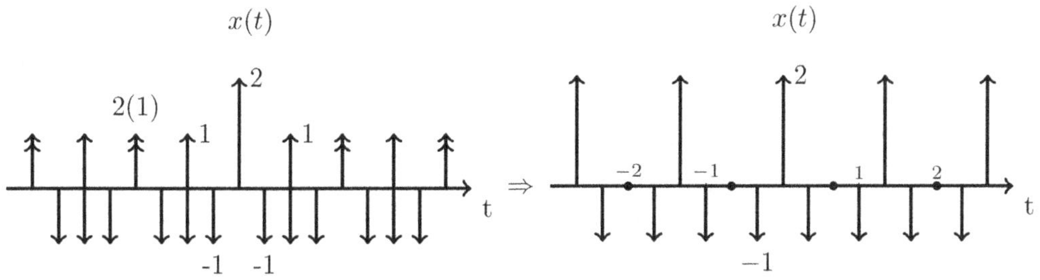

Figure 4.7: Signal $x(t)$ considered in Problem 65.

Signal $x(t)$ is composed of two parts $x_1(t)$ ans $x_2(t)$ that are equal to $\sum_{m=-\infty}^{\infty}(-1)^m\delta(t-\frac{1}{3}m)$ and $\sum_{m=-\infty}^{\infty}(-1)^m\delta(t+\frac{2}{3}m)$, respectively. The fundamental period T_0 of $x_1(t)$ and $x_2(t)$ is $\frac{2}{3}$ and $\frac{4}{3}$, respectively. Also the fundamental angular frequency w_0 of $x_1(t)$ and $x_2(t)$ is 3π and $\frac{3\pi}{2}$, respectively. Therefore, the fundamental period T_0 of $x(t)$ is $\frac{4}{3}$ and the fundamental angular frequency w_0 is $\frac{3\pi}{2}$. Thus,

$$\begin{aligned}
a_k &= \frac{1}{T}\int_T x(t)e^{-jkw_0t}dt \\
&= \frac{3}{4}\int_{-\frac{2}{3}}^{\frac{2}{3}} x(t)e^{-jk(\frac{3\pi}{2})t}dt \\
&= \frac{3}{4}\int_{-\frac{2}{3}}^{\frac{2}{3}} [2\delta(t)-\delta(t+\frac{1}{3})-\delta(t-\frac{1}{3})]e^{-jk(\frac{3\pi}{2})t}dt \\
&= \frac{3}{4}(2-e^{-jk(\frac{\pi}{2})}-e^{jk(\frac{\pi}{2})}) \\
&= \frac{3}{4}(2-2\cos(\frac{k\pi}{2})) = \frac{3}{2}(1-\cos(\frac{k\pi}{2})).
\end{aligned}$$

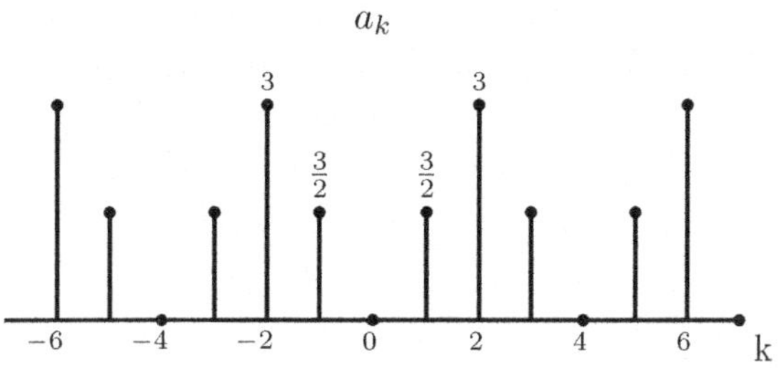

Figure 4.8: The Fourier series coefficients of the signal $x(t)$ in Problem 65.

Problem 66. Determine the complex exponential Fourier series representation of the following signals.

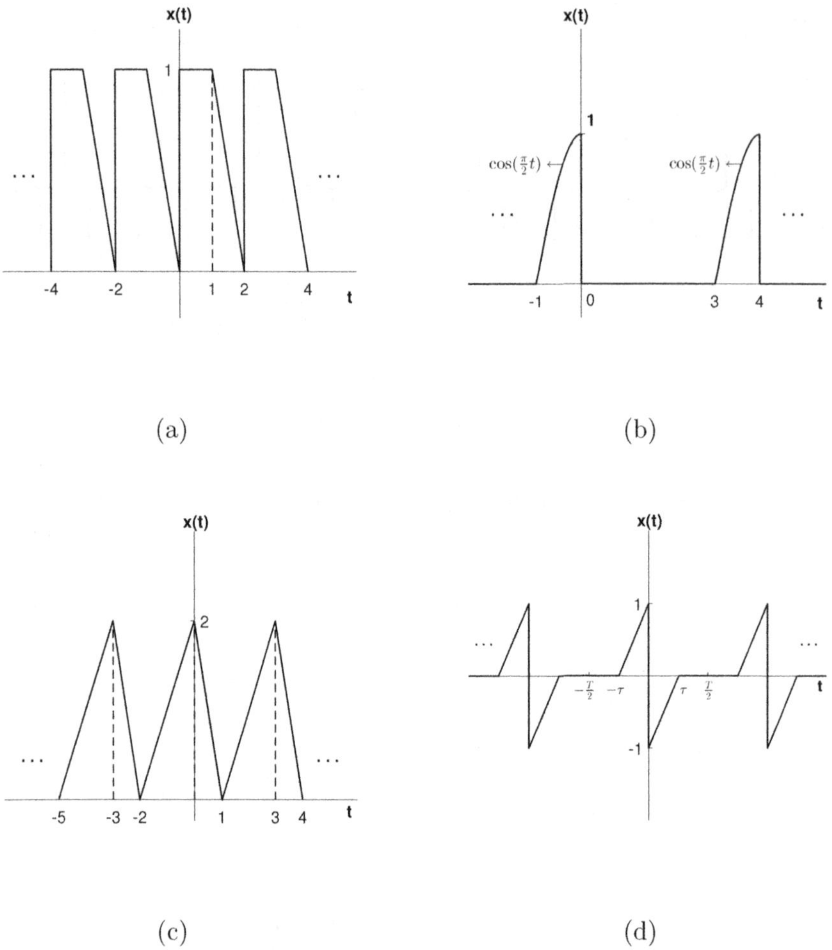

Figure 4.13: Problem 66.

Solution.

(a) The fundamental period T_0 of $x(t)$ is 2 and $w_0 = \pi$. Take the period for $0 < t < 2$. For $0 < t < 1$, $x(t) = 1$ and for $1 < t < 2$, $x(t) = 2 - t$. Thus,

$$a_0 = \frac{1}{2}\int_0^2 x(t)dt = \frac{3}{4},$$

and

$$
\begin{aligned}
a_k &= \frac{1}{T}\int_0^T x(t)e^{-jw_0kt}dt = \frac{1}{2}\int_0^2 x(t)e^{-j\pi kt}dt \\
&= \frac{1}{2}\int_0^1 e^{-jw_0kt}dt + \frac{1}{2}\int_1^2 (2-t)e^{-jw_0kt}dt \\
&= \frac{1}{2}\left[\frac{je^{-jk\pi t}}{\pi k}\right]_0^1 + \frac{1}{2}\left[\frac{2je^{-jk\pi t}}{\pi k} + \frac{(jk\pi t+1)e^{-jk\pi t}}{\pi^2k^2}\right]_1^2 \\
&= \frac{1}{2k\pi}(je^{-jk\pi} - j) + \frac{1}{2k\pi}(2je^{-2jk\pi} + \frac{(2jk\pi+1)e^{-2jk\pi}}{k\pi} - 2je^{-jk\pi} - \frac{(jk\pi+1)e^{-jk\pi}}{k\pi}) \\
&= \frac{j(e^{-jk\pi}-1)}{2k\pi} + \frac{e^{-2jk\pi}(4jk\pi+1) - e^{-jk\pi}(3jk\pi+1)}{2\pi^2k^2}.
\end{aligned}
$$

Thus, the complex Fourier coefficients for $x(t)$ are

$$
\begin{cases}
a_k = \frac{jk\pi}{2\pi^2k^2} = \frac{j}{2\pi k} \quad , & \text{if k is even,} \\
a_k = -\frac{2j}{2k\pi} + \frac{7j}{2k\pi} + \frac{1}{\pi^2k^2} = \frac{5j}{2k\pi} + \frac{1}{\pi^2k^2} \quad , & \text{if k is odd.}
\end{cases}
$$

(b) The fundamental period T_0 of $x(t)$ is 4 and $w_0 = \frac{\pi}{2}$. Take the period for $-1 < t < 3$. For $-1 < t < 0$, $x(t) = \cos(\frac{\pi t}{2})$ and for $0 < t < 3$, $x(t) = 0$. Thus,

$$
\begin{aligned}
a_k &= \frac{1}{T}\int_0^T x(t)e^{-jw_0kt}dt \\
&= \frac{1}{4}\int_{-1}^0 \cos(\frac{\pi t}{2})e^{-j\frac{k\pi t}{2}}dt + \frac{1}{2}\int_0^3 0\, e^{-j\frac{k\pi t}{2}}dt \\
&= \frac{1}{8}\int_{-1}^0 (e^{j\frac{\pi t}{2}} + e^{-j\frac{\pi t}{2}})e^{-j\frac{k\pi t}{2}}dt \\
&= \frac{1}{8}\left[\frac{2}{j\pi(1-k)}e^{j\pi t\frac{(1-k)}{2}} + \frac{2}{-j\pi(1+k)}e^{-j\pi t\frac{(1+k)}{2}}\right]_{-1}^0 \\
&= \frac{1}{4}\left(\frac{2k+2je^{j\frac{k\pi}{2}}}{j\pi(1-k^2)}\right) \\
&= \frac{1}{2}j\frac{k+je^{j\frac{k\pi}{2}}}{\pi(k^2-1)}.
\end{aligned}
$$

(c) The fundamental period T_0 of $x(t)$ is 3 and $w_0 = \frac{2\pi}{3}$. Take the period for $0 < t < 3$. For $0 < t < 1$, $x(t) = 2 - 2t$ and for $1 < t < 3$, $x(t) = t - 1$. Thus,

$$
a_0 = \frac{1}{3}\int_0^3 x(t)dt = 1,
$$

and

$$\begin{aligned}
a_k &= \frac{1}{T}\int_0^T x(t)e^{-jw_0kt}dt \\
&= \frac{1}{3}\int_0^3 x(t)e^{-j\frac{2k\pi t}{3}}dt \\
&= \frac{1}{3}\int_0^1 (2-2t)e^{-j\frac{2k\pi t}{3}}dt + \frac{1}{3}\int_1^3 (t-1)e^{-j\frac{2k\pi t}{3}}dt \\
&= \frac{1}{2}\left[\frac{2j\pi k - 2j\pi kt - 3}{\pi^2k^2}e^{-j\frac{2k\pi t}{3}}\right]_0^1 + \frac{1}{4}\left[\frac{3 - 2j\pi k + 2j\pi kt}{\pi^2k^2}e^{-j\frac{2k\pi t}{3}}\right]_1^3 \\
&= \frac{9}{4\pi^2k^2}(1 - e^{-j\frac{2k\pi}{3}}).
\end{aligned}$$

(d) The fundamental period T_0 of $x(t)$ is T and $w_0 = \frac{2\pi}{T}$. Take the period for $-T/2 < t < T/2$. For $-T/2 < t < -\tau$ and $\tau < t < T/2$, we have $x(t) = 0$. For $-\tau < t < 0$, $x(t) = t/\tau + 1$ and for $0 < t < \tau$, $x(t) = t/\tau - 1$. Thus,

Thus,

$$a_0 = \frac{1}{T}\int_0^T x(t)dt = 0,$$

and

$$\begin{aligned}
a_k &= \frac{1}{T}\int_0^T x(t)e^{-jw_0kt}dt \\
&= \frac{1}{T}\int_{-\tau}^0 (\frac{t}{\tau}+1)e^{-jw_0kt}dt + \frac{1}{T}\int_0^\tau (\frac{t}{\tau}-1)e^{-jw_0kt}dt \\
&= \frac{1}{T}\left[\frac{t}{\tau}(-\frac{1}{jkw_0})e^{-jw_0kt} + \frac{1}{\tau}\frac{1}{k^2w_0^2}e^{-jw_0kt} - \frac{1}{jkw_0}e^{-jw_0kt}\right]_{-\tau}^0 \\
&+ \frac{1}{T}\left[\frac{t}{\tau}(-\frac{1}{jkw_0})e^{-jw_0kt} + \frac{1}{\tau}\frac{1}{k^2w_0^2}e^{-jw_0kt} + \frac{1}{jkw_0}e^{-jw_0kt}\right]_0^\tau \\
&= \frac{2j}{kw_0T}\left[1 - \frac{\sin(kw_0\tau)}{kw_0\tau}\right].
\end{aligned}$$

Problem 67. A continuous-time periodic signal $x(t)$ is real valued and has fundamental period $T = 2$ and $x(t) = e^{-t}$ for $-1 < t < 1$.

(a) Determine the Fourier series coefficients of $x(t)$ through through direct evaluation.

(b) Is it possible to obtain the result in Part (a) without direct evaluatin?

Solution.

(a)

$$T_0 = 2 \Rightarrow w_0 = \frac{2\pi}{T} = \frac{2\pi}{2} = \pi,$$

then,

$$\begin{aligned}a_k &= \frac{1}{T}\int_0^T x(t)e^{-jw_0kt}dt\\&= \frac{1}{2}\int_{-1}^{1} e^{-t}e^{-jkw_0t}dt\\&= \frac{1}{2}\int_{-1}^{1} e^{-(1+jk\pi)t}dt\\&= \frac{-1}{2(1+jk\pi)}\left[e^{-(1+jk\pi)t}\right]_{-1}^{1}\\&= \frac{1}{2(1+jk\pi)}\left[e^{(1+jk\pi)} - e^{-(1+jk\pi)}\right],\end{aligned}$$

thus,

$$\begin{cases} a_k = \frac{-(e^{-1}-e)}{2(1+jk\pi)} & , \quad \text{if k is even,} \\ a_k = \frac{-(-e^{-1}+e)}{2(1+jk\pi)} & , \quad \text{if k is odd,} \end{cases}$$

finally,

$$a_k = \frac{-(e^{-1}-e)}{2(1+jk\pi)}(-1)^k.$$

(b) First we plot the signal $x(t)$ and $x'(t)$:

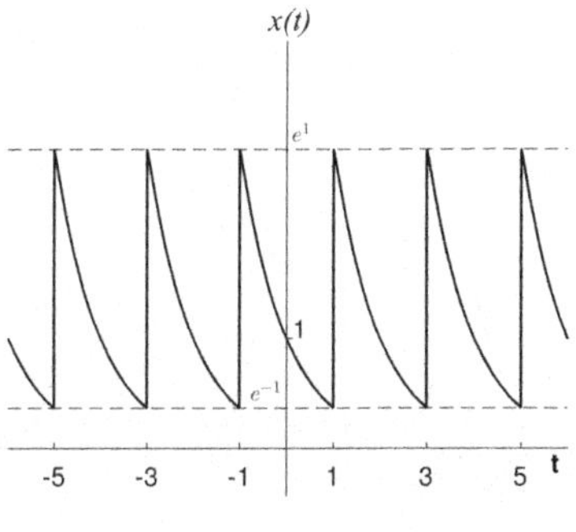

Figure 4.14: Problem 67, Signal $x(t)$.

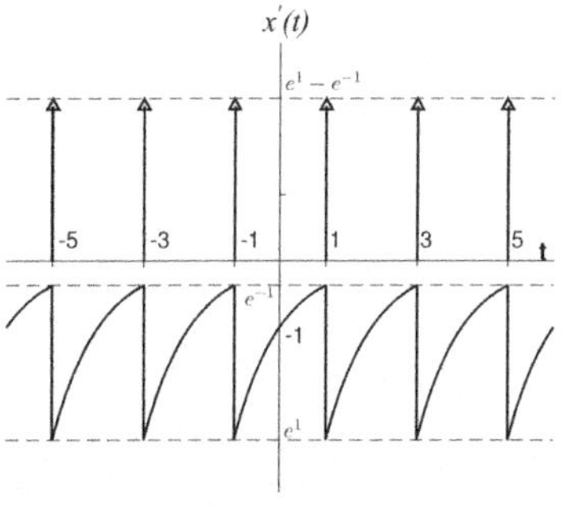

Figure 4.15: Problem 67, Signal $x'(t)$.

From figures 4.14 and 4.15, the relationship between $x(t)$ and $x'(t)$ is as

$$\frac{dx(t)}{dt} = -x(t) + z(t) \tag{4.31}$$

where

$$z(t) = (e - e^{-1}) \sum_{n=-\infty}^{\infty} \delta(t - 2n + 1).$$

Using (4.17), the Fourier series coefficients of the periodic impulse train with fundamental period

T is equal to $\frac{1}{T}$. Using the Time Shifting property of Section 4.5, the Fourier series coefficients of $z(t)$ are equal to

$$b_k = \frac{(e - e^{-1})}{2} e^{jk\pi}.$$

Using time-differentiation property of Section 4.5,

$$x(t) \xrightarrow{\mathcal{F.S.}} a_k \quad \Rrightarrow \quad \frac{dx(t)}{dt} \xrightarrow{\mathcal{F.S.}} jkw_0 a_k.$$

As such, based on the relationship(4.31), we can write:

$$jk\pi a_k = -a_k + b_k = -a_k + \frac{(e - e^{-1})}{2} e^{jk\pi}.$$

Thus we have

$$\rightarrow \quad a_k = \frac{(e - e^{-1})e^{jk\pi}}{2(1 + jk\pi)}$$

$$\xrightarrow{e^{jk\pi} = (-1)^k} \quad a_k = \frac{-(e^{-1} - e)}{2(1 + jk\pi)} (-1)^k.$$

Problem 68. Use the Fourier series properties to calculate the coefficients a_k for the following continuous-time periodic signals.

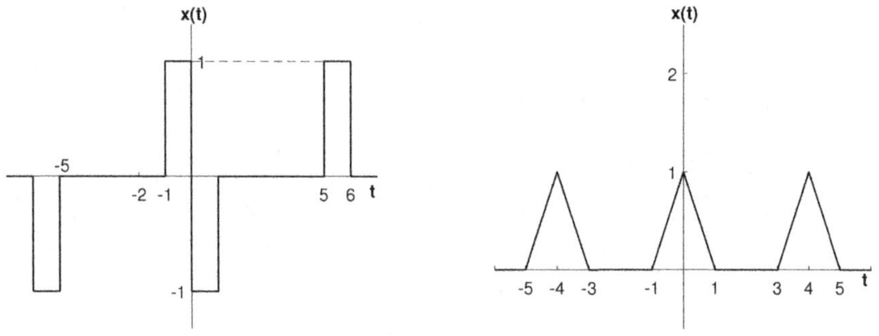

(a) (b)

Figure 4.18: Problem 68.

Solution. To solve this problem, we review the Fourier series formulation for Impulse Train and Pulse Train signals discussed in Section 4.3.4:
- The coefficients a_k for the rectangular pulse train:

$$x(t) = 1 \quad \text{for} \quad -T_1 < t < T_1,$$

with fundamental frequency $w_0 = \frac{2\pi}{T}$ is equal to,

$$a_k = \frac{\sin(k\pi\frac{2T_1}{T})}{k\pi} \quad \text{and} \quad a_0 = \frac{2T_1}{T}.$$

- The coefficients a_k for the impulse train:

$$x(t) = \sum_{n=-\infty}^{\infty} \delta(t - nT),$$

is equal to

$$a_k = \frac{1}{T}.$$

Also, we review the Time Shifting property as well as Time-Differentiation property discussed in Section 4.5:
- Time Shifting property:

$$x(t) \overset{\mathcal{F.S.}}{\longleftrightarrow} a_k$$

$$x(t - t_0) \overset{\mathcal{F.S.}}{\longleftrightarrow} a_k e^{-jkw_0 t_0}$$

- Time differentiation property:

$$x(t) \overset{\mathcal{F.S.}}{\longleftrightarrow} a_k$$

$$\frac{dx(t)}{dt} \overset{\mathcal{F.S.}}{\longleftrightarrow} jkw_0 a_k,$$

where w_0 is fundamental angular frequency of signal $x(t)$.

(a) Signal $x(t)$ can be seen as the sum of two periodic signals $x_1(t)$ and $x_2(t)$ as seen in Fig. 4.19:

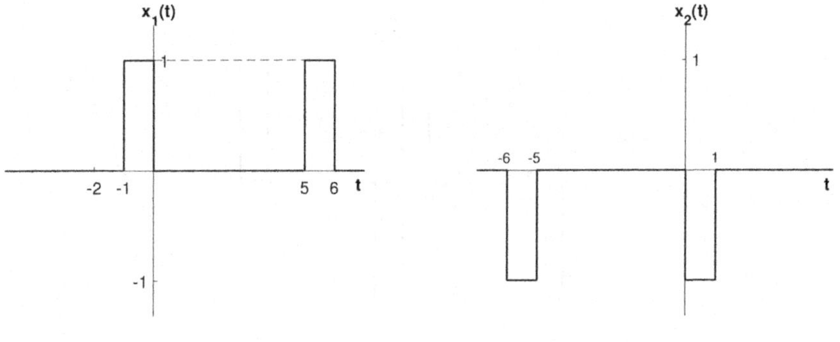

Figure 4.19: Problem 68 - Part (a), Signals $x_1(t)$ and $x_2(t)$.

Signals $x_1(t)$ ans $x_2(t)$ are shifted versions of the rectangular pulse $w(t)$ signal with $T_1 = 1/2$ and $T = 6$:

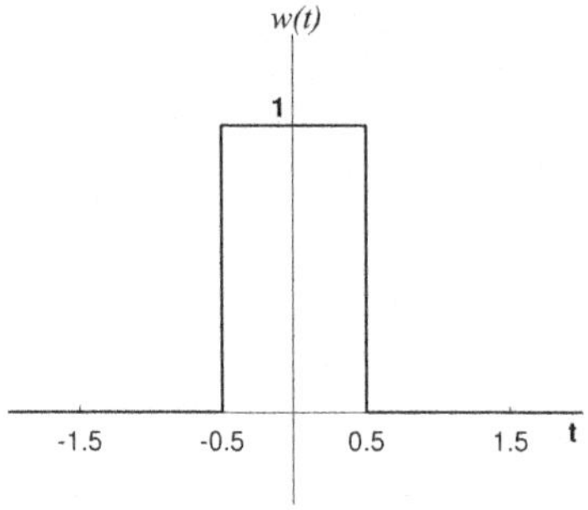

$$\xrightarrow[T_1=\frac{1}{2} \quad \text{and} \quad T=6]{\mathcal{F.S.}} \quad a_k = \frac{\sin(k\pi\frac{2\times\frac{1}{2}}{6})}{k\pi} = \frac{\sin(\frac{k\pi}{6})}{k\pi}.$$

Now we have:

$$x_1(t) = w(t+\frac{1}{2}) \quad \xrightarrow{\mathcal{F.S.}} \quad a_{k1} = a_k e^{jk(\frac{2\pi}{6})(\frac{1}{2})} = \frac{\sin(\frac{k\pi}{6})}{k\pi} e^{j\frac{k\pi}{6}},$$

and,

$$x_2(t) = -w(t-\frac{1}{2}) \quad \xrightarrow{\mathcal{F.S.}} \quad a_{k2} = -a_k e^{-jk(\frac{2\pi}{6})(\frac{1}{2})} = -\frac{\sin(\frac{k\pi}{6})}{k\pi} e^{-j\frac{k\pi}{6}}.$$

Finally according to the linearity property we can write

$$x(t) = x_1(t) + x_2(t) \quad \xrightarrow{\mathcal{F.S.}} \quad c_k = a_{k1} + a_{k2} = \frac{\sin(\frac{k\pi}{6})}{k\pi} e^{j\frac{k\pi}{6}} - \frac{\sin(\frac{k\pi}{6})}{k\pi} e^{-j\frac{k\pi}{6}}.$$

(b) It will not be easy to obtain the Fourier series directly. So we differentiate $x(t)$ and sketch $x'(t)$ as shown in the Fig. 4.20. As can be seen, $x'(t)$ is the same signal as Part (a).

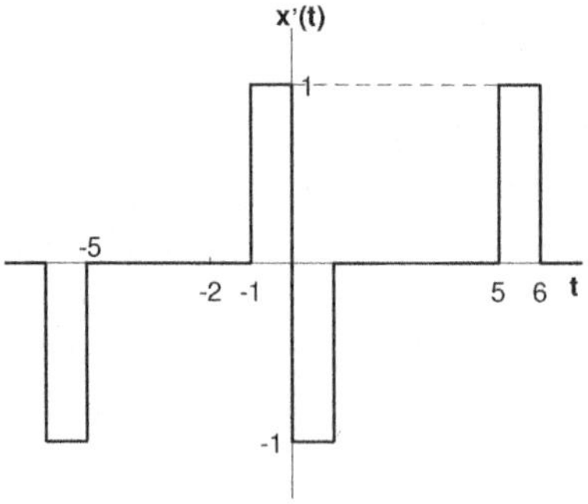

Figure 4.20: Problem 68 - Part (b), The derivative of signal $x(t)$.

So we have

$$x'(t) \quad \xrightarrow{\mathcal{F.S.}} \quad \frac{\sin(\frac{k\pi}{6})}{k\pi} e^{j\frac{k\pi}{6}} - \frac{\sin(\frac{k\pi}{6})}{k\pi} e^{-j\frac{k\pi}{6}}.$$

$$x(t) \xrightarrow{\mathcal{F.S.}} c_k = \frac{a_k}{jkw_0} = \frac{\sin(\frac{k\pi}{6})}{jk^2w_0\pi}e^{j\frac{k\pi}{6}} - \frac{\sin(\frac{k\pi}{6})}{jk^2w_0\pi}e^{-j\frac{k\pi}{6}}.$$

$$\xrightarrow[w_0=\frac{\pi}{3}]{\mathcal{F.S.}} c_k = 2\frac{\sin(\frac{k\pi}{6})}{k^2w_0\pi}\sin(\frac{k\pi}{6}) = \frac{6\sin^2(\frac{k\pi}{6})}{(k\pi)^2}.$$

Problem 69. A continuous-time periodic signal $x(t)$ is real valued and has a fundamental period $T = 4$ ($w = \frac{\pi}{2}$). Determine $x(t)$ if the Fourier series coefficients of $x(t)$ are

$$a_k = \frac{\sin(\frac{k\pi}{4})}{k\pi}(2\cos(\frac{k\pi}{4}) + 2e^{-\frac{jk\pi}{4}}).$$

Solution. We have to see if we can rewrite a_k as the summation of Fourier series coefficient of some known signals. Due to the existence of the sine function in numerator and k in the denominator, it is likely that the signal includes some elements of the pulse train. The complex exponential term also represents some kind of time shift. To get rid of the cosine term, we can also write it in terms of complex exponential functions using the Euler's formula. We write:

$$\begin{aligned} a_k &= \frac{\sin(\frac{k\pi}{4})}{k\pi}(2\cos(\frac{k\pi}{4}) + 2e^{-\frac{jk\pi}{4}}) \\ &= \frac{\sin(\frac{k\pi}{4})}{k\pi}(e^{\frac{jk\pi}{4}} + e^{-\frac{jk\pi}{4}} + 2e^{-\frac{jk\pi}{4}}) \\ &= \left[\frac{\sin(\frac{k\pi}{4})}{k\pi}e^{\frac{jk\pi}{4}}\right] + \left[\frac{3\sin(\frac{k\pi}{4})}{k\pi}e^{-\frac{jk\pi}{4}}\right]. \end{aligned}$$

Comparing the formulation of the Fourier series of a pulse train $w(t)$ and the above equation, we see that the coefficient of $k\pi$ inside the sinus function is $\frac{1}{4}$ which has to be set equal to $\frac{2T_1}{T}$ where $T = 4$ and T_1 is half of the pulse width. Thus we obtain $T_1 = \frac{1}{2}$. In other words:

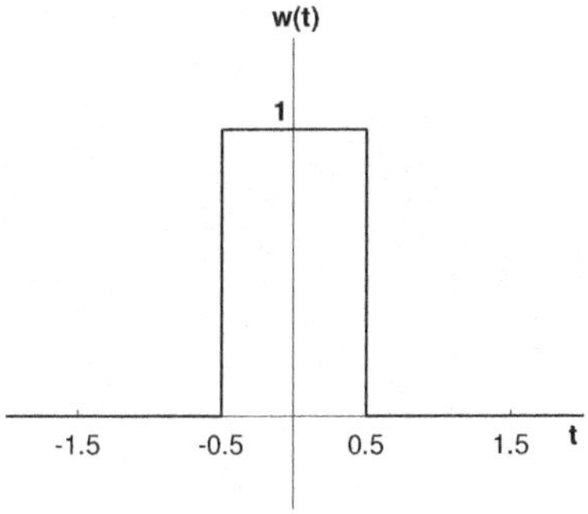

$$\xrightarrow[T_1=\frac{1}{2} \quad \text{and} \quad T=4]{\mathcal{F.S.}} \quad c_k = \frac{\sin(k\pi\frac{2\times\frac{1}{2}}{4})}{k\pi} = \frac{\sin(\frac{k\pi}{4})}{k\pi}.$$

On the other hand, we can see that terms in a_k represent the time shifts of $w(t)$ as follows:

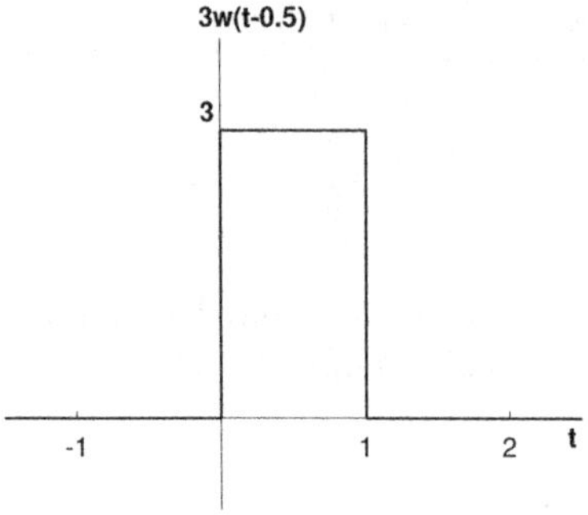

$$\xrightarrow[T=4]{\mathcal{F.S.}} \quad c_k^{'} = 3\frac{\sin(\frac{k\pi}{4})}{k\pi}e^{jkw(-\frac{1}{2})},$$

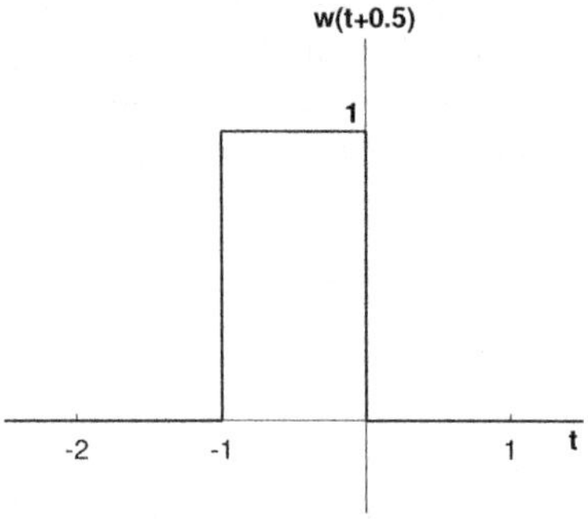

$$\xrightarrow[T=4]{\mathcal{F.S.}} \quad c_k^{''} = \frac{\sin(\frac{k\pi}{4})}{k\pi}e^{jkw(\frac{1}{2})}.$$

Now using the fact that $a_k = c_k^{''} + c_k^{'}$, we have

$$x(t) = w(t+\frac{1}{2}) + 3w(t-\frac{1}{2}).$$

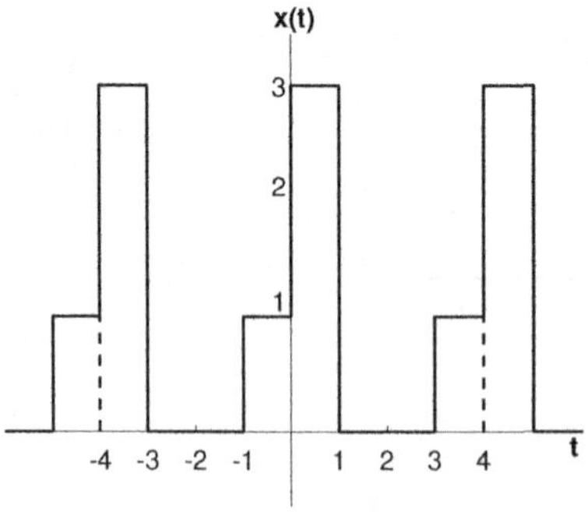

Figure 4.21: Signal $x(t)$ of Problem 69.

Problem 70. Let $x(t)$ be a periodic signal with period $T = 5$ where the Fourier series coefficients of the signal are given as follows:

$$a_{-2} = 3,\ a_{-1} = 1,\ a_0 = 3,\ a_1 = 1,\ a_2 = 3.$$

Evaluate the integrals in Parts (a) and (b).

(a) $A = \int_{-1}^{4} x(t)dt$ (b) $B = \int_{0}^{5} x(t)e^{j\frac{4\pi}{5}t}dt$

Solution. Since $T = 5$, w_0 is equal to $\frac{2\pi}{5}$. As can be seen, both integrals are taken over one period. Therefore, we may be able to use the values of a_k. By definition,

$$a_k = \frac{1}{T}\int_T x(t)e^{-jkw_0t}dt = \frac{1}{T}\int_T x(t)e^{-jk\frac{2\pi}{5}t}dt. \tag{4.32}$$

(a) Setting $k = 0$ in (4.32), we get

$$a_0 = \frac{1}{T}\int_T x(t)dt \rightarrow A = \int_{-1}^{4} x(t)dt = 5a_0 = 15$$

(b) Setting $k = -2$ in (4.32), we get

$$a_{-2} = \frac{1}{T}\int_T x(t)e^{jk\frac{4\pi}{5}t}dt \rightarrow B = \int_0^5 x(t)e^{j\frac{4\pi}{5}t}dt = 5a_{-2} = 15.$$

Problem 71. In each of the following, we specify the Fourier series coefficients of a signal that is periodic. Determine $x(t)$ in each case.

(a) $a_1 = j,\ a_{-1} = -j,\ a_3 = 2,\ a_{-3} = 2,\ a_k = 0$, otherwise, $w_0 = 2\pi$

(b) $a_k = (-\frac{1}{5})^{|k|}\ ,\ w_0 = 1$

(c) $a_k = \begin{cases} ke^{j\pi/2} & ,|k| \leq 3 \\ 0 & ,o.w. \end{cases}\ ,\ w_0 = \frac{2\pi}{T}$

Solution.

(a) The Fourier series coefficients are zero except for $k \in \{-3, -1, 1, 3\}$. According to the definition we have:

$$\begin{aligned} x(t) &= \sum_{k=-\infty}^{\infty} a_k e^{j2k\pi t} \\ &= je^{j(1)2\pi t} - je^{j(-1)2\pi t} + 2e^{j(3)2\pi t} + 2e^{j(-3)2\pi t} \\ &= -2\sin(2\pi t) + 4\cos(6\pi t). \end{aligned}$$

(b) We simply go by the definition and then try to reach a closed form if possible:

$$\begin{aligned}
x(t) &= \sum_{k=-\infty}^{\infty} a_k e^{jkt} \\
&= \sum_{k=-\infty}^{\infty} (-\frac{1}{5})^{|k|} e^{jkt} \\
&= \sum_{k=-\infty}^{-1} (-\frac{1}{5})^{|k|} e^{jkt} + \sum_{k=0}^{\infty} (-\frac{1}{5})^{|k|} e^{jkt} \\
&= \sum_{k=-\infty}^{-1} (-\frac{1}{5})^{(-k)} e^{jkt} + \sum_{k=0}^{\infty} (-\frac{1}{5})^{k} e^{jkt} \\
&= \sum_{k=1}^{\infty} (-\frac{1}{5})^{(k)} e^{-jkt} + \sum_{k=0}^{\infty} (-\frac{1}{5})^{k} e^{jkt} \\
&= \sum_{k=1}^{\infty} (-\frac{1}{5} e^{-jt})^{(k)} + \sum_{k=0}^{\infty} (-\frac{1}{5} e^{jt})^{k}.
\end{aligned}$$

Both terms are infinite sums of geometric series, so we write:

$$\begin{aligned}
x(t) &= \sum_{k=1}^{\infty} (-\frac{1}{5} e^{-jt})^{(k)} + \sum_{k=0}^{\infty} (-\frac{1}{5} e^{jt})^{k} \\
&= \frac{1}{1+\frac{1}{5}e^{jt}} - \frac{\frac{1}{5}e^{-jt}}{1+\frac{1}{5}e^{-jt}} \\
&= \frac{5}{5+e^{jt}} - \frac{e^{-jt}}{5+e^{-jt}} \\
&= \frac{5(5+e^{-jt}) - (5+e^{jt})e^{-jt}}{(5+e^{jt})(5+e^{-jt})} \\
&= \frac{24}{25+5e^{jt}+5e^{-jt}+1} \\
&= \frac{24}{10\cos(t)+26}.
\end{aligned}$$

(c) We only have 6 nonzero Fourier series coefficients at: a_{-3}, a_{-2} , a_{-1}, a_1, a_2 and a_3. Since

$x(t)$ has fundamental frequency of $w_0 = \frac{2\pi}{T}$ and $e^{j\pi/2} = j$, it follows that

$$\begin{aligned}
x(t) &= \sum_{k=-\infty}^{\infty} a_k e^{jk\frac{2\pi}{T}t} \\
&= \sum_{k=-3}^{3} (jk) e^{jk\frac{2\pi}{T}t} \\
&= (-3j)e^{-3j\frac{2\pi}{T}t} + (-2j)e^{-2j\frac{2\pi}{T}t} + (-j)e^{-1j\frac{2\pi}{T}t} + 0 + (j)e^{j\frac{2\pi}{T}t} + (2j)e^{2j\frac{2\pi}{T}t} + (3j)e^{3j\frac{2\pi}{T}t} \\
&= 3j\left(e^{j\frac{6\pi}{T}t} - e^{-j\frac{6\pi}{T}t}\right) + 2j\left(e^{j\frac{4\pi}{T}t} - e^{-j\frac{4\pi}{T}t}\right) + j\left(e^{j\frac{2\pi}{T}t} - e^{-j\frac{2\pi}{T}t}\right) \\
&= -6\sin(\frac{6\pi}{T}t) - 4\sin(\frac{4\pi}{T}t) - 2\sin(\frac{2\pi}{T}t).
\end{aligned}$$

Problem 72. Suppose that $x(t)$ is a periodic signal with period T and Fourier series coefficients a_k.

(a) Let $T = 7$. Now let periodic signal $y(t)$ with period $T' = 7$ have Fourier series coefficients $b_k = a_{k-5}e^{jk\frac{4\pi}{7}}$. Write $y(t)$ based on $x(t)$.

(b) Let $T = 3$. Determine the periodic signal $y(t)$ with period $T' = 3$ and Fourier series coefficients $b_k = \text{Re}\{a_k\}$.

Solution.

(a) We have to use time-shifting and frequency-shifting properties in Section 4.5. We have:

$$T = 7 \rightarrow w_0 = \frac{2\pi}{7} \quad \text{(fundamental frequency)}$$

$$\rightarrow b_k = a_{k-5}e^{jk\frac{4\pi}{7}} = a_{k-5}e^{jkw_0(2)}.$$

So the value of time shift is 2. The following 2 lines show the changes that happen in parallel to a_k and $x(t)$ during time-shift and frequency-shift:

$$a_k \xrightarrow{k \longrightarrow k-5} a_{k-5} \longrightarrow a_{k-5}e^{jkw_0(2)}$$

$$x(t) \longrightarrow x(t)e^{j(5)w_0 t} \xrightarrow{t \rightarrow t+2} x(t+2)e^{j(5)w_0(t+2)}$$

As such, we have:

$$\begin{aligned}
y(t) &= x(t+2)e^{j(5)w_0(t+2)} \\
&= x(t+2)e^{j(5)\frac{2\pi}{7}(t+2)} \\
&= x(t+2)e^{\frac{10\pi}{7}(t+2)}
\end{aligned}$$

(b) We know that $\text{Re}\{z\} = \dfrac{z + z^*}{2}$. Using the linearity, conjugation, and time reversal properties of Section 4.5, we see that if the Fourier series coefficient of $x(t)$ are a_k's, then

$$x(t) + x^*(-t) \quad \xrightarrow{\mathcal{F.S.}} \quad a_k + a_k^*,$$

where $w_0 = 2\pi/3$.
As such,

$$y(t) = \frac{x(t) + x^*(-t)}{2}.$$

Problem 73. Let $x(t)$ be a periodic signal with period $T = 6$ and Fourier series coefficients a_k. Determine signal $y(t)$ with Fourier series coefficients $b_k = (-1)^{3k}a_k + (-1)^{3k}a_{-k}$.

Solution. Since $T = 6$, we have $w_0 = \frac{\pi}{3}$. We try to manipulate the relationship between a_k and b_k so that we can use the properties of Section 4.5. We have:

$$\begin{aligned} b_k &= (-1)^{3k}a_k + (-1)^{3k}a_{-k} \\ &= e^{j3k\pi}a_k + e^{j3k\pi}a_{-k} \\ &= e^{jk\frac{\pi}{3}(9)}a_k + e^{jk\frac{\pi}{3}(9)}a_{-k} \\ &= e^{jkw_0(9)}a_k + e^{jkw_0(9)}a_{-k}, \qquad w_0 = \frac{\pi}{3}. \end{aligned}$$

Now if a_k's represent the Fourier series coefficients of $x(t)$, then according to the linearity and time reversal property we have:

$$a_k + a_{-k} \xrightarrow{Inverse\,\mathcal{F.S.}} x(t) + x(-t).$$

Based on time-shifting property we can write:

$$b_k = e^{jkw_0(9)}(a_k + a_{-k}) \xrightarrow[t \to t+9]{Inverse\,\mathcal{F.S.}} x(t+9) + x(-(t+9)).$$

Thus:

$$y(t) = x(t+9) + x(-(t+9)).$$

Problem 74. Let $x(t) = |\cos(2t)|$ with Fourier series coefficients a_k. Compute the following expression:

$$A = \sum_{k=-\infty}^{\infty} |a_k|^2.$$

Solution. Note that we do not really need to obtain a_k's to evaluate the summation. Using Parseval's identity of Section 4.5 and noting that the period of the signal is $T = \frac{\pi}{2}$, we have

$$\begin{aligned} A &= \sum_{k=-\infty}^{\infty} |a_k|^2 = \frac{1}{T}\int_T |x(t)|^2 dt \\ &= \frac{2}{\pi}\int_{-\frac{\pi}{4}}^{\frac{\pi}{4}} \cos^2(2t)dt \\ &= \frac{2}{\pi}\int_{-\frac{\pi}{4}}^{\frac{\pi}{4}} \frac{1+\cos(4t)}{2}dt \\ &= \frac{2}{\pi} \times \frac{\pi}{4} = \frac{1}{2}. \end{aligned}$$

Problem 75. As shown in the Fig. 4.22, $x(t)$ is a periodic signal with period $T = 4$ and Fourier series coefficients a_k. Compute the following expression:

$$A = \sum_{k=-\infty}^{\infty} |ka_k|^2$$

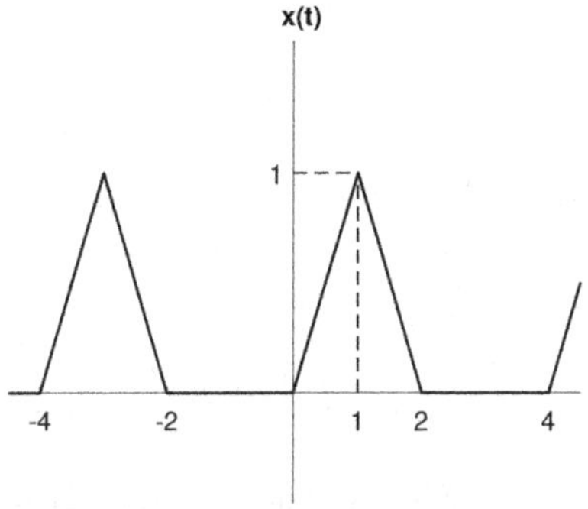

Figure 4.22: Problem 75, Periodic signal $x(t)$.

Solution. Similar to Problem 74, we should find out a way to deploy the Parseval's identity. We know from time-differentiation property that:

$$x(t) \xrightarrow{\mathcal{F.S.}} a_k \quad \Rrightarrow \quad x^{'}(t) \xrightarrow{\mathcal{F.S.}} jkw_0a_k.$$

Applying the Parseval's identity, we have:

$$\sum_{k=-\infty}^{\infty} |jkw_0a_k|^2 = \frac{1}{T}\int_T |x^{'}(t)|^2dt.$$

Manipulating the left hand side,

$$\sum_{k=-\infty}^{\infty} |jkw_0a_k|^2 = \sum_{k=-\infty}^{\infty} w_0^2|ka_k|^2.$$

As such,

$$\Rrightarrow \quad \sum_{k=-\infty}^{\infty} |ka_k|^2 = \frac{1}{Tw_0^2}\int_T |x^{'}(t)|^2dt.$$

Therefore, we have to find the derivative of the signal:

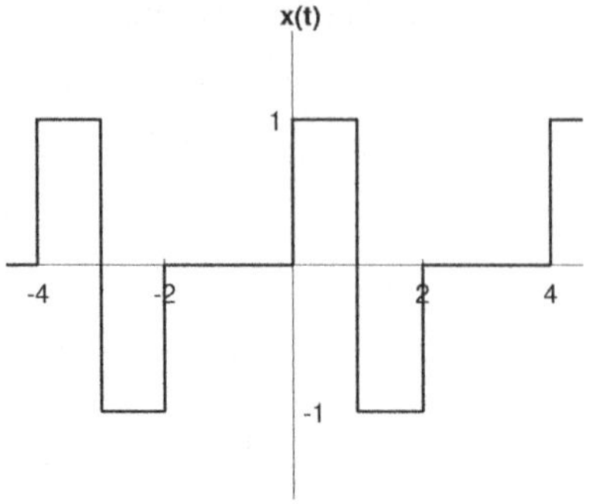

Figure 4.23: Problem 75, Signal $x'(t)$.

Finally,

$$A = \sum_{k=-\infty}^{\infty} |ka_k|^2 = \frac{1}{Tw_0^2} \int_T |x^{'}(t)|^2 dt = \frac{1}{4 \times \frac{\pi^2}{4}} \times 2 = \frac{2}{\pi^2}.$$

Problem 76. Consider signal $x(t)$ with a fundamental period of T_0. Determine the Fourier series coefficients for each of the following cases.

(a) $y(t) = x(1 + 3t)$

(b) $y(t) = x^*(5t - 3)$

(c) $y(t) = x(t) + x(-t) \quad , \quad x(t)$ is real

(d) $y(t) = x(t). \cos(\frac{2\pi t}{T_0})$

Solution. For each case, let w denote the fundamental frequency. For $x(t)$ we have $w = w_0 = 2\pi/T_0$. Also let $z(t)$ be an arbitrary signal with Fourier series coefficients c_k. When necessary, this signal is used to show how each system operate on the input, in both time and frequency domains.

(a) $y(t) = x(1 + 3t)$
In order to have $x(\alpha t + \beta)$, we first perform the time shifting and then apply the time scaling:

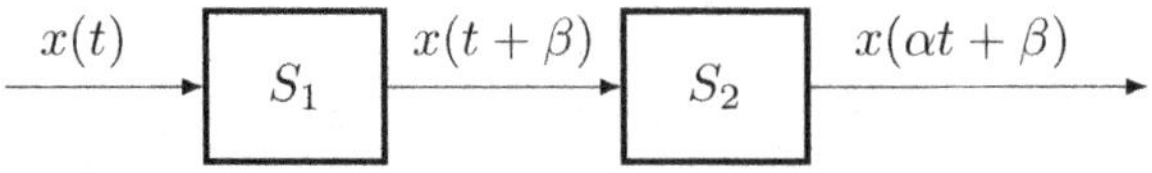

$$\begin{cases} S_1 : & z(t) \to z(t+\beta) \overset{\mathcal{F.S.}}{\longleftrightarrow} c_k \to c_k e^{jkw_0\beta}, \ w = w_0 \\ S_2 : & z(t) \to z(\alpha t) \overset{\mathcal{F.S.}}{\longleftrightarrow} c_k \to c_k, \ w = \alpha w_0 \end{cases}.$$

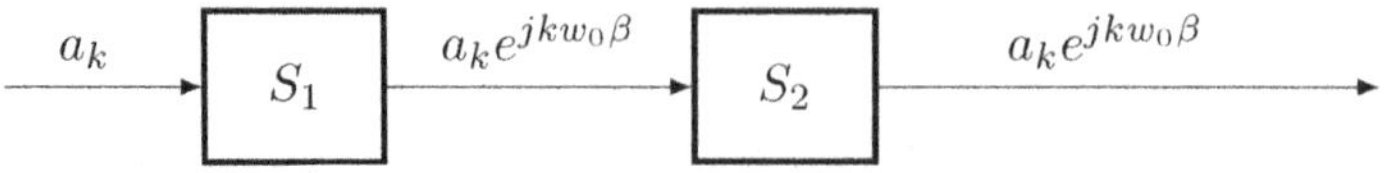

Thus,

$$b_k = a_k e^{jkw_0}, \quad w = 3w_0.$$

(b) $y(t) = x^*(5t - 3)$
The following block diagram demonstrates the steps to come up with $y(t)$:

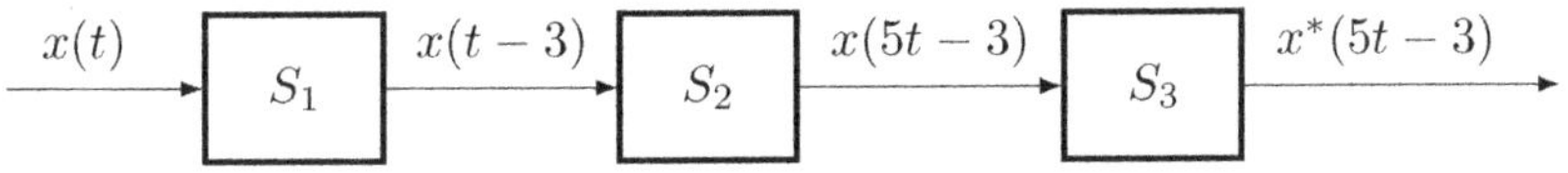

$$\begin{cases} S_1: & z(t) \rightarrow z(t+\beta) \overset{\mathcal{F.S.}}{\longleftrightarrow} c_k \rightarrow c_k e^{jkw_0\beta}, \; w = w_0 \\ S_2: & z(t) \rightarrow z(\alpha t) \overset{\mathcal{F.S.}}{\longleftrightarrow} c_k \rightarrow c_k, \; w = \alpha w_0 \\ S_3: & z(t) \rightarrow z^*(t) \overset{\mathcal{F.S.}}{\longleftrightarrow} c_k \rightarrow c^*_{-k}, \; w = w_0 \end{cases}$$

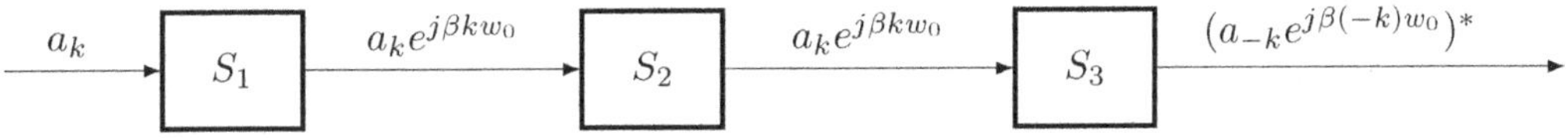

Thus,

$$b_k = (a_{-k}e^{-j\beta k w_0})^* = a^*_{-k}e^{j\beta k w_0},$$

substituting $\beta = -3$ into above equation, we have

$$b_k = a^*_{-k}e^{-j3kw_0}, \; w = 5w_0.$$

(c) $y(t) = x(t) + x(-t)$, $x(t)$ is real.

$$x(-t) \longleftrightarrow a_{-k}, \; w = w_0.$$

Thus,

$$x(t) + x(-t) \longleftrightarrow a_k + a_{-k}, \; w = w_0.$$

Furthermore, as $x(t)$ is real, the Hermitian symmetry holds:

$$a_{-k} = a_k^*.$$

As such:

$$b_k = a_k + a_{-k} = a_k + a_k^* = 2\mathrm{Re}\{a_k\}.$$

(d) $y(t) = x(t).\cos(\frac{2\pi t}{T_0})$

$$\begin{aligned} y(t) &= x(t).\cos(\frac{2\pi t}{T_0}) \\ &= x(t) \times \frac{1}{2}(e^{jw_0 t} + e^{-jw_0 t}) \\ &= \frac{1}{2}x(t)e^{jw_0 t} + \frac{1}{2}x(t)e^{-jw_0 t}, \\ \Rightarrow b_k &= \frac{1}{2}a_{k-1} + \frac{1}{2}a_{k+1}. \end{aligned}$$

Problem 77. Consider signal $x(t)$ with a fundamental period of $T = 4$ and the Fourier series coefficients $F_k = \mathrm{sinc}(k\frac{\pi}{2})$. Determine the Fourier series coefficients for each of the following cases.

(a) $y(t) = x(2t)$

(b) $y(t) = x^{'}(t)$

(c) $y(t) = x(t - \frac{1}{5})$

(d) $y(t) = \text{Re}\{x(t)\}$

(e) $y(t) = x(t).\cos(2\pi t)$

Solution. The fundamental frequency w of $x(t)$ is $w_0 = \frac{2\pi}{T} = \frac{\pi}{2}$.

(a) $y(t) = x(2t)$
Time scaling is an operation that changes the period of the underlying signal. It, however, generally does not change the Fourier coefficients:

$$y(t) = x(2t) \leftrightarrow a_k = F_k = \text{sinc}(k\frac{\pi}{2}), \; w = 2w_0.$$

(b) $y(t) = x^{'}(t)$
Using the time-differentiation property we write:

$$a_k = jkw_0F_k \quad \xrightarrow{w_0=\pi/2} \quad a_k = jk\frac{\pi}{2}\text{sinc}(k\frac{\pi}{2}), \; w = w_0.$$

(c) $y(t) = x(t - \frac{1}{5})$
Using the time shifting property we have:

$$a_k = e^{-\frac{jkw_0}{5}}F_k \quad \xrightarrow{w_0=\pi/2} \quad a_k = e^{-\frac{jk\pi}{10}}\text{sinc}(k\frac{\pi}{2}), \; w = w_0.$$

(d) $y(t) = \text{Re}\{x(t)\}$

$$y(t) = \text{Re}\{x(t)\} = \frac{1}{2}\{x(t) + x^*(t)\}$$

$$a_k = \frac{1}{2}F_k + \frac{1}{2}F^*_{-k} = \frac{1}{2}\text{sinc}(k\frac{\pi}{2}) + \frac{1}{2}\text{sinc}(-k\frac{\pi}{2}), \; w = w_0.$$

(e) $y(t) = x(t).\cos(2\pi t)$

$$y(t) = x(t).\cos(2\pi t) = \frac{1}{2}\left[e^{j2\pi t} + e^{-j2\pi t}\right]x(t) = \frac{1}{2}\left[e^{j4w_0t} + e^{-j4w_0t}\right]x(t).$$

Using the frequency-shifting property,

$$a_k = \frac{1}{2}F_{k-4} + \frac{1}{2}F_{k+4}$$

As such,

$$a_k = \text{sinc}((k-4)\frac{\pi}{2}) + \text{sinc}((k+4)\frac{\pi}{2}).$$

Problem 78. Suppose we are given the following facts about a signal $x(t)$:

(a) $x(t)$ is real and periodic signal with period $T = 4$, and it has Fourier series coefficients a_k.

(b) $a_k = 0$ for $k \geq 3$.

(c) the DC Fourier coefficient a_0 is zero.

(d) $x(t) = -x(t-2)$

(e) $\int_0^4 |x(t)|^2 dt = 2$

(f) a_1 is a negative real number.

Determine $x(t)$.

Solution.
First note that $w_0 = 2\pi/T = \pi/2$.
To find $x(t)$, we may want to first find out its Fourier series coefficients. Since the signal is real, a_k is conjugate symmetric, i.e, $a_k^* = a_{-k}$ and since for $k \leq -3$, $a_k = 0$, for $k \geq 3$, we also have $a_k = 0$. We also know from Part (c) that $a_0 = 0$. As such, we just have nonzero coefficients a_2, a_1, a_{-1} and a_{-2}.
Based on the time-shifting property, the coefficients b_k for signal $-x(t-2)$ are

$$b_k = -e^{-jkw_0(2)} a_k.$$

Based on Part (d), $b_k = a_k, \forall k$ and so

$$-e^{-jk\pi} a_k = a_k.$$

Now for odd values of k, the identity changes to the trivial identity $a_k = a_k$. For even values of k, we have $-a_k = a_k$ which means that $a_k = 0$. Therefore, $a_2 = a_{-2} = 0$.

Now we end up with only two nonzero components, a_1 and a_{-1}. Using Parseval's identity and Part (e) we have

$$\sum_{k=-\infty}^{\infty} |a_k|^2 = \frac{1}{T} \int_T |x(t)|^2 dt = \frac{1}{4} \int_0^4 |x(t)|^2 dt = \frac{1}{4} \times 2.$$

Thus:

$$|a_1|^2 + |a_{-1}|^2 = \frac{1}{2}.$$

Using the conjugate symmetry property again we have

$$\xrightarrow{a_1^* = a_{-1}} |a_1|^2 + |a_{-1}|^2 = |a_1|^2 + |a_1|^2 = \frac{1}{2}$$

$$\rightarrow |a_1|^2 = \frac{1}{4} \rightarrow |a_1| = \frac{1}{2},$$

Using Part (f), we can say that

$$a_1 = -\frac{1}{2}$$

and according to conjugate symmetry property,

$$a_k^* = a_{-k} \xrightarrow{k=1} a_1^* = a_{-1} \Rightarrow a_{-1} = a_1^* = -\frac{1}{2}.$$

Now we can use Fourier series definition to write:

$$\begin{aligned}x(t) &= \sum_{k=-\infty}^{\infty} a_k e^{jkw_0 t} \\ &= a_{-1}e^{j(-1)\frac{2\pi}{4}t} + a_1 e^{j(1)\frac{2\pi}{4}t} \\ &= -\frac{1}{2}e^{j(-1)\frac{2\pi}{4}t} - \frac{1}{2}e^{j(1)\frac{2\pi}{4}t} \\ &= -\cos(\frac{\pi}{2}t).\end{aligned}$$

Problem 79. Suppose we are given a signal $x(t)$ with period $T = 2$, and it has Fourier series coefficients C_n where,

$$C_n = \begin{cases} 0, & n \text{ is even} \\ 1, & n \text{ is odd} \end{cases}.$$

The signal is entered into a filter with frequency response $H(jw)$, shown in the following figure. Find the output signal $y(t)$.

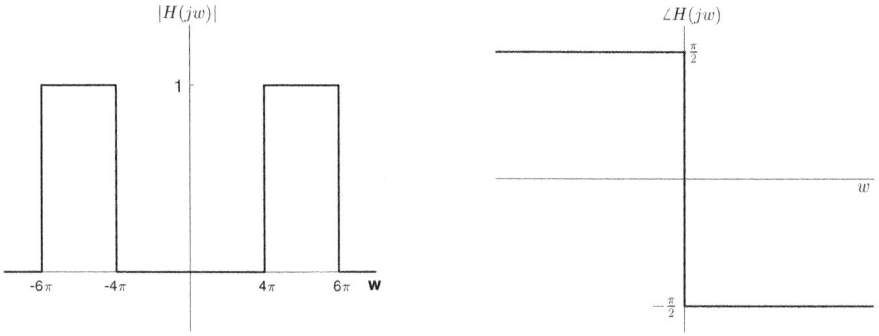

Figure 4.24: Problem 79, Magnitude and Phase of the considered filter $H(jw)$.

Solution.
For this signal, we have $w_0 = \frac{2\pi}{2} = \pi$. We expand $x(t)$ as

$$\begin{aligned}x(t) &= \sum_{n=-\infty}^{\infty} C_n e^{jkw_0 t} \\ &= \overbrace{2\cos(\pi t)}^{n=1,-1} + \overbrace{2\cos(3\pi t)}^{n=3,-3} + \overbrace{2\cos(5\pi t)}^{n=5,-5} + \ldots\end{aligned},$$

Now according to the frequency response $H(jw)$, we see that only harmonic frequency 5π can pass and other components are removed by the filter. We now rewrite the cosine function as the sum of 2 exponentials to also capture the phase change due to the filter:

$$2\cos(5\pi t) = e^{j5\pi t} + e^{-j5\pi t}.$$

The filter does not change the magnitude of the exponential functions and so we have:

$$\begin{aligned}
y(t) &= (1\angle 90^\circ)e^{-j5\pi t} + (1\angle -90^\circ)e^{j5\pi t} \\
&= e^{\frac{j\pi}{2}}e^{-5\pi t} + e^{-\frac{j\pi}{2}}e^{5\pi t} \\
&= j(e^{-j5\pi t} - e^{j5\pi t}) \\
&= 2\sin(5\pi t) = 2\cos(5\pi t - \frac{\pi}{2}).
\end{aligned}$$

Problem 80. Consider signals $x(t)$ and $y(t) = x^*(t) + x(-\frac{t}{3})$ with the Fourier series coefficients a_k and b_k, respectively. Find the value of b_k based on a_k.

Solution.

Suppose that the fundamental frequency of $x(t)$ is w_0, thus, the fundamental frequencies of $x^*(t)$ and $x(-\frac{t}{3})$ are w_0 and $\frac{w_0}{3}$, respectively. The fundamental frequency of $y(t)$ is equal to $\frac{w_0}{3}$ because that is the greatest common divisor of w_0 and $\frac{w_0}{3}$.

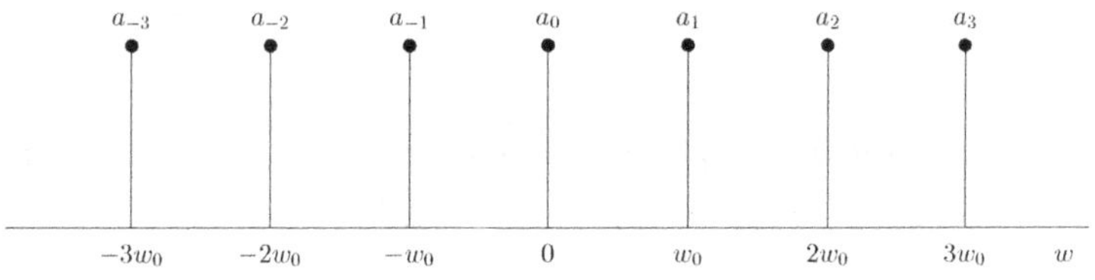

Figure 4.25: Problem 80, Fourier series coefficients of $x(t)$.

Now according to the plot of a_k's, we obtain the plot for coefficients of $x(t/3), x(-t/3)$ and $x^*(t)$.

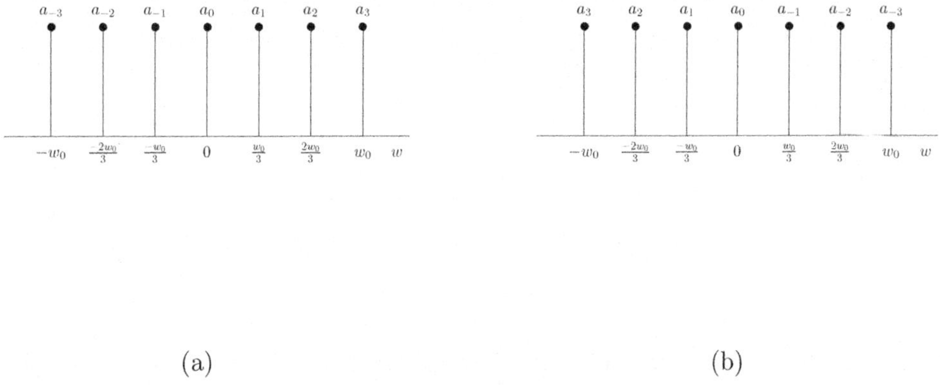

Figure 4.28: Problem 80: Fourier coefficients for (a) $x(t/3)$; (b) $x(-t/3)$.

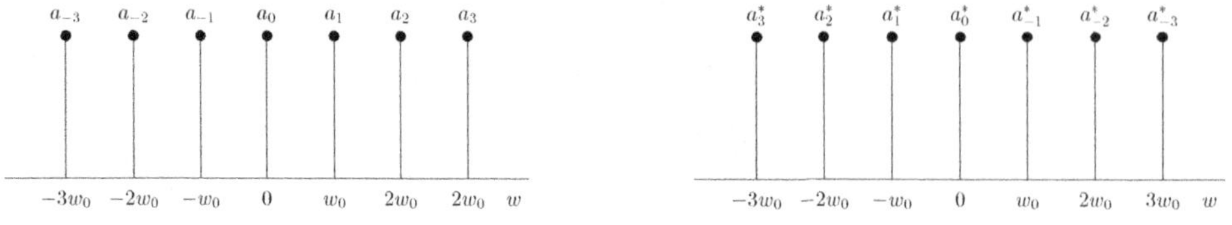

(a) (b)

Figure 4.31: Problem 80, Fourier series coefficients for (a) $x(t)$; (b) $x^*(t)$.

Using the plots for $x^*(t)$ and $x(-t/3)$, we obtain the sketch of b_k's as in the figure below.

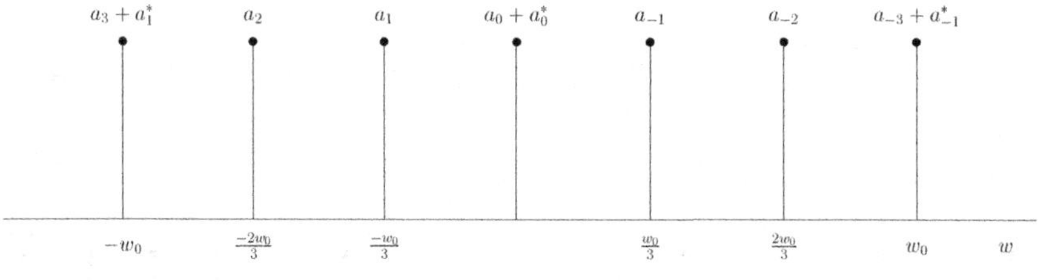

Figure 4.32: Problem 80, Fourier series coefficients of $y(t)$.

Problem 81. Suppose that periodic signal $x(t) = \cos(t)$ is the input signal to an LTI system with frequency response $H(jw)$. $H(jw)$ is demonstrated by its magnitude $|H(jw)|$ and phase $\angle H(jw)$ in figures below. Determine $y(t)$, the output of the system.

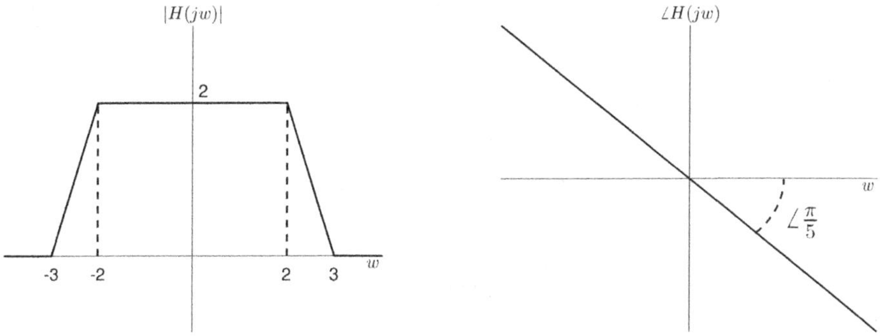

Figure 4.33: Problem 81, Magnitude and Phase of the frequency response $H(jw)$.

Solution.
We have to obtain b_k's, the Fourier series coefficients of $y(t)$ based on a_k's, the Fourier series coefficients of $x(t)$, using (4.27). The value of a_k's are shown in the figure below:

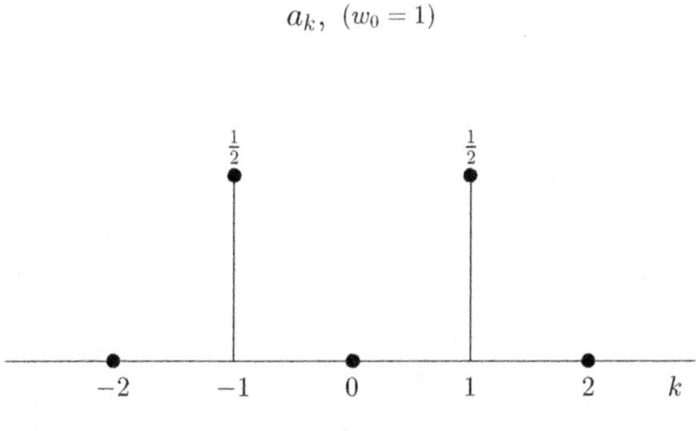

Figure 4.34: Problem 81, Fourier series coefficients of $x(t)$.

Using (4.27),

$$k = 1 \Rightarrow b_1 = a_1 H(jw_0) \xrightarrow{w_0=1} b_1 = \frac{1}{2}|H(j)|e^{j\angle H(j)} = \frac{1}{2} \times 2e^{-j\frac{\pi}{5}} = e^{-j\frac{\pi}{5}}$$

$$k = -1 \Rightarrow b_{-1} = a_{-1} H(-jw_0) \xrightarrow{w_0=1} b_{-1} = \frac{1}{2}|H(-j)|e^{j\angle H(-j)} = \frac{1}{2} \times 2e^{j\frac{\pi}{5}} = e^{j\frac{\pi}{5}}$$

Thus,

$$\begin{aligned}
y(t) &= \sum_{k=-\infty}^{\infty} b_k e^{jkw_0 t} \\
&= b_{-1}e^{-jt} + b_1 e^{jt} \\
&= e^{j\frac{\pi}{5}} e^{-jt} + e^{-j\frac{\pi}{5}} e^{jt} \\
&= e^{j(\frac{\pi}{5}-t)} + e^{-j(\frac{\pi}{5}-t)} \\
&= 2\cos(t - \frac{\pi}{5}).
\end{aligned}$$

Chapter 5

Fourier Series Representation of Discrete-time Signals

5.1 Introduction

In the previous chapter, we introduced the continuous-time Fourier Series. In this chapter, we will introduce the Fourier Series representation of discrete-time periodic signals. Expectedly, there are many similarities between the properties of discrete-time Fourier analysis and its continuous-time counterpart. Nevertheless, there are also some important differences.

5.2 The Response of the Discrete-Time LTI Systems to Complex Exponentials

Consider a discrete-time LTI system with impulse response $h[n]$ and the input $x[n]$. The response of an LTI system to the input is given by the convolution sum. With $x[n] = z^n$,

$$\begin{aligned} y[n] = x[n] * h[n] &= \sum_{k=-\infty}^{\infty} h[k]x[n-k] \\ &= \sum_{k=-\infty}^{\infty} h[k]z^{(n-k)} \\ &= z^n \sum_{k=-\infty}^{\infty} h[k]z^{-k} = z^n H(z). \end{aligned} \tag{5.1}$$

where z can be a complex variable.

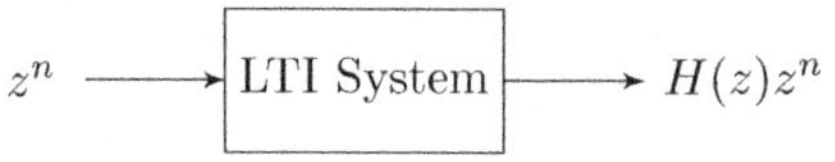

Figure 5.1: The response of LTI systems to complex exponentials.

The input signal $x[n] = z^n$ has a nice advantage: the very same version of the signal is observed at the output multiplied by $H(z)$. Using the linearity, if the input to an LTI system can be represented as a linear combination of complex exponentials, the output can also be represented as a linear combination of the same complex exponential signals, that is, if

$$x[n] = a_1 z_1^n + a_2 z_2^n + \cdots \tag{5.2}$$

then

$$y[n] = a_1 H(z_1) z_1^n + a_2 H(z_2) z_2^n + \cdots \tag{5.3}$$

For the discrete-time Fourier analysis, we suppose that $|z| = 1$ and then

$$z = |z|e^{\angle z} = |z|e^{jw} \quad \Rightarrow \quad z = e^{jw}, \quad -\pi < w < \pi$$

In general, we have 2 types of signal representation based on complex exponentials:

1. **The Discrete-Time Fourier Series**

   If a signal can be expanded solely based on linear sum of z^n with $z = e^{jkw_0}$ and coefficients a_k where
   $$k = 1, 2, 3, \cdots, N-1,$$
   then we call the resulting equivalence the Fourier series expansion of the signal.

2. **The Discrete-Time Fourier Transform**

   If the signal can not be expanded solely based on linear sum of z^n with $z = e^{jkw_0}$, the signal can still be presented based on a linear combination of signals e^{jwn} using the discrete-time Fourier transform which will be discussed in next chapters.

5.3 Fourier Series Representation of Discrete-Time Periodic Signals

In this section we will introduce the discrete-time Fourier series and in the next sections we will introduce its applications in the analysis of discrete-time signals and systems.

From Chapter 1, we know that a periodic signal is defined as

$$x[n] = x[n+N] \quad \text{for all } n. \tag{5.4}$$

The fundamental period of $x[n]$ is the smallest positive, nonzero value of N for which (5.4) is satisfied. The value f_0 is referred to as the *fundamental frequency* for which we have

$$f_0 = \frac{1}{N}, \qquad w_0 = 2\pi f_0 = \frac{2\pi}{N}.$$

5.3.1 Obtaining the Fourier Series

The pair of the Fourier series of a periodic discrete-time signal is defined as

$$x[n] = \sum_{k=0}^{N-1} a_k e^{jkw_0 n} = \sum_{k=<N>} a_k e^{jk(2\pi/N)n} \tag{5.5}$$

$$a_k = \frac{1}{N}\sum_{n=0}^{N-1} x[n]e^{-jkw_0 n} = \sum_{n=<N>} x[n]e^{-jk(2\pi/N)n} \tag{5.6}$$

where a_k's are the Fourier coefficients. Note that by $\sum_{k=<N>}$ or $\sum_{n=<N>}$, we mean the summation can be done over any consecutive range of integer numbers respectively k or n with length N. Note that for (5.6), this is equivalent to the fact that in the continuous case, we can obtain the coefficients by integrating over any arbitrary period.

The coefficient a_0 is the DC or the average value of $x[n]$ which is indicated by setting $k = 0$ in (5.6), i.e.,

$$a_0 = \frac{1}{N}\sum_{n=<N>} x[n]. \tag{5.7}$$

According to the definition of a_k, the coefficients a_k are complex in general. So the coefficients are shown by the *amplitude spectrum*, $|a_k|$, and the *phase spectrum*, $\angle a_k$, which are plotted versus the angular frequency w.

From (5.5), we see that the Fourier series representation of a discrete-time periodic signal is a finite series, in contrast to the continuous-time case. As a result, we are not worried about summation convergence as is the case for the infinite sum corresponding to continuous-time Fourier series.

5.3.2 Periodicity of Fourier Coefficients

The Fourier series coefficients of discrete-time periodic signal are periodic with period N. This is another major difference with the continuous-time case, i.e., if

$$a_k = \frac{1}{N}\sum_{n=<N>} x[n]e^{-jkw_0 n} \tag{5.8}$$

then we have

$$\begin{aligned} a_{k+N} &= \frac{1}{N}\sum_{n=<N>} x[n]e^{-j(k+N)w_0 n} \\ &= \frac{1}{N}\sum_{n=<N>} x[n]e^{-jkw_0 n}\, e^{-jN(2\pi/N)n} \\ &= \frac{1}{N}\sum_{n=<N>} x[n]e^{-jkw_0 n}\, e^{-j2\pi n} = a_k \end{aligned} \tag{5.9}$$

Note that $e^{-j2\pi n} = 1$ for all n.

5.3.3 Fourier Series of Important Signals

Some common signals and their Fourier Series are:

(a) $x[n] = e^{jw_0 n}$, $w_0 = \frac{2m\pi}{N}$

$$\Rightarrow \quad a_k = \begin{cases} 1 & k = m,\ m \pm N,\ m \pm 2N, \cdots \\ 0 & \text{otherwise} \end{cases}. \tag{5.10}$$

(b) $x[n] = \cos w_0 n$, $w_0 = \frac{2m\pi}{N}$

$$\Rightarrow \quad a_k = \begin{cases} \frac{1}{2} & k = \pm m,\ \pm m \pm N,\ \pm m \pm 2N, \cdots \\ 0 & \text{otherwise} \end{cases}. \tag{5.11}$$

(c) $x[n] = \sin w_0 n$, $w_0 = \frac{2m\pi}{N}$

$$\Rightarrow \quad a_k = \begin{cases} \frac{1}{2j} & k = m,\ m \pm N,\ m \pm 2N, \cdots \\ -\frac{1}{2j} & k = -m,\ -m \pm N,\ -m \pm 2N, \cdots \\ 0 & \text{otherwise} \end{cases}. \tag{5.12}$$

(d) $x[n] = 1$,

$$\Rightarrow \quad a_k = \begin{cases} 1 & k = 0,\ \pm N,\ \pm 2N \\ 0 & \text{otherwise} \end{cases}. \tag{5.13}$$

(e) Impulse Train with period N:

$$x[n] = \sum_{k=-\infty}^{+\infty} \delta[n - kN] \quad \Rightarrow \quad a_k = \frac{1}{N}. \tag{5.14}$$

(f) $x[n] = \begin{cases} 1 & |n| \le N_1 \\ 0 & N_1 < |n| \le \frac{N}{2} \end{cases}$, $\quad x[n + N] = x[n]$,

$$a_k = \begin{cases} \dfrac{\sin\left(\frac{2k\pi(N_1+1/2)}{N}\right)}{N \sin\left(\frac{2k\pi}{2N}\right)}, & k \ne 0,\ \pm N, \pm 2N,\ \cdots \\ \frac{2N_1+1}{N}, & k = 0,\ \pm N,\ \pm 2N,\ \cdots \end{cases} \tag{5.15}$$

5.4 Properties of the Discrete-Time Fourier Series

Let $x[n]$ and $y[n]$ be periodic signals with the Fourier series coefficients a_k and b_k, respectively. Basic properties of the Fourier series are presented in the following.

(a) Linearity

$$Ax[n] + By[n] \quad \overset{\mathcal{F.S.}}{\longleftrightarrow} \quad Aa_k + Bb_k. \tag{5.16}$$

(b) Time Shifting

The Fourier series coefficients b_k of the signal $y[n] = x[n - n_0]$ is given by

$$y[n] = x[n - n_0] \quad \stackrel{\mathcal{F.S.}}{\longleftrightarrow} \quad b_k = e^{-jkw_0 n_0} a_k. \tag{5.17}$$

From (5.17), we can see that

$$|b_k| = |a_k|, \qquad \angle b_k = -jkw_0 n_0 + \angle a_k.$$

(c) Frequency Shifting

The frequency shifting property indicates that the multiplication of $x[n]$ by a complex exponential signal $e^{jMw_0 n}$ results in an index shift of M for a_k:

$$e^{jMw_0 n}\, x[n] \quad \stackrel{\mathcal{F.S.}}{\longleftrightarrow} \quad a_{k-M}. \tag{5.18}$$

Note that

$$(-1)^n = e^{j\pi n} = e^{j(N/2)w_0 n},$$

so the Fourier series coefficients of $(-1)^n\, x[n]$ are $a_{k-N/2}$.

(d) Time Reversal

If a periodic signal undergoes time reversal, its Fourier series coefficients will see a sign change in the index:

$$x[-n] \quad \stackrel{\mathcal{F.S.}}{\longleftrightarrow} \quad a_{-k}.$$

From this property, if signal $x[n]$ is even, then we have

$$x[-n] = x[n] \quad \Rightarrow \quad a_{-k} = a_k.$$

If signal $x(t)$ is odd, then we have

$$x[-n] = -x[n] \quad \Rightarrow \quad a_{-k} = -a_k.$$

(e) Conjugation

The complex conjugation of a periodic signal $x[n]$ causes the complex conjugation and time reversal on the corresponding Fourier series coefficients. That is,

$$x^*[n] \quad \stackrel{\mathcal{F.S.}}{\longleftrightarrow} \quad a^*_{-k}.$$

(f) Time Scaling

Generally, time scaling operation causes a change in the signal period and the fundamental frequency will be changed to $w_0 = \frac{2\pi}{mN}$. Then the Fourier series coefficients are given by

$$x_m[n] = \begin{cases} x[\frac{n}{m}], & \text{if } n \text{ is a multiple of } m \\ 0, & \text{if } n \text{ is not a multiple of } m \end{cases}, \tag{5.19}$$

so,

$$x_m[n] \overset{\mathcal{F.S.}}{\longleftrightarrow} \frac{1}{m} a_k.$$

(g) Multiplication

$$x[n].y[n] \overset{\mathcal{F.S.}}{\longleftrightarrow} \sum_{l=<N>} a_k b_{k-l}. \tag{5.20}$$

(h) First Difference

$$x[n] - x[n-1] \overset{\mathcal{F.S.}}{\longleftrightarrow} (1 - e^{-jkw_0}) a_k. \tag{5.21}$$

(i) Running Sum

$$\sum_{k=-\infty}^{n} x[k] \overset{\mathcal{F.S.}}{\longleftrightarrow} \frac{1}{(1 - e^{-jkw_0})} a_k. \tag{5.22}$$

which holds only when $a_0 = 0$.

(j) Periodic Convolution

$$\sum_{r=<N>} x[r]y[n-r] \overset{\mathcal{F.S.}}{\longleftrightarrow} N a_k\, b_k. \tag{5.23}$$

(k) Parseval's Relation for Periodic Signals

Parseval's relation for discrete-time periodic signals is

$$\frac{1}{N} \sum_{n=<N>} |x[n]|^2 = \sum_{k=n=<N>} |a_k|^2. \tag{5.24}$$

(l) Conjugate Symmetry for Real Signals

As in the continuous-time case, if a signal $x[n]$ is real, then its Fourier series coefficients are *conjugate symmetric*, i.e.,

$$x[n] = x^*[n] \;\Rightarrow\; a_k = a^*_{-k} \;\Rightarrow\; a_{-k} = a^*_k,$$

then we have

$$\begin{aligned}
&|a_k| \text{ is even} \quad : \quad |a_k| = |a_{-k}| \\
&\angle a_k \text{ is odd} \quad : \quad \angle a_k = -\angle a_{-k} \\
&\Re\{a_k\} \text{ is even} \quad \Re\{a_k\} = \Re\{a_{-k}\} \\
&\Im\{a_k\} \text{ is odd} \quad : \quad \Im\{a_k\} = -\Im\{a_{-k}\} \\
&x_e[n] \overset{\mathcal{F.S.}}{\longleftrightarrow} \Re\{a_k\} \\
&x_o[n] \overset{\mathcal{F.S.}}{\longleftrightarrow} j\Im\{a_k\}
\end{aligned}$$

Note that if $x[n]$ is real and even, so are its Fourier series coefficients, and if $x[n]$ is real and odd, then its Fourier series coefficients are imaginary and odd. Also if $x[n]$ is even or odd, then its Fourier series coefficients are even or odd, respectively.

(m) Duality Property for Discrete Fourier Series

Let $x[n]$ be a periodic signal with period N and the Fourier series coefficients a_k. If the fundamental period of a_k is the same as $x[n]$, then the Fourier series coefficients of signal a_n are given by

$$a_n \xleftrightarrow{\mathcal{F.S.}} \frac{1}{N} x[-k]. \tag{5.25}$$

5.5 Fourier Series and LTI Systems

Consider an LTI system with the impulse response $h[n]$. Let the input be a periodic signal $x[n]$ with period N. Based on the properties of LTI systems, the output signal $y[n]$ is a periodic signal with the same period as $x[n]$. The Fourier series coefficients of $y[n]$, and eventually $y[n]$ itself, can be obtained based on the Fourier coefficients of $x[n]$ and the frequency response of the system. Let

$$x[n] = \sum_{k=-\infty}^{+\infty} a_k e^{jkw_0 n}, \tag{5.26}$$

and

$$y[n] = \sum_{k=-\infty}^{+\infty} b_k e^{jkw_0 n}. \tag{5.27}$$

Since the system is LTI we have

$$y[n] = x[n] * h[n] \quad \Rightarrow \quad b_k = a_k . H(e^{jkw_0}), \tag{5.28}$$

where $H(e^{jkw_0})$ is the frequency response of the system which is given by

$$H(e^{jkw_0}) = \sum_{n=-\infty}^{+\infty} h[n] e^{-jkw_0 n}. \tag{5.29}$$

So we obtain

$$\begin{aligned} |b_k| &= |a_k| . |H(e^{jkw_0})|, \\ \angle b_k &= \angle a_k + \angle H(e^{jkw_0}). \end{aligned} \tag{5.30}$$

5.6 Problems

Problem 82. Determine the fundamental period N_0 and frequency w_0 of the following signals. Then, find their Fourier series coefficients.

(a) $x[n] = A\cos(\frac{\pi n}{2} + \varphi)$

(b) $x[n] = \cos(\frac{\pi n}{2})\cos(\frac{3\pi n}{7})$

(c) $x[n] = 1 + 2\cos(\frac{\pi n}{3} + \frac{\pi}{4}) + \sin(\frac{\pi n}{6})$

(d) $x[n] = \sum_{m=-\infty}^{+\infty}(-1)^m\left[\delta[n-m] + \delta[n+3m]\right]$

(e) $x[n] = e^{j\frac{3\pi}{5}n} + \sum_{m=-\infty}^{+\infty}\delta[n-10m]$

Solution.

(a) $x[n] = A\cos(\frac{\pi n}{2} + \varphi)$

The fundamental period N_0 of $x[n]$ is given by

$$N_0 = \frac{2\pi}{\frac{\pi}{2}}m \quad \xrightarrow{m=1} \quad N_0 = 4,$$

then

$$w_0 = \frac{2\pi}{N_0} = \frac{\pi}{2}.$$

Thus

$$x[n] = \sum_{k=<N>} a_k e^{jkw_0 n} = \sum_{k=<N>} a_k e^{jk\frac{\pi}{2}n}.$$

Now, using Euler's formula, we get

$$\begin{aligned} x[n] &= \frac{A}{2}\left(e^{j(\frac{\pi n}{2}+\varphi)} + e^{-j(\frac{\pi n}{2}+\varphi)}\right) \\ &= \frac{A}{2}e^{j\varphi}e^{j\frac{\pi n}{2}} + \frac{A}{2}e^{-j\varphi}e^{-j\frac{\pi n}{2}}. \end{aligned}$$

Thus the Fourier series coefficients for $x[n]$ are

$$a_1 = \frac{A}{2}e^{j\varphi}, \qquad a_{-1} = \frac{A}{2}e^{-j\varphi}.$$

and $a_k = 0$ for other values of k over the corresponding period.

(b) $x[n] = \cos(\frac{\pi n}{2})\cos(\frac{3\pi n}{7})$

Hint: $\cos a . \cos b = \frac{1}{2}\cos(a+b) + \frac{1}{2}\cos(a-b)$.

$$\begin{aligned} x[n] &= \frac{1}{2}\cos(\frac{\pi n}{2} + \frac{3\pi n}{7}) + \frac{1}{2}\cos(\frac{\pi n}{2} - \frac{3\pi n}{7}) \\ &= \frac{1}{2}\cos(\frac{13\pi n}{14}) + \frac{1}{2}\cos(\frac{\pi n}{14}). \end{aligned}$$

The signal can be seen as the summation of signals $x_1[n]$ and $x_2[n]$ where

$$x_1[n] = \frac{1}{2}\cos(\frac{13\pi n}{14}) \quad \Rightarrow \quad N_1 = \frac{2\pi}{\frac{13\pi}{14}}m = \frac{28}{13}m \xrightarrow{m=13} N_1 = 28,$$

$$x_2[n] = \frac{1}{2}\cos(\frac{n\pi}{14}) \quad \Rightarrow \quad N_2 = \frac{2\pi}{\frac{\pi}{14}}m = 28m \xrightarrow{m=1} N_2 = 28.$$

Thus fundamental period of $x[n]$ is $N_0 = 28$ and $w_0 = \frac{2\pi}{N_0} = \frac{\pi}{14}$. As such

$$x[n] = \sum_{k=<N>} a_k e^{jkw_0 n} = \sum_{k=<N>} a_k e^{jk\frac{\pi}{14}n}.$$

Again by Euler's formula we have

$$x[n] = \frac{1}{4}e^{j13n(\pi/14)} + \frac{1}{4}e^{-j13n(\pi/14)} + \frac{1}{4}e^{jn(\pi/14)} + \frac{1}{4}e^{-jn(\pi/14)}.$$

Thus the Fourier coefficients for $x[n]$ are

$$a_1 = \frac{1}{4}, \quad a_{-1} = \frac{1}{4}, \quad a_{13} = \frac{1}{4}, \quad a_{-13} = \frac{1}{4},$$

and $a_k = 0$ for other values of k over the corresponding period.

(c) $x[n] = 1 + 2\cos(\frac{\pi n}{3} + \frac{\pi}{4}) + \sin(\frac{\pi n}{6})$

Putting the DC term aside, the signal can be seen as the summation of signals $x_1[n]$ and $x_2[n]$ where

$$x_1[n] = 2\cos(\frac{\pi n}{3} + \frac{\pi}{4}) \quad \Rightarrow \quad N_1 = \frac{2\pi}{\frac{\pi}{3}}m \xrightarrow{m=1} N_1 = 6,$$

$$x_2[n] = \sin(\frac{\pi n}{6}) \quad \Rightarrow \quad N_2 = \frac{2\pi}{\frac{\pi}{6}}m \xrightarrow{m=1} N_2 = 12.$$

Thus the fundamental period of $x[n]$ is $N_0 = 12$ and $w_0 = \frac{2\pi}{N_0} = \frac{\pi}{6}$. So

$$x[n] = \sum_{k=<N>} a_k e^{jkw_0 n} = \sum_{k=<N>} a_k e^{jk\frac{\pi}{6}n}.$$

Then we have

$$x[n] = 1 + e^{j(\pi/4)}e^{j2w_0 n} + e^{-j(\pi/4)}e^{-j2w_0 n} + \frac{1}{2j}e^{jw_0 n} - \frac{1}{2j}e^{-jw_0 n}.$$

Thus the Fourier coefficients for $x[n]$ are

$$a_0 = 1, \quad a_1 = \frac{1}{2j}, \quad a_{-1} = -\frac{1}{2j}, \quad a_2 = e^{j(\pi/4)}, \quad a_{-2} = e^{-j(\pi/4)},$$

and $a_k = 0$ for other values of k over the corresponding period.

(d) $x[n] = \sum_{m=-\infty}^{+\infty}(-1)^m\left[\delta[n-m] + \delta[n+3m]\right]$

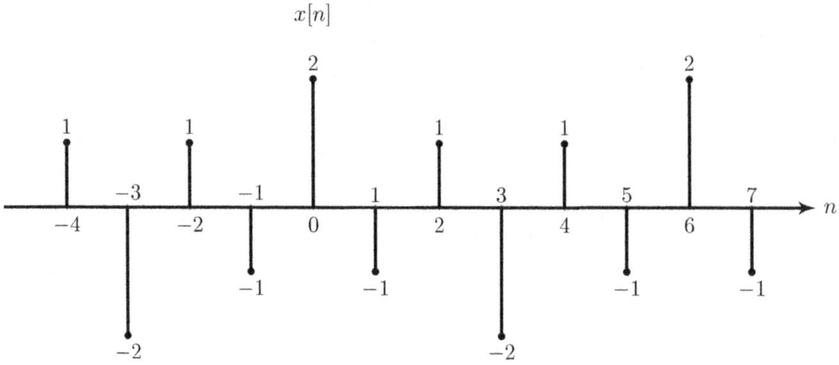

Figure 5.2: Problem 82 - Part (d), Signal $x_2[n]$.

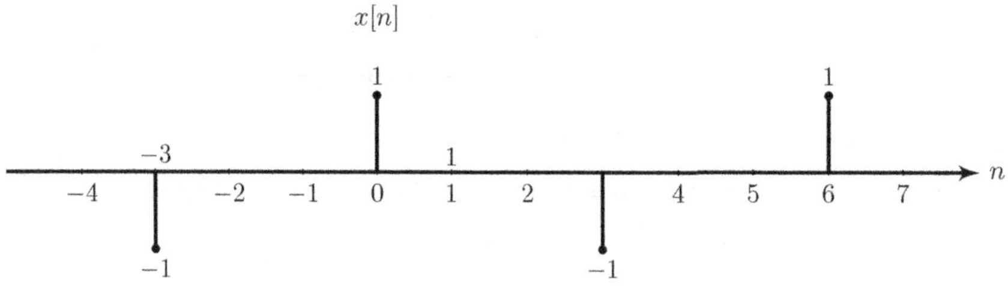

Figure 5.3: Problem 82 - Part (d), Signal $x_1[n]$.

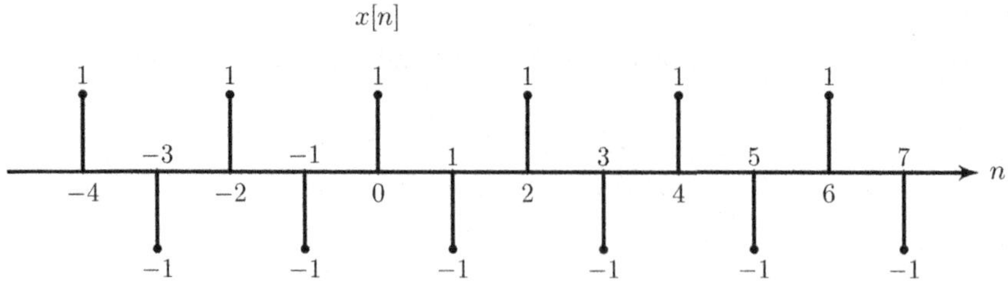

Figure 5.4: Problem 82 - Part (d), Signal $x_2[n]$.

The signal can be seen as the sum of signals $x_1[n]$ and $x_2[n]$ that are equal to

$$x_1[n] = \sum_{m=-\infty}^{\infty}(-1)^m\delta[n+3m]$$

and

$$x_2[n] = \sum_{m=-\infty}^{\infty} (-1)^m \delta[n-m]$$

The fundamental period N of $x_1[n]$ and $x_2[n]$ are 6 and 2, respectively. Thus, the fundamental period of $x[n]$ is $N_0 = 6$ and its fundamental frequency is $w_0 = \frac{\pi}{3}$. As such by choosing the period from $n = -2$ to $n = 3$, the Fourier series coefficients are given by

$$\begin{aligned}
a_k &= \frac{1}{N} \sum_{n=<N>} x[n]e^{-jkw_0 n} \\
&= \frac{1}{6} \sum_{n=-2}^{3} x[n]e^{-jk(\pi/3)n} \\
&= \frac{1}{6} \big[x[-2]e^{-jk(\pi/3)(-2)} + x[-1]e^{-jk(\pi/3)(-1)} + x[0] \\
&\quad + x[1]e^{-jk(\pi/3)} + x[2]e^{-j2k(\pi/3)} + x[3]e^{-j3k(\pi/3)}\big] \\
&= \frac{1}{6}\cos(\frac{2\pi}{3}k) - \frac{1}{6}\cos(\frac{\pi}{3}k) + \frac{2}{3}.
\end{aligned}$$

(e) $x[n] = e^{j\frac{3\pi}{5}n} + \sum_{m=-\infty}^{+\infty} \delta[n-10m]$

The signal can be seen as the sum of signals $x_1[n]$ and $x_2[n]$ where

$$x_1[n] = e^{j\frac{3\pi}{5}n} \;\Rightarrow\; N_1 = \frac{2\pi}{\frac{3\pi}{5}}m \xrightarrow{m=3} N_1 = 10,$$

$$x_2[n] = \sum_{m=-\infty}^{+\infty} \delta[n-10m] \;\Rightarrow\; N_2 = 10.$$

The fundamental period of $x[n]$ is $N_0 = 10$ and $w_0 = \frac{2\pi}{N_0} = \frac{\pi}{5}$. Let a_k^{xi} denote the Fourier series coefficients of $x_i[n]$ for $i = 1, 2, 3$. Then we have

$$x_1[n] = e^{j3w_0 n} = e^{j\frac{3\pi}{5}n} \xrightarrow{\mathcal{F.S.}} a_3^{x1} = 1, a_k^{x1} = 0 \text{ for } k \neq 3.$$

To determine the Fourier series coefficients a_k of $x_2[n]$, we select the period $-N_0/2 + 1 \leq n \leq N_0/2$. Within this interval, $x[n]$ is equal to $\delta[n]$ and it follows that

$$a_k^{x2} = \frac{1}{N_0} \sum_{n=-4}^{5} \delta[n]e^{-jk(\pi/5)n} = \frac{1}{10} \quad \forall k.$$

Thus, the Fourier series coefficients of $x[n]$ are given by

$$a_k = \begin{cases} 1 + \frac{1}{10}, & k = 3 \\ \frac{1}{10}, & \text{else} \end{cases}.$$

Problem 83. $x[n]$ is a periodic signal with fundamental period N where

$$x[n] = 1 - \cos(\frac{n\pi}{4} + \frac{\pi}{4}) \qquad 0 \leq n \leq N-1.$$

Determine its Fourier series coefficients for $N = 4$.

Solution. For $N = 4$, the fundamental frequency of $x[n]$ is $w_0 = \frac{2\pi}{N} = \frac{\pi}{2}$. Then, we should find the values of $x[n]$ over a period where

$$x[0] = 1 - \frac{\sqrt{2}}{2}, \qquad x[1] = 1 \qquad x[2] = 1 + \frac{\sqrt{2}}{2}, \qquad x[3] = 2.$$

We have

$$\begin{aligned} a_k &= \frac{1}{N} \sum_{n=<N>} x[n] e^{-jk(\pi/2)n} \\ &= \frac{1}{4} x[0] + \frac{1}{4} x[1] e^{-jk(\pi/2)} + \frac{1}{4} x[2] e^{-jk(2\pi/2)} + \frac{1}{4} x[3] e^{-jk(3\pi/2)} \\ &= \frac{1}{4}(1 - \frac{\sqrt{2}}{2}) + \frac{1}{4} e^{-jk(\pi/2)} + \frac{1}{4}(1 + \frac{\sqrt{2}}{2}) e^{-jk\pi} + \frac{1}{2} e^{-jk(3\pi/2)} \end{aligned}$$

Problem 84. Determine the discrete Fourier series representation for each of the following signals.

(a) $x[n] = \cos(\frac{\pi}{3}n) + (-1)^n$

(b) $x[n] = (-1)^n + e^{jn(\pi/2)}$

(c) $x[n] = (j)^n$

Solution.

(a) $x[n] = \cos(\frac{\pi}{3}n) + (-1)^n$

The signal can be seen as the summation of signals $x_1[n]$ and $x_2[n]$ where

$$\begin{aligned} &x_1[n] = \cos(\frac{\pi}{3}n) \;\Rightarrow\; N_1 = \frac{2\pi}{\frac{\pi}{3}} m \xrightarrow{m=1} N_1 = 6, \\ &x_2[n] = (-1)^n \;\Rightarrow\; N_2 = 2. \end{aligned}$$

Thus the fundamental period of $x[n]$ is $N_0 = 6$ and $w_0 = \frac{2\pi}{N_0} = \frac{\pi}{3}$. Now, using Euler's formula we get

$$\begin{aligned} x[n] &= \frac{1}{2}e^{j(\pi/3)n} + \frac{1}{2}e^{-j(\pi/3)n} + e^{j\pi n} \\ &= \sum_{k=<N>} a_k e^{jk\frac{\pi}{3}n}. \end{aligned}$$

Thus the Fourier coefficients for $x[n]$ are

$$a_1 = \frac{1}{2}, \quad a_{-1} = \frac{1}{2}, \quad a_3 = 1.$$

and $a_k = 0$ over a period for other values of k.

(b) $x[n] = (-1)^n + e^{jn(\pi/2)}$

The signal can be seen as the summation of a constant, and signals $x_1[n]$ and $x_2[n]$ where

$$\begin{aligned} x_1[n] = (-1)^n = e^{jn\pi} \quad &\Rightarrow \quad N_1 = \frac{2\pi}{\pi}m \xrightarrow{m=1} N_1 = 2, \\ x_2[n] = e^{jn(\pi/2)} \quad &\Rightarrow \quad N_2 = \frac{2\pi}{\frac{\pi}{2}}m \xrightarrow{m=1} N_2 = 4. \end{aligned}$$

So the fundamental period of $x[n]$ is $N_0 = 4$ and $w_0 = \frac{2\pi}{N_0} = \frac{\pi}{2}$. Thus,

$$x[n] = e^{jn\pi} + e^{jn(\pi/2)} = \sum_{k=<N_0>} a_k e^{jk\frac{\pi}{2}n}.$$

The Fourier coefficients of $x[n]$ over a period are

$$a_0 = 0, \quad a_1 = 1, \quad a_2 = 1, \quad a_4 = 0.$$

(c) $x[n] = (j)^n$

$$x[n] = (j)^n = e^{jn(\pi/2)} \quad \Rightarrow \quad N = \frac{2\pi}{\frac{\pi}{2}}m \xrightarrow{m=1} N = 4.$$

Thus $w_0 = \frac{\pi}{2}$. To determine the Fourier series coefficients, we see that over a period the values of $x[n]$ are

$$x[0] = 1, \quad x[1] = j, \quad x[2] = -1, \quad x[3] = -j.$$

So the fundamental period of $x[n]$ is $N_0 = 4$ and $w_0 = \frac{2\pi}{N_0} = \frac{\pi}{2}$. Thus,

$$a_0 = 0, \quad a_1 = 1, \quad a_2 = 0, \quad a_3 = 0.$$

Or we obtain

$$
\begin{aligned}
a_k &= \frac{1}{4}\sum_{n=0}^{3} x[n]e^{-jk(\pi/2)n} \\
&= \frac{1}{4}\left[x[0] + x[1]e^{-jk(\pi/2)} + x[2]e^{-jk(2\pi/2)} + x[3]e^{-jk(3\pi/2)}\right] \\
&= \frac{1}{4}\left[1 + je^{-jk(\pi/2)} - e^{-jk(2\pi/2)} - je^{-jk(3\pi/2)}\right],
\end{aligned}
$$

and then over a period, the Fourier series coefficients are calculated as follows:

$$
\begin{aligned}
a_0 &= \frac{1}{4}(1 + j - 1 - j) = 0 \\
a_1 &= \frac{1}{4}(1 + j(-j) - (-1) - j(+j)) = 1 \\
a_2 &= \frac{1}{4}(1 + j(-1) - (1) - j(-1)) = 0 \\
a_3 &= \frac{1}{4}(1 + j(j) - 1(-1) - j(-j)) = 0.
\end{aligned}
$$

Problem 85. Find the discrete Fourier series for each of the following periodic signals shown in Fig. 5.5.

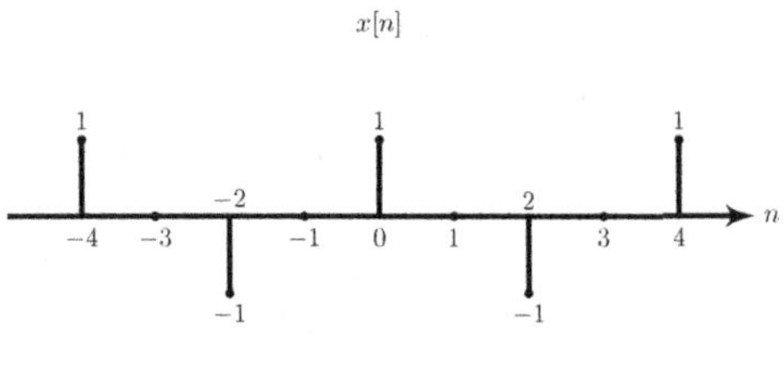

(a)

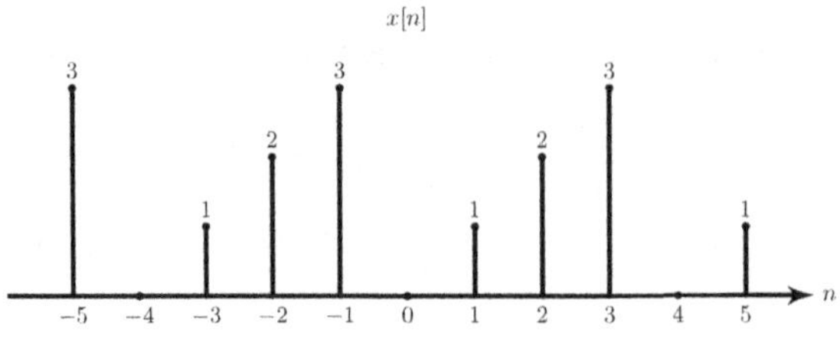

(b)

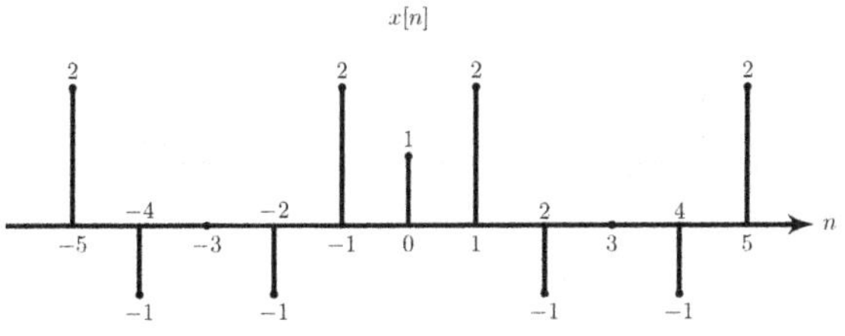

(c)

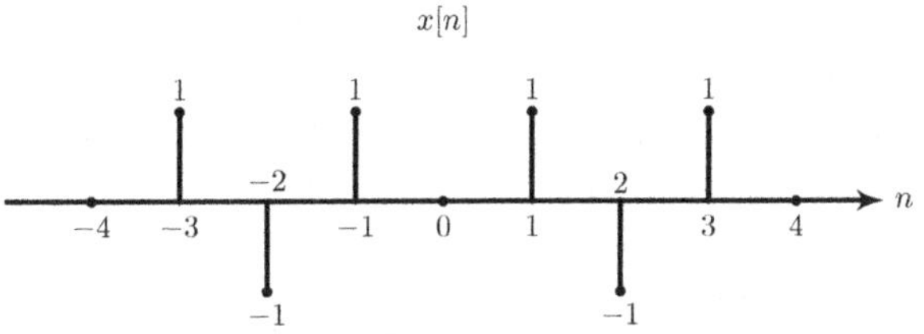

(d)

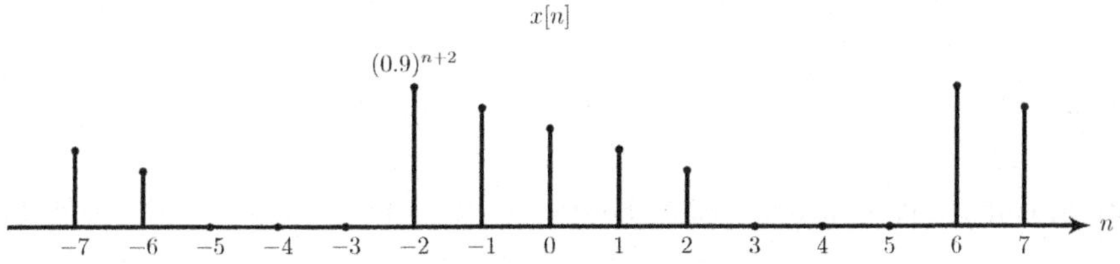

(e)

Figure 5.5: Problem 85

Solution.

(a) The signal is composed of two impulse trains that are shown in Fig. 5.6.

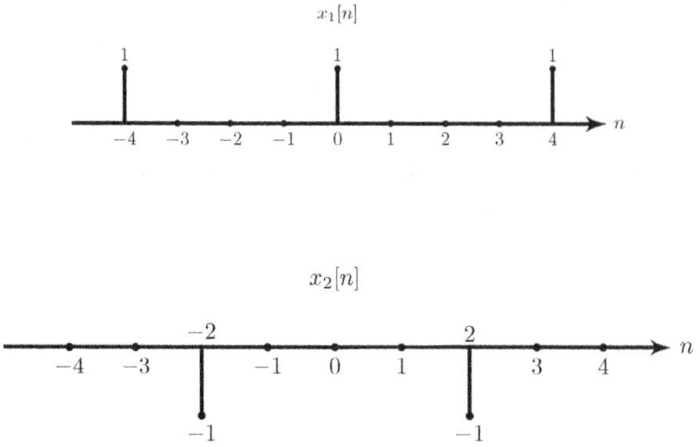

Figure 5.6: Problem 85 - Part (a), Two impulse trains corresponding to $x[n]$,

The first part, $x_1[n]$, is a periodic train of impulses with period $N = 4$ and $w_0 = \frac{\pi}{2}$. Thus the Fourier series coefficients are $a_k = \frac{1}{N}$. So we have

$$x_1[n] \quad \xrightarrow{\mathcal{F.S.}} \quad b_k = \frac{1}{4}.$$

From Fig. 5.6, we see that

$$x_2[n] = -x_1[n-2],$$

and using time shifting properties of Fourier series, the Fourier series coefficients of $x_2[n]$ are given by

$$x_2[n] \quad \xrightarrow{\mathcal{F.S.}} \quad c_k = -b_k e^{-jk(\pi/2)2} = -\frac{1}{4}e^{-jk\pi} = -\frac{1}{4}(-1)^k.$$

Finally, the Fourier series coefficients of $x[n]$ are

$$a_k = b_k + c_k = \frac{1}{4}\big(1 - (-1)^k\big) = \begin{cases} 0 & k \text{ even} \\ \frac{1}{2} & k \text{ odd} \end{cases}.$$

(b) The signal period is $N = 4$ and $w_0 = \frac{\pi}{2}$. Based on the definition we have

$$\begin{aligned} a_k &= \frac{1}{4}\sum_{n=0}^{3} x[n]e^{-jkn(\pi/2)} \\ &= \frac{1}{4}\Big(x[0] + x[1]e^{-jk(\pi/2)} + x[2]e^{-jk\pi} + x[3]e^{-jk(3\pi/2)}\Big). \end{aligned}$$

We also know that

$$e^{-jk\pi} = (-1)^k, \quad e^{-jk(\pi/2)} = (-j)^k.$$

Then over a period, the Fourier series coefficients are calculated as follows

$$\begin{aligned} a_0 &= \frac{1}{4}(0+1+2+3) = \frac{3}{2} \\ a_1 &= \frac{1}{4}(0-j-2+3j) = \frac{1}{2}(-1+j) \\ a_2 &= \frac{1}{4}(0-1+2-3) = -\frac{1}{2} \\ a_3 &= \frac{1}{4}(0+j-2-3j) = -\frac{1}{2}(1+j). \end{aligned}$$

(c) The signal is periodic with period $N = 6$ and $w_0 = \frac{\pi}{3}$. Based on the definition we have

$$\begin{aligned} a_k &= \frac{1}{N} \sum_{n=<N>} x[n]e^{-jkw_0 n} \\ &= \frac{1}{6} \sum_{n=0}^{5} x[n]e^{-jkn(\pi/3)} \\ &= \frac{1}{6}\left[1 + 2e^{-jk(\pi/3)} - e^{-j2k(\pi/3)} - e^{-j4k(\pi/3)} + 2e^{-j5k(\pi/3)}\right] \\ &= \frac{1}{6}\left[1 + 2e^{-jk(\pi/3)} - e^{-j2k(\pi/3)} - e^{j2k(\pi/3)} + 2e^{jk(\pi/3)}\right] \\ &= \frac{1}{6}\left[1 + 4\cos(\frac{k\pi}{3}) - 2\cos(\frac{2k\pi}{3})\right]. \end{aligned}$$

(d) Let's define signal $x_1[n]$ as an impulse train with period $N = 4$ and $w_0 = \frac{\pi}{2}$. This signal is shown in Fig. 5.7:

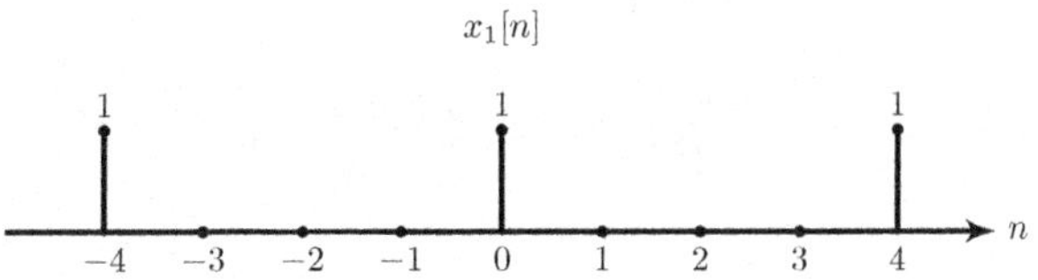

Figure 5.7: Problem 85 - Part (d), Signal $x_1[n]$.

Thus, we have

$$x_1[n] \quad \xrightarrow{\mathcal{F.S.}} \quad b_k = \frac{1}{4}.$$

Now signal $x[n]$ can be written as follows

$$x[n] = x_1[n+1] + x_1[n-1] - x_1[n-2]. \tag{5.31}$$

From (5.31) and using time shifting properties of Fourier series, we have

$$\begin{aligned} a_k &= b_k e^{jk(\pi/2)} + b_k e^{-jk(\pi/2)} - b_k e^{-j2k(\pi/2)} \\ &= \frac{1}{4} e^{jk(\pi/2)} + \frac{1}{4} e^{-jk(\pi/2)} - \frac{1}{4} e^{-jk\pi} \\ &= \frac{1}{4}\left(e^{jk(\pi/2)} + e^{-jk(\pi/2)}\right) - \frac{1}{4}(-1)^k \\ &= \frac{1}{2}\cos(\frac{k\pi}{2}) - \frac{1}{4}(-1)^k. \end{aligned}$$

(e) The signal has period $N = 8$ and $w_0 = \frac{\pi}{4}$. Based on the definition we have

$$\begin{aligned} a_k &= \frac{1}{8}\sum_{n=-2}^{5} x[n] e^{-jkn(\pi/4)} \\ &= \frac{1}{8}\sum_{n=-2}^{2} (0.9)^{n+2} e^{-jkn(\pi/4)}. \end{aligned}$$

Substituting $r = n + 2$ and interchanging the order of summation, we have

$$\begin{aligned} a_k &= \frac{1}{8}\sum_{r=0}^{4} (0.9)^r e^{jk(\pi/2)} e^{-jkr(\pi/4)} \\ &= \frac{e^{jk(\pi/2)}}{8} \frac{1 - (0.9e^{-jk(\pi/4)})^5}{1 - 0.9e^{-jk(\pi/4)}}. \end{aligned}$$

Problem 86. Consider a discrete-time LTI system. Find the corresponding system output $y[n]$, if

(a) The frequency response of this system is

$$H(e^{jw}) = \frac{\cos(\frac{3}{2}w)}{\cos(\frac{w}{2})} + 1$$

and the input is

$$x[n] = \sum_{k=-2}^{2} |k| e^{jk(2\pi/3)n}.$$

(b) The frequency response is

$$H(e^{jw}) = 4\cos(2w)$$

and the input to this system is

$$x[n] = \begin{cases} 0 & n \text{ even} \\ 1 & n \text{ odd} \end{cases}.$$

Solution.

Let b_k be the Fourier series coefficients of $y[n]$, the output of the discrete-time LTI system with frequency response $H(e^{jw})$, to a periodic input $x[n]$ with Fourier series coefficients a_k. Then we have

$$b_k = a_k.H(e^{jkw_0}).$$

(a) Based on the definition we know that

$$x[n] = \sum_{k<N>} a_k e^{jk(2\pi/N)n}. \tag{5.32}$$

Comparing the input signal with above equation, we see that the Fourier coefficients of $x[n]$ with period $N = 3$ are

$$a_k = |k| = \begin{cases} 0 & k = 0 \\ 1 & k = \pm 1 \\ 2 & k = \pm 2 \end{cases}.$$

Hence, the Fourier series coefficients b_k of $y[n]$ are given by

$$\begin{aligned} b_k &= \begin{cases} a_0 H(e^{j0}) & k = 0 \\ a_{\pm 1} H(e^{j(\pm 1)(2\pi/3)}) & k = \pm 1 \\ a_{\pm 2} H(e^{j(\pm 2)(2\pi/3)}) & k = \pm 2 \end{cases} \\ &= \begin{cases} 0 & k = 0 \\ -1 & k = \pm 1 \\ -2 & k = \pm 2 \end{cases}, \end{aligned}$$

then, the output $y[n]$ is

$$\begin{aligned} y[n] &= -e^{-j(2\pi/3)n} - e^{j(2\pi/3)n} - 2e^{-j(4\pi/3)n} - 2e^{j(4\pi/3)n} \\ &= -2\cos(\frac{2\pi}{3}n) - 4\cos(\frac{4\pi}{3}n). \end{aligned}$$

(b) The input signal is an impulse train with the period $N = 2$ and $w_0 = \pi$.

Therefore, the Fourier series coefficients a_k of $x[n]$ are given by

$$a_k = \frac{1}{N} e^{-jk\pi} = \frac{1}{2}(-1)^k.$$

Then, we have

$$\begin{aligned} b_0 &= a_0 H(e^{j0}) \xrightarrow{w=0} b_0 = a_0 H(0) = \frac{1}{2} \times 4 = 2 \\ b_1 &= a_1 H(e^{j\pi}) \xrightarrow{w=\pi} b_1 = a_1 H(\pi) = -\frac{1}{2} \times 4 = -2 \\ b_2 &= a_2 H(e^{j2\pi}) \xrightarrow{w=2\pi} b_2 = a_2 H(2\pi) = \frac{1}{2} \times 4 = 2. \end{aligned}$$

It can be concluded that in general, $b_k = 2(-1)^k = 4a_k$, and thus $y[n] = 4x[n]$.

Problem 87. As shown in the following figure, $x[n]$ is a periodic signal with period $N = 5$ and the Fourier series coefficients a_k.

(a) Find the value of

$$\sum_{k=0}^{4} |a_k|^2.$$

(b) Find the value of

$$\sum_{k=0}^{4} (Re\{a_k\})^2.$$

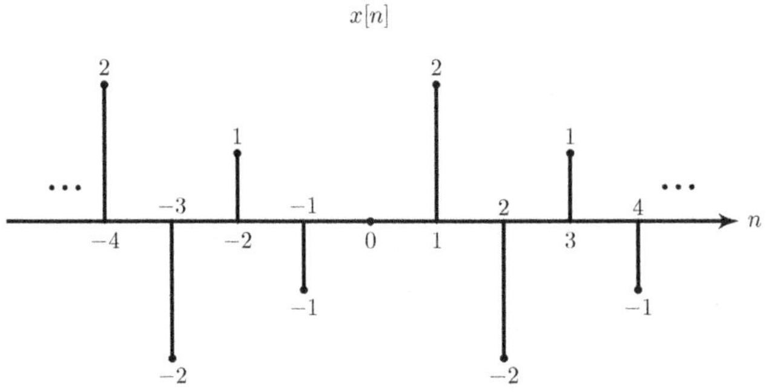

Figure 5.8: Problem 87, Discrete-time periodic signal $x[n]$.

Solution.

(a) According to the Parseval's relation for discrete-time periodic signals, we have

$$\begin{aligned}
\sum_{k < N >} |a_k|^2 &= \frac{1}{N} \sum_{n < N >} |x[n]|^2 \\
&= \frac{1}{5} \sum_{n=0}^{4} |x[n]|^2 \\
&= \frac{1}{5}\left(0 + (2)^2 + (-2)^2 + (1)^2 + (-1)^2\right) = \frac{10}{5} = 2.
\end{aligned}$$

(b) As we know, if signal $x[n]$ is real, that is, when $x[n] = x^*[n]$, the Fourier series coefficients will be conjugate symmetric, i.e.,

$$x[n] = x^*[n] \quad \Leftrightarrow \quad a_k = a^*_{-k} \quad \text{or} \quad a_{-k} = a^*_k$$

and from even-odd decomposition of real signal, we have

$$a_k = a^*_{-k} \Leftrightarrow \begin{cases} x_e[n] \xrightarrow{\mathcal{F.S.}} Re\{a_k\} \\ \\ x_o[n] \xrightarrow{\mathcal{F.S.}} jIm\{a_k\} \end{cases}.$$

If we write Parseval's relation for the even part of the signal, we have

$$\frac{1}{N}\sum_{n<N>}|x_e[n]|^2 = \sum_{k<N>}|Re\{a_k\}|^2 = \sum_{k=0}^{4}(Re\{a_k\})^2.$$

Now we should find the even part of the signal as follows:

$$x_e[n] = \frac{x[n]+x[-n]}{2},$$

which is depicted in Fig. 5.9.

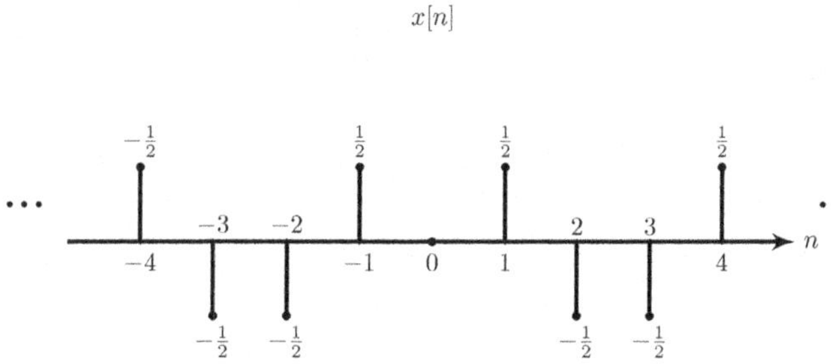

Figure 5.9: Problem 87 - Part (b), The even part of signal $x[n]$.

Thus we obtain

$$\sum_{k=0}^{4}(Re\{a_k\})^2 = \frac{1}{5}(0+\frac{1}{4}+\frac{1}{4}+\frac{1}{4}+\frac{1}{4}) = \frac{1}{5}.$$

Problem 88. Consider the discrete-time signal $x[n]$ with period N and the Fourier series coefficients a_k. Determine the Fourier series coefficients for each of the following cases based on a_k.

(a) $y[n] = x[n]\cos(\frac{2\pi}{N}n)$

(b) $y[n] = (-1)^n x[n], \quad N = \text{even}$

(c) $y[n] = x_2[n-1]$

(d) $y[n] = x^*[-n+1], \quad N = 8$

Solution.

(a) $y[n] = x[n]\cos(\frac{2\pi}{N}n)$

Using Euler's formula, $y[n]$ can be expressed as

$$\begin{aligned} y[n] &= x[n]\left(\frac{e^{j(2\pi/N)n} + e^{-j(2\pi/N)n}}{2}\right) \\ &= \frac{1}{2}e^{j(2\pi/N)n}x[n] + \frac{1}{2}e^{-j(2\pi/N)n}x[n] \end{aligned}$$

and using frequency shifting properties of Fourier series, i.e.,

$$e^{jmw_0 n}x[n] \quad \xrightarrow{\mathcal{F.S.}} \quad a_{k-m},$$

and setting $w_0 = \frac{2\pi}{N}$, we have

$$y[n] \xrightarrow{\mathcal{F.S.}} b_k = \frac{1}{2}a_{k-1} + \frac{1}{2}a_{k+1}.$$

(b) $y[n] = (-1)^n x[n], \quad N = \text{even}$

The period of $(-1)^n$ is 2, and since N is even, the period of $y[n]$ is also N. Let $w_0 = \frac{2\pi}{N}$. We rewrite $y[n]$ as

$$y[n] = (-1)^n x[n] = e^{j\pi n}x[n] = e^{j(N/2)(2\pi/N)n}x[n].$$

Using frequency shifting properties of Fourier series, we obtain

$$y[n] \xrightarrow{\mathcal{F.S.}} b_k = a_{k-\frac{N}{2}}.$$

(c) $y[n] = x_2[n-1]$

We are dealing with a time-scaled and time-shifted version of $x[n]$ here. In order to generate $x_2[n-1]$, let's first perform the time shifting and then apply the time scaling. Let $z[n]$ be an arbitrary signal with Fourier series coefficients c_k. This signal is used to show how each system operates on the input, in both time and frequency domains.

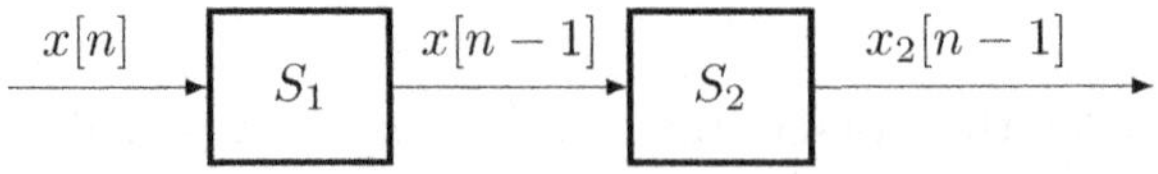

$$\begin{cases} S_1: & z[n] \rightarrow z[n-1] \xleftrightarrow{\mathcal{F.S.}.} c_k \rightarrow c_k e^{-jkw_0},\ w_0 \\ \\ S_2: & z[n] \rightarrow z_2[n] \xleftrightarrow{\mathcal{F.S.}} c_k \rightarrow \frac{1}{2}c_k,\ \frac{w_0}{2} \end{cases}$$

Note that the fundamental frequency of $x_2[n]$ is half the fundamental frequency of $x[n]$, i.e, $\frac{w_0}{2}$.

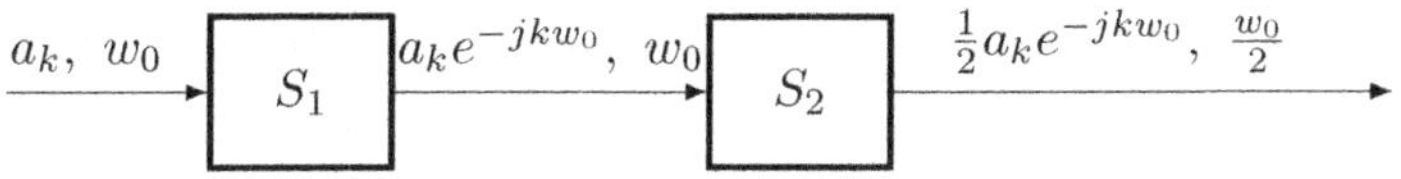

Thus,

$$y[n] \xrightarrow{\mathcal{F.S.}} b_k = \frac{1}{2}a_k e^{-jkw_0}, \quad \frac{w_0}{2}.$$

(d) $y[n] = x^*[-n+1], \quad N = 8$
Similar to above, we use $z[n]$ to to show how each system operates on the input, in both time and frequency domains. The following block diagram denotes the steps to make $y[n]$:

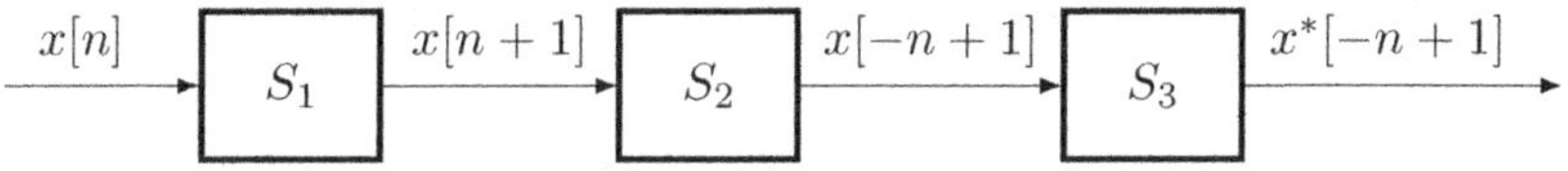

$$\begin{cases} S_1: & z[n] \to z[n+1] \xleftrightarrow{\mathcal{F.S.}} c_k \to c_k e^{jk(\pi/4)},\ w_0 = \frac{\pi}{4} \\ S_2: & z[n] \to z[-n] \xleftrightarrow{\mathcal{F.S.}} c_k \to c_{-k},\ w_0 \\ S_3: & z[n] \to z^*[n] \xleftrightarrow{\mathcal{F.S.}} c_k \to (c_{-k})^*,\ w_0 \end{cases}$$

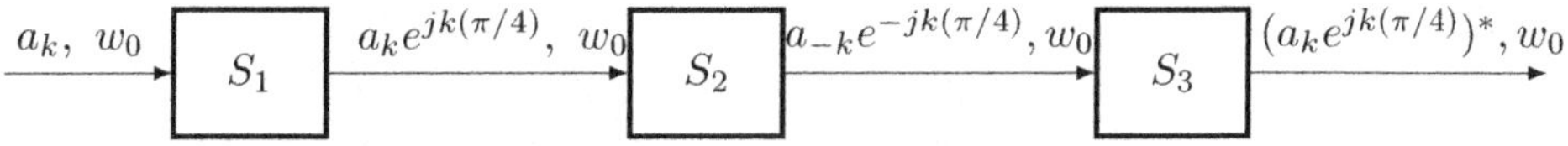

Thus, we have

$$y[n] \xrightarrow{\mathcal{F.S.}} b_k = {a_k}^* e^{-jk(\pi/4)},\ w_0 = \frac{\pi}{4}.$$

Problem 89. Let $x[n]$ be a discrete-time periodic signal with period N and Fourier coefficients a_k.

(a) If the period is $N = 3$ and the Fourier series coefficients are $a_k = j^k$ where $j = \sqrt{-1}$, determine the power of the signal $y[n] = 2x[n] + 2$.

(b) Now assume that $x[n]$ is another signal with period $N = 4$ which is real and has the average value of zero. Given that

$$a_1 = -1 + j, \qquad a_2 = -2,$$

determine the power of $x[n]$.

Solution.

(a) Let b_k be the Fourier coefficients of the signal $y[n]$. Since the period of $x[n]$ is 3, the period of $y[n]$ is also 3 and we only need to find b_0, b_1 and b_2.

We have

$$b_k = 2a_k$$

for $k \neq 0$. Note that the constant value of 2 only affects b_0 or the average value of $y[n]$, in other words,

$$b_0 = 2a_0 + 2.$$

Thus

$$\begin{aligned} b_0 &= 2a_0 + 2 = 2 \times 1 + 2 = 4, \\ b_1 &= 2a_1 = 2(j), \\ b_2 &= 2a_2 = 2(-1) = -2. \end{aligned}$$

Now using Parseval's relation for discrete-time periodic signals, we have

$$\begin{aligned} P &= \frac{1}{N} \sum_{n=0}^{2} |y[n]|^2 = \sum_{k=0}^{2} |b_k|^2 \\ &= |b_0|^2 + |b_1|^2 + |b_3|^2 = 16 + 4 + 4 = 24. \end{aligned}$$

(b) Let's determine the Fourier coefficients of $x[n]$ over period $k = -1$ to $k = 2$. Since signal $x[n]$ is real, its Fourier series coefficients will be conjugate symmetric. The value of a_{-1} is given by

$$a_k^* = a_{-k} \xrightarrow{k=1} a_{-1} = a_1^* = -1 - j$$

and since the average value of the signal is zero, then $a_0 = 0$. Now according to Parseval's relation, we have

$$\begin{aligned} P &= \frac{1}{N} \sum_{n<N>} |x[n]|^2 = \sum_{k<N>} |a_k|^2 = \sum_{k=-1}^{2} |a_k|^2 \\ &= |a_{-1}|^2 + |a_0|^2 + |a_1|^2 + |a_2|^2 = 8. \end{aligned}$$

Problem 90. Suppose we are given the following facts about a signal $x[n]$:

1. $x[n]$ is a real and even signal.

2. $x[n]$ is periodic with period $N = 6$ and has Fourier coefficients a_k.

3. $a_7 = 2$.

4. $\frac{1}{6} \sum_{n=0}^{5} |x[n]|^2 = 8$.

Using the above information, find $x[n]$.

Solution.

First, we try to find the Fourier series coefficients and then use them to determine $x[n]$. According to Fact 1, the Fourier coefficients are also real and even. Since $x[n]$ is periodic with period $N = 6$ (Fact 2), its Fourier coefficients a_k has period $N = 6$. From Fact 3, we can write

$$a_7 = a_1 = 2$$

and also

$$a_1 = a_{-1} = 2.$$

Now Using Parseval's relation and Fact 4, we get

$$\begin{aligned}\frac{1}{6}\sum_{n=0}^{5} |x[n]|^2 &= \sum_{k=-2}^{3} |a_k|^2 \\ &= |a_0|^2 + 2|2|^2 + 2|a_2|^2 + |a_3|^2 = 8.\end{aligned}$$

So,

$$a_0 = a_2 = a_3 = 0.$$

Thus by using Fourier series definition,

$$\begin{aligned}x[n] &= \sum_{k=-2}^{3} a_k e^{jk(\pi/3)n} \\ &= a_{-1}e^{-j(\pi/3)n} + a_1 e^{j(\pi/3)n} \\ &= 2e^{-j(\pi/3)n} + 2e^{j(\pi/3)n} \\ &= 4\cos(\frac{\pi}{3}n).\end{aligned}$$

Problem 91. Suppose a discrete-time periodic signal $x[n]$ has fundamental period N and Fourier coefficients a_k.

(a) Let us define $y[n]$ as

$$y[n] = 2x^*[n] + 2x_2[n]$$

with the Fourier coefficients b_k, where $x_2[n]$ is the time-scaled version of $x[n]$. If the period is $N = 3$, obtain the Fourier series coefficients of $y[n]$ based on a_k's.

(b) Signal $y[n]$ is defined as

$$y[n] = x[n]\cos(\frac{n\pi}{2})$$

If the period of $x[n]$ is $N = 4$, obtain the Fourier series coefficients of $y[n]$ based on a_k's.

Solution.

(a)

$$y[n] = 2x^*[n] + 2x_2[n]$$

First we should find the fundamental period of $y[n]$ as follows:

$$x^*[n] \quad \Rightarrow \quad N = 3,$$

$$x_2[n] \quad \Rightarrow \quad N = 6.$$

Thus, the period of $y[n]$ is $N = 6$ and $w_0 = \frac{\pi}{3}$. Let us demonstrate typical values of the Fourier coefficients of $x[n]$ through Fig. 5.10.

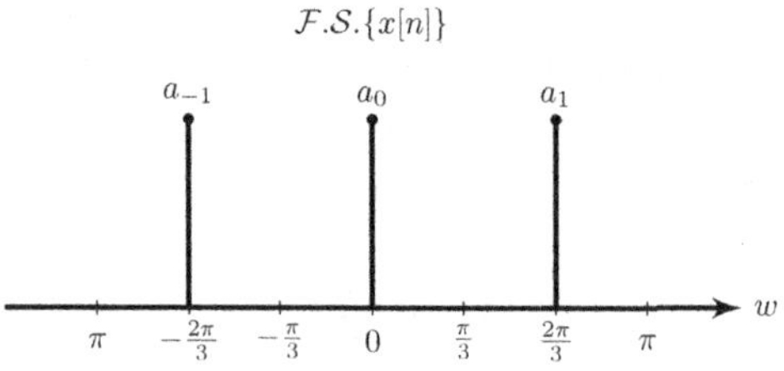

Figure 5.10: Problem 91 - Part (a), Fourier coefficients for $x[n]$.

Now according to this figure, we demonstrate the resulting Fourier series coefficients in Fig. 5.11.

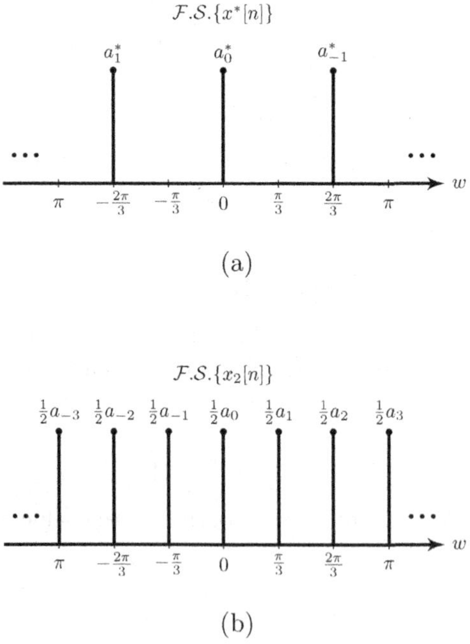

Figure 5.11: Problem 91 - Part (a), Fourier coefficients for (a) $x^*[n]$; (b) $x_2[n]$.

Now, according to the linearity property, we demonstrate the Fourier series coefficients of $y[n]$ in Fig. 5.12.

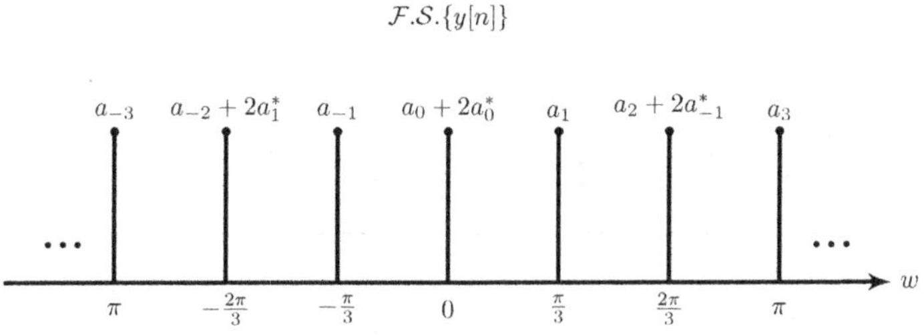

Figure 5.12: Problem 91 - Part (a), Fourier coefficients for $y[n]$.

(b)

$$y[n] = x[n]\cos(\frac{n\pi}{2})$$

First, let's find the overall period:

$$x[n] \Rightarrow N = 4$$
$$\cos(\frac{n\pi}{2}) \Rightarrow N = 4$$

Thus, the period of $y[n]$ is $N = 4$ and $w_0 = \frac{\pi}{2}$. We see that signal $y[n]$ is the multiplication of two signals $x[n]$ and $\cos(\frac{n\pi}{2})$, so

$$x[n]\cos(\frac{n\pi}{2}) \overset{\mathcal{F.S.}}{\longleftrightarrow} \mathcal{F.S.}\{x[n]\} * \mathcal{F.S.}\{\cos(\frac{n\pi}{2})\}.$$

The Fourier series coefficients of $x[n]$, $\cos(\frac{n\pi}{2})$ and $x[n]\cos(\frac{n\pi}{2})$ are depicted in Fig. 5.14.

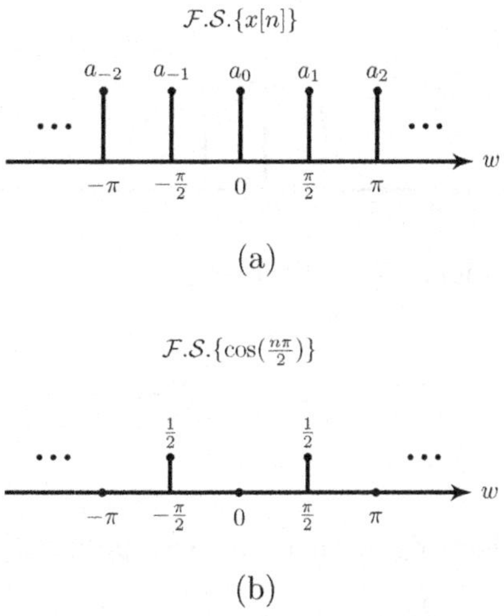

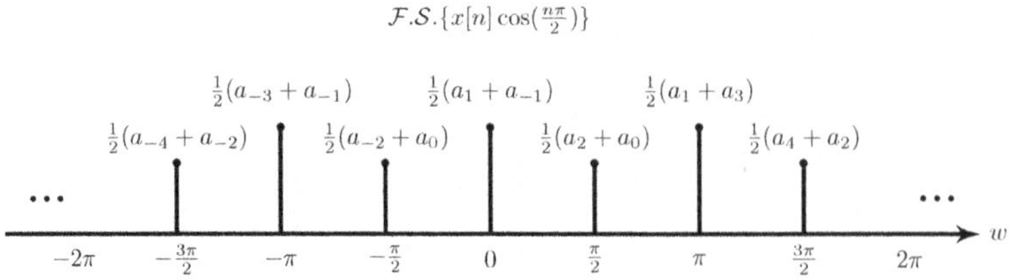

(c)

Figure 5.13: Problem 91 - Part (b), Fourier coefficients for (a) $x[n]$; (b) $\cos(\frac{n\pi}{2})$; (c) $x[n]\cos(\frac{n\pi}{2})$.

Finally, we obtain

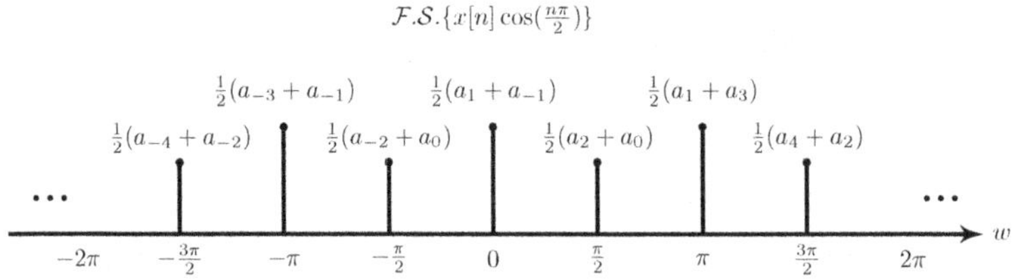

Figure 5.14: Problem 91 - Part (b), Fourier series coefficients of $y[n]$.

Problem 92. In each of the following, we specify the Fourier series coefficients of a periodic discrete-time signal. Determine signal $x[n]$ in each case.

(a) $a_k = \cos(\frac{4\pi}{23}k)$

(b) $a_k = \sum_{m=-\infty}^{+\infty}(-1)^m\big(\delta[k-2m]+\delta[k+5m]\big)$

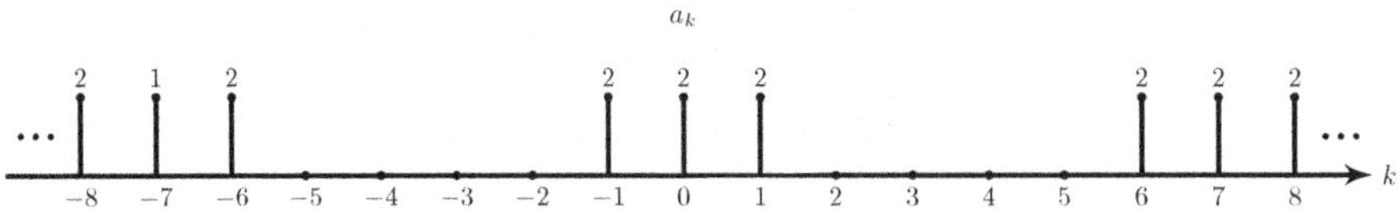

Figure 5.15: Problem 92 - Part (c)

Solution.

(a) $a_k = \cos(\frac{4\pi}{23}k)$

The Fourier series coefficients a_k are periodic with period N where

$$N = \frac{2\pi}{\frac{4\pi}{23}}m \xrightarrow{m=2} N = 23, \qquad w_0 = \frac{2\pi}{23}$$

Using Euler's formula we have

$$\begin{aligned} a_k &= \frac{1}{2}e^{jk(4\pi/23)} + \frac{1}{2}e^{-jk(4\pi/23)} \\ &= \frac{1}{2}e^{j2k(2\pi/23)} + \frac{1}{2}e^{-j2k(2\pi/23)} \\ &= \frac{1}{23}\sum_{n=-11}^{11} x[n]e^{-jkn(2\pi/23)}. \end{aligned}$$

Comparing the sides of the equality, signal $x[n]$ over one period is obtained as

$$x[n] = \begin{cases} \frac{23}{2} & n = \pm 2 \\ 0 & \text{otherwise} \end{cases}.$$

(b) $a_k = \sum_{m=-\infty}^{+\infty}(-1)^m(\delta[k-2m] + \delta[k+5m])$
Let's define sequences

$$c_k = \sum_{m=-\infty}^{+\infty} (\delta[k-2m])$$

and

$$d_k = \sum_{m=-\infty}^{+\infty} (\delta[k+5m]).$$

It is easy to see that the period of c_k is $N_c = 2$ and the period of d_k is $N_d = 5$. Consequently, the periods corresponding to sequences $(-1)^n c_k$ and $(-1)^n d_k$ is 4 and 10, respectively. On the other hand, we have

$$a_k = (-1)^n c_k + (-1)^n d_k.$$

Thus the period of a_k is $N_0 = 20$ and $w_0 = \frac{\pi}{10}$.

We choose the period from -9 to 10 and based on definition we have

$$\begin{aligned} x[n] &= \sum_{k=-9}^{10} a_k e^{jk(\pi/10)n} \\ &= 2\cos(\frac{4\pi}{5}n) - 2\cos(\frac{3\pi}{5}n) + 2\cos(\frac{2\pi}{5}n) - 2\cos(\frac{\pi}{5}n) - 2\cos(\frac{\pi}{2}n) + 1. \end{aligned}$$

(c) According to Fig. 5.15 (c), $x[n]$ has nonzero Fourier series coefficients for $k = \{-1, 0, 1\}$ and period $N_0 = 7$. Thus, we have $w_0 = \frac{2\pi}{7}$ and we write

$$\begin{aligned} x[n] &= \sum_{k=-3}^{3} a_k e^{jk(2\pi/7)n} \\ &= 2e^{-j(2\pi/7)n} + 2e^{j(2\pi/7)n} + 2 \end{aligned}$$

$$= 4\cos(\frac{2\pi}{7}n) + 2.$$

Problem 93. Consider signal $x[n]$ depicted in Fig. 5.16. This signal is periodic with period $N = 4$ and has the Fourier series coefficients a_k. Determine signal $z[n]$ corresponding to each of the following Fourier series coefficients.

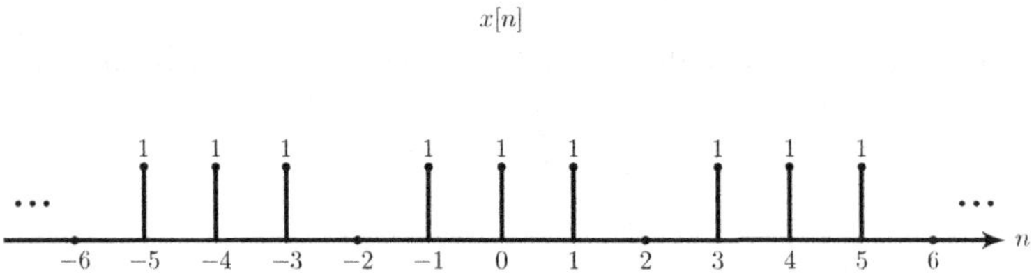

Figure 5.16: Problem 93, Periodic Signal $x[n]$.

(a) $(-1)^k a_k$ **(b)** a_{k-2}

Solution.

(a)

$$b_k = (-1)^k a_k \tag{5.33}$$

Substituting $(-1)^k = e^{-jk\pi}$ into (5.33), we have

$$b_k = e^{-jk\pi} a_k = e^{-jk(2\pi/4)2} a_k \tag{5.34}$$

Thus, the signal corresponds to b_k is

$$z[n] = x[n-2]$$

that is depicted in Fig. 5.17.

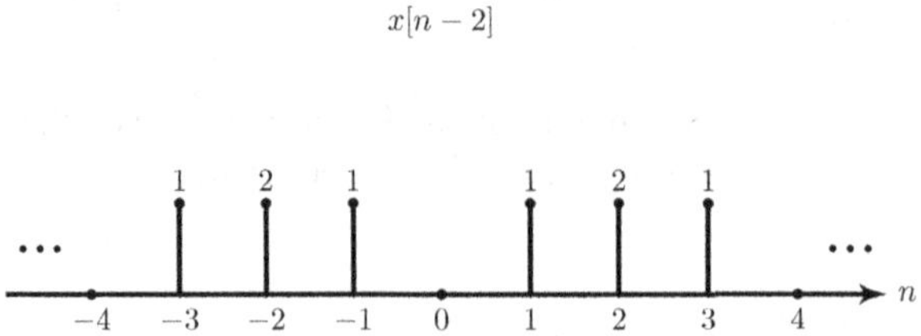

Figure 5.17: Problem 93 - Part (a), Periodic signal $x[n-2]$.

(b) $b_k = a_{k-2}$

Using frequency shifting properties of Fourier series, the signal corresponding to the b_k is

$$z[n] = e^{j2(2\pi/4)n}x[n] = e^{jn\pi}x[n] = (-1)^n x[n],$$

that is depicted in Fig. 5.18.

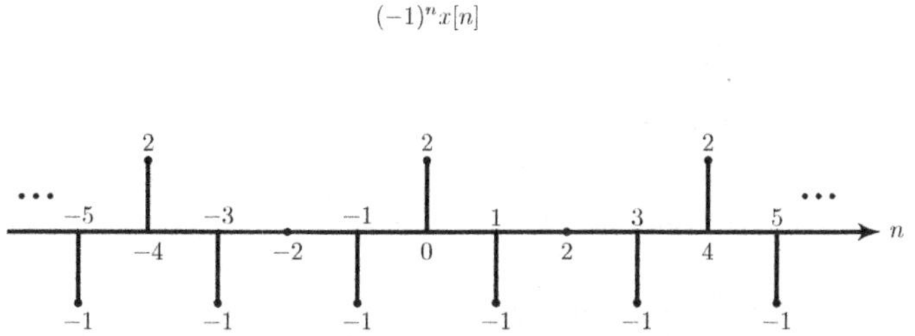

Figure 5.18: Problem 93 - Part (b), Periodic signal $(-1)^n x[n]$.

Problem 94. Let $x[n]$ be a real and even periodic signal with fundamental period $N = 10$ ($w_0 = \frac{\pi}{5}$) and power $P = 3/2$. We also know that the maximum value of $x[n]$ is equal to 2. This signal enters a filter with frequency response $H(jw)$, shown in Fig. 5.19. The output signal has the average power equal to the power of $x[n]$. Find signal $x[n]$.

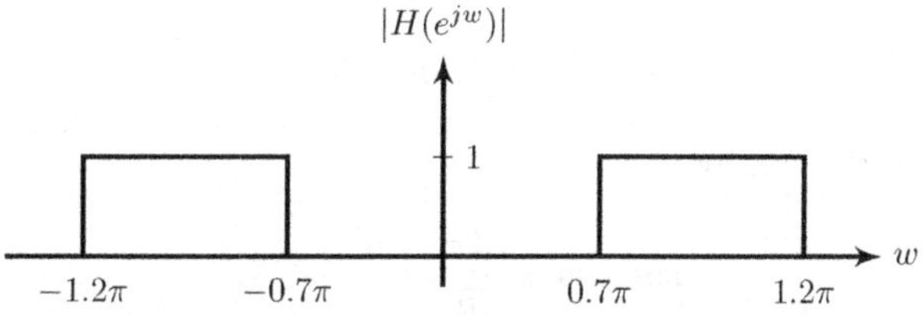

Figure 5.19: Problem 94, Frequency response of the ideal band-pass filter.

Solution.

Using the fact that the signal period is 10, we write the Fourier series representation of $x[n]$ as

$$x[n] = \sum_{k=0}^{9} a_k e^{jk(\pi/5)n}, \quad w_0 = \frac{\pi}{5}.$$

The band-pass filter shown in Fig. 5.19 passes 3 harmonic components ($4\pi/5, \pi, 6\pi/5$). Since this filter does not change the amplitude of passing harmonic components and since the power of the input and output signals are equal, signal $x[n]$ can only be composed of these harmonics. As such

$$\begin{aligned} x[n] &= a_4 e^{j4(\pi/5)n} + a_5 e^{j5(\pi/5)n} + a_6 e^{j6(\pi/5)n} \\ &= a_4 e^{j(4\pi/5)n} + a_5 e^{j\pi n} + a_6 e^{j6(\pi/5)n} \end{aligned}$$

Since $x[n]$ is even, we have

$$a_k = a_{-k} \quad \Rightarrow \quad a_4 = a_{-4}.$$

and so

$$a_{-4} = a_{-4+10} = a_6 \quad \Rightarrow \quad a_4 = a_6$$

Now using the Euler's formula we have

$$x[n] = 2a_4 \cos(\frac{4\pi}{5}n) + a_5 e^{j\pi n}. \tag{5.35}$$

Now note that based on the above equation, the second term takes values $\pm a_5$ depending on n being even or odd. Let's assume that a_5 is positive (if at the end, we come up with a negative value, we have to change this assumption and redo the solution). Then, $x[n]$ takes its maximum when $n = 0$. By setting $n = 0$ in (5.35) we obtain

$$2a_4 + a_5 = 2. \tag{5.36}$$

Now since the signal power is assumed to be $\frac{3}{2}$, using the Parseval's relationship, we have

$$a_5^2 + 2a_4^2 = \frac{3}{2}. \tag{5.37}$$

Combining (5.36) and (5.37), we can write

$$a_4 = 0.5, \, a_5 = 1.$$

Finally, the signal $x[n]$ is given by

$$x[n] = \cos(\frac{4\pi}{5}n) + (-1)^n.$$

Problem 95. A discrete-time periodic real signal $x[n]$ has period $N = 5$ and Fourier coefficients a_k. Let

$$a_0 = 2, \quad a_2 = e^{j(\pi/5)}, \quad a_4 = e^{j(\pi/4)}.$$

(a) Determine the values of a_1, a_{-1} and a_{-2}.

(b) Determine signal $x[n]$ using the result of Part (a).

Solution.

(a) We know that the Fourier coefficients a_k are periodic with period $N = 5$. So we have

$$a_k = a_{k+N} \quad \Rightarrow \quad a_{-1} = a_4 = e^{j(\pi/4)}.$$

Since $x[n]$ is real, the Fourier series coefficients will be conjugate symmetric. So we have

$$a_{-k} = a_k^*$$

$$\xrightarrow{k=-1} \quad a_1 = a_{-1}^* = (e^{j(\pi/4)})^* = e^{-j(\pi/4)},$$
$$\xrightarrow{k=2} \quad a_{-2} = a_2^* = (e^{j(\pi/5)})^* = e^{-j(\pi/5)}.$$

(b) From the result of Part (a), the Fourier series coefficients over a period are

$$a_{-2} = e^{-j(\pi/5)}, \quad a_{-1} = a_4 = e^{j(\pi/4)}, \quad a_0 = 2, \quad a_1 = e^{-j(\pi/4)}, \quad a_2 = e^{j(\pi/5)}.$$

Now we simply use the definition to obtain $x[n]$:

$$\begin{aligned} x[n] &= \sum_{k=-2}^{2} a_k e^{jk(2\pi/5)n} \\ &= a_{-2}e^{-j2(2\pi/5)n} + a_{-1}e^{-j(2\pi/5)n} + a_0 + a_1 e^{j(2\pi/5)n} + a_2 e^{j2(2\pi/5)n} \\ &= e^{-j(\frac{4\pi}{5}n+\frac{\pi}{5})} + e^{-j(\frac{2\pi}{5}n-\frac{\pi}{4})} + 2 + e^{j(\frac{2\pi}{5}n-\frac{\pi}{4})} + e^{j(\frac{4\pi}{5}n+\frac{\pi}{4})} \\ &= 2\cos(\frac{4\pi}{5}n + \frac{\pi}{5}) + 2\cos(\frac{2\pi}{5}n - \frac{\pi}{4}) + 2. \end{aligned}$$

Problem 96. As shown in Fig. 5.20, $x[n]$ is a periodic signal with the Fourier series coefficients a_k. Find the value of

$$I = \sum_{k=-6}^{5} a_k.$$

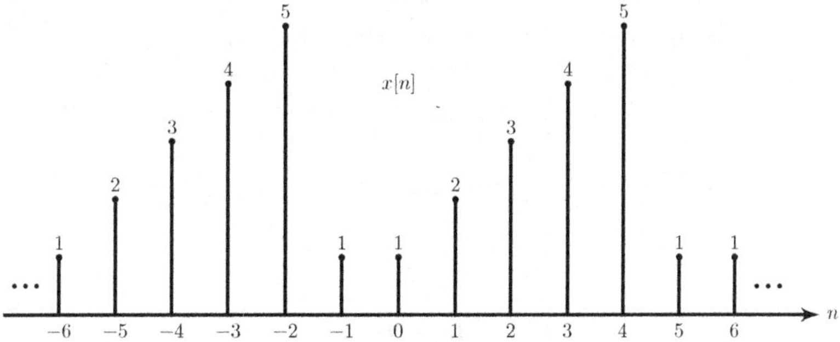

Figure 5.20: Problem 96, Periodic signal $x[n]$.

Solution.

From Fig. 5.20, we see that the period of signal is $N = 6$ and $w_0 = \frac{\pi}{3}$. By definition,

$$x[n] = \sum_{k<N>} a_k e^{jkw_0 n}. \tag{5.38}$$

Substituting $n = 0$ in (5.38) and from Fig. 5.20 (summing over 1 period), we obtain

$$x[0] = \sum_{k<N>} a_k = 1. \tag{5.39}$$

On the other hand,

$$I = \sum_{k=-6}^{5} a_k = \sum_{k=-6}^{-1} a_k + \sum_{k=0}^{5} a_k = 2 \sum_{k<N>} a_k = 2.$$

Problem 97. Consider a discrete-time LTI system whose frequency response is

$$H(e^{jw}) = \cos(4w).$$

If the input to this system is the following periodic signal

$$x[n] = \sum_{k=-\infty}^{+\infty} \delta[n - 4k],$$

find the corresponding system output $y[n]$.

Solution.
The signal is an impulse train with period $N = 4$ and $w_0 = \pi/2$. We know that the Fourier series coefficients of the impulse train are equal to $\frac{1}{N}$:

$$a_k = \frac{1}{N} = \frac{1}{4}.$$

The Fourier series coefficients of the system output are obtained as

$$b_k = a_k H(kw_0) = \frac{1}{4} H(k\pi/2). \tag{5.40}$$

Substituting $k = 0, 1, 2, 3$, in (5.40), we obtain

$$\begin{aligned} b_0 &= \frac{1}{4} H(0) = \frac{1}{4}, \\ b_1 &= \frac{1}{4} H(\pi/2) = \frac{1}{4}, \\ b_2 &= \frac{1}{4} H(\pi) = \frac{1}{4}, \\ b_3 &= \frac{1}{4} H(3\pi/2) = \frac{1}{4}. \end{aligned}$$

and using the definition, we get

$$y[n] = \sum_{k=0}^{3} b_k e^{jk\frac{\pi}{2}n} = b_0 + b_1 e^{j1\frac{\pi}{2}n} + b_2 e^{j2\frac{\pi}{2}n} + b_3 e^{j3\frac{\pi}{2}n}$$
$$= \frac{1}{4} + \frac{1}{4}e^{j\frac{\pi}{2}n} + \frac{1}{4}e^{j\pi n} + \frac{1}{4}e^{j\frac{3\pi}{2}n}$$
$$= \frac{1}{4} + \frac{1}{4}(-1)^n + \frac{1}{2}\cos(\pi n/2).$$

Problem 98. Consider a discrete-time periodic signal $x[n]$ with period N and Fourier series coefficients a_k.

(a) If $x[n]$ is as shown in Fig. 5.21 with period $N = 4$, determine signal $y[n]$ at time $n = 2$ with Fourier series coefficients $b_k = (a_k)^2$.

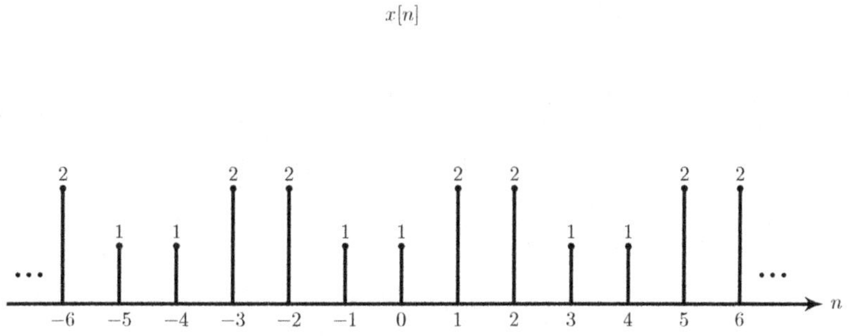

Figure 5.21: Problem 98 - Part (a), Periodic signal $x[n]$.

(b) If a_k as shown in Fig. 5.22 with period $N = 8$ and we let $y[n] = (x[n])^2$ with Fourier series coefficients b_k, determine the value of b_0.

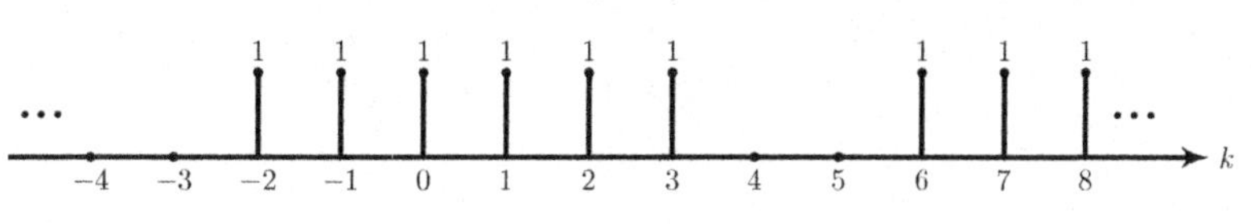

Figure 5.22: Problem 98 - Part (b), Fourier series coefficients a_k.

Solution.

(a) According to the periodic convolution property of the discrete-time Fourier series, we have

$$b_k = (a_k)^2 = a_k \cdot a_k \quad \xrightarrow{\mathcal{F.S.}^{-1}} \quad y[n] = \frac{1}{N} x[n] \circledast x[n]$$

Then

$$\begin{aligned} y[n] &= \frac{1}{N} x[n] \circledast x[n] \\ &= \frac{1}{N} \sum_{l<N>} x[l]x[n-l] \\ &= \frac{1}{4} \sum_{l=0}^{3} x[l]x[n-l] \end{aligned} \tag{5.41}$$

Substituting $n = 2$ in (5.41) and from Fig. 5.21, we obtain

$$\begin{aligned} y[2] &= \frac{1}{4} \sum_{l=0}^{3} x[l]x[2-l] \\ &= \frac{1}{4} \big(x[0]x[2] + x[1]x[1] + x[2]x[0] + x[3]x[-1] \big) = \frac{9}{4}. \end{aligned}$$

(b) According to the multiplication property of the discrete-time Fourier series, we have

$$y[n] = x[n] \cdot x[n] \quad \xrightarrow{\mathcal{F.S.}} \quad b_k = a_k \circledast a_k.$$

Thus

$$\begin{aligned} b_k &= a_k \circledast a_k \\ &= \sum_{l<N>} a_l a_{k-l} = \sum_{l=0}^{7} a_l a_{k-l}. \end{aligned} \tag{5.42}$$

Substituting $k = 0$ in (5.42) and from Fig. 5.22, we obtain

$$\begin{aligned} b_0 &= \sum_{l=0}^{7} a_l a_{-l} \\ &= a_0 a_0 + a_1 a_{-1} + \cdots + a_7 a_{-7} \\ &= 1 + 1 + 1 + 0 + 0 + 0 + 1 + 1 = 5. \end{aligned}$$

Chapter 6

The Continuous-time Fourier Transform

6.1 Introduction

In the Chapter 4 we reviewed the continuous-time Fourier series which is mainly defined for periodic signals. We also observed an important aspect of LTI systems when the input is the complex exponential e^{jwt} and the corresponding Fourier analysis. To be able to extend such an analysis to other signals which are not necessarily periodic, the Continuous-Time Fourier transform is used which is the focus of this chapter.

6.2 The Continuous-Time Fourier Transform

Let $x(t)$ be an aperiodic signal. The Fourier transform of $x(t)$ is denoted by $X(jw)$ (or $X(w)$). The Fourier transform pair of $x(t)$ and $X(jw)$ is defined as

$$X(jw) = \int_{-\infty}^{+\infty} x(t)e^{-jwt}dt, \tag{6.1}$$

$$x(t) = \frac{1}{2\pi}\int_{-\infty}^{+\infty} X(jw)e^{jwt}dw. \tag{6.2}$$

According to the definition of $X(jw)$, the function $X(jw)$ is generally complex and is shown by the *amplitude spectrum*, $|X(jw)|$, and the *phase spectrum*, $\angle X(jw)$, which are plotted versus the angular frequency w:

$$X(jw) = |X(jw)|\, e^{\angle X(jw)}. \tag{6.3}$$

By setting $w = 0$ in (6.1), we see that $X(0)$ represents the DC component of the aperiodic signal $x(t)$, i.e.,

$$X(0) = \int_{-\infty}^{+\infty} x(t)dt. \tag{6.4}$$

6.3 Convergence of Fourier Transforms

An aperiodic signal $x(t)$ has a Fourier transform representation if it satisfies the following Dirichlet conditions:

1. $x(t)$ must be *absolutely integrable*; that is,

$$\int_{-\infty}^{+\infty} |x(t)|dt < \infty, \tag{6.5}$$

2. In any finite interval of t, $x(t)$ has no more than a finite number of maxima and minima.

3. In any finite interval of t, $x(t)$ has only a finite number of discontinuities, and each of them is finite.

6.4 Properties of the Continuous-Time Fourier Transform

(a) Linearity

Consider two given signals $x(t)$ and $y(t)$. If

$$x(t) \overset{\mathcal{F.T.}}{\longleftrightarrow} X(jw),$$
$$y(t) \overset{\mathcal{F.T.}}{\longleftrightarrow} Y(jw).$$

Then we have

$$z(t) = ax(t) + by(t) \overset{\mathcal{F.T.}}{\longleftrightarrow} Z(jw) = aX(jw) + bY(jw). \tag{6.6}$$

(b) Time Shifting

Consider signal $x(t)$ with the Fourier transform $X(jw)$. The Fourier transform $x(t - t_0)$ is given by

$$x(t - t_0) \overset{\mathcal{F.T.}}{\longleftrightarrow} e^{-jwt_0}X(jw). \tag{6.7}$$

Equation (6.7) indicates that when an aperiodic signal is shifted in time domain, the magnitude of its Fourier transform remain unchanged but the phase is rotated by $-wt_0$.

(c) Frequency Shifting

Consider signal $x(t)$ with the Fourier transform $X(jw)$. The frequency shifting property indicates that the multiplication of $x(t)$ by a complex exponential signal e^{jw_0t} results in a shift of $X(jw)$ in the frequency domain:

$$e^{jw_0t}\, x(t) \overset{F.}{\longleftrightarrow} X\left(j(w - w_0)\right). \tag{6.8}$$

(d) Conjugation

The complex conjugation of signal $x(t)$ causes complex conjugation and frequency reversal in the corresponding Fourier transform. That is,

$$x(t) \xleftrightarrow{\mathcal{F.T.}} X(jw),$$
$$x^*(t) \xleftrightarrow{\mathcal{F.T.}} X^*(-jw).$$

(e) Multiplication

Consider two signals $x(t)$ and $y(t)$ with the Fourier transforms $X(jw)$ and $Y(jw)$, respectively. Then the Fourier transform of $x(t).y(t)$ is given as

$$x(t).y(t) \xleftrightarrow{F.} \frac{1}{2\pi} X(jw) * Y(jw). \tag{6.9}$$

(f) Time and Frequency Scaling

The time scaling operation by factor a causes a frequency scaling of $1/a$ and vice versa. If $a > 1$, time compression of a signal results in its spectral expansion and if $a < 1$, time expansion of the signal results in its spectral compression:

$$x(at) \xleftrightarrow{\mathcal{F.T.}} \frac{1}{|a|} X\left(\frac{jw}{a}\right), \tag{6.10}$$

where a is a real constant.

(g) Duality

Consider a signal $x(t)$ with Fourier transform $X(jw)$, then the Fourier transform of signal $X(t)$ will be $2\pi x(-jw)$:

$$x(t) \xleftrightarrow{\mathcal{F.T.}} X(jw)$$
$$X(t) \xleftrightarrow{\mathcal{F.T.}} 2\pi x(-jw).$$

For example

$$x_1(t) = \begin{cases} 1 & |t| < T \\ 0 & |t| > T \end{cases} \xleftrightarrow{\mathcal{F.T.}} X_1(jw) = \frac{2 \sin wT}{w},$$

$$x_2(t) = \frac{2 \sin Tt}{t} \xleftrightarrow{F.} X_2(jw) = \begin{cases} 2\pi & |w| < T \\ 0 & |w| > T \end{cases}.$$

(h) Parseval's Relation

Consider a signal $x(t)$ with the Fourier transform $X(jw)$. The Parseval's relation is given by

$$\int_{-\infty}^{+\infty} |x(t)|^2 dt = \frac{1}{2\pi} \int_{-\infty}^{+\infty} |X(jw)|^2 dw. \tag{6.11}$$

Note: This identity can be generalized to 2 signals, i.e.,

$$\int_{-\infty}^{\infty} x(t)y^*(t)\ dt = \int_{-\infty}^{\infty} X(f)Y^*(f)\ df. \tag{6.12}$$

Proof:

$$\begin{aligned}\int_{-\infty}^{+\infty} x(t)y^*(t)dt &= \int_{-\infty}^{+\infty} x(t)\left[\frac{1}{2\pi}\int_{-\infty}^{+\infty} Y^*(jw)e^{-jwt}dw\right]dt\\ &= \frac{1}{2\pi}\int_{-\infty}^{+\infty} Y^*(jw)\left[\int_{-\infty}^{+\infty} x(t)e^{-jwt}dt\right]dw\\ &= \frac{1}{2\pi}\int_{-\infty}^{+\infty} Y^*(jw)X(jw)dw\\ &= \int_{-\infty}^{+\infty} X(f)Y^*(f)df.\end{aligned}$$

(i) Conjugate Symmetry for Real Signals

If $x(t)$ is real, then its Fourier transform is *conjugate symmetric*, i.e., if $x(t) = x^*(t)$,

$$X(jw) = X^*(-jw) \quad \text{or} \quad X(-jw) = X^*(jw).$$

As such we have

$$\begin{aligned}&|X(jw)| \text{ is even} &&: \ |X(jw)| = |X(-jw)|\\ &\angle X(jw) \text{ is odd} &&: \ \angle X(jw) = -\angle X(-jw)\\ &\Re\{X(jw)\} \text{ is even} &&: \ \Re\{X(jw)\} = \Re\{X(-jw)\}\\ &\Im\{X(jw)\} \text{ is odd} &&: \ \Im\{X(jw)\} = -\Im\{X(-jw)\}\\ &x_e(t) \overset{\mathcal{F.T.}}{\longleftrightarrow} \Re\{X(jw)\}\\ &x_o(t) \overset{\mathcal{F.T.}}{\longleftrightarrow} j\Im\{X(jw)\}\end{aligned}$$

Note that if $x(t)$ is real and even, so is its Fourier transform, and if $x(t)$ is real and odd, then its Fourier transform is imaginary and odd.

6.5 The Fourier Transform of Periodic Signals

Consider a periodic signal $x(t)$ with period T and the Fourier series coefficients a_k. From the definition of the Fourier series coefficients, we have

$$x(t) = \sum_{k=-\infty}^{+\infty} a_k e^{jkw_0 t}. \tag{6.13}$$

We now apply the Fourier transform to both sides of (6.13):

$$X(jw) = \mathcal{F.T.}\left\{\sum_{k=-\infty}^{+\infty} a_k e^{jkw_0t}\right\}$$
$$= \sum_{k=-\infty}^{+\infty} a_k\ \mathcal{F.T.}\left\{e^{jkw_0t}\right\}. \quad (6.14)$$

From Table 6.2, we have

$$e^{jkw_0t} \overset{\mathcal{F.T.}}{\longleftrightarrow} 2\pi\delta(w - kw_0)$$

Therefore, we obtain

$$X(jw) = \sum_{k=-\infty}^{+\infty} 2\pi a_k \delta(w - kw_0). \quad (6.15)$$

In other words, the Fourier transform of a periodic signal $x(t)$ is an impulse train with magnitude $2\pi a_k$ for $w = kw_0$.

6.6 Tables of the Fourier Trasform Properties and Some Fourier Transform Pairs

Some of the important properties of the Fourier transform are summarized in Table 6.1. In particular, please see the time differentiation and integration properties. In Table 6.2, we have summarized some frequently-used Fourier transform pairs.

Table 6.1: Properties of the Fourier Transform

Property	**Aperiodic signal**	**Fourier transform**
	$x(t)$	$X(jw)$
	$y(t)$	$Y(jw)$
Linearity	$ax(t)+by(t)$	$aX(jw)+bY(jw)$
Time Shifting	$x(t-t_0)$	$e^{-jwt_0}X(jw)$
Frequency Shifting	$e^{jw_0t}x(t)$	$X(j(w-w_0))$
Conjugation	$x^*(t)$	$X^*(-jw)$
Time Reversal	$x(-t)$	$X(-jw)$
Time scaling	$x(at)$	$\frac{1}{\|a\|}X\left(\frac{jw}{a}\right)$
Convolution	$x(t)*y(t)$	$X(jw)Y(jw)$
Multiplication	$x(t)y(t)$	$\frac{1}{2\pi}X(jw)*Y(jw)$
Differentiation in Time	$\frac{d}{dt}x(t)$	$jwX(jw)$
Integration	$\int_{-\infty}^{t}x(\tau)d\tau$	$\frac{1}{jw}X(jw)+\pi X(0)\delta(w)$
Differentiation in Frequency	$tx(t)$	$j\frac{d}{dw}X(jw)$
Real signal	$x(t)=x_e(t)+x_o(t)$	$X(-jw)=X^*(jw)$
Even component	$x_e(t)$	$\Re\{X(jw)\}$
Odd component	$x_o(t)$	$j\Im\{X(jw)\}$

Parseval's relations

$$\int_{-\infty}^{+\infty}|x(t)|^2dt=\frac{1}{2\pi}\int_{-\infty}^{+\infty}|X(jw)|^2dw$$

Table 6.2: Some Fourier Transforms Pairs

Signal	**Fourier Transform**
$\delta(t)$	1
$\delta(t-t_0)$	e^{-jwt_0}
1	$2\pi\delta(w)$
e^{jw_0t}	$2\pi\delta(w-w_0)$
$\cos w_0t$	$\pi[\delta(w-w_0)+\delta(w+w_0)]$
$\sin w_0t$	$\frac{\pi}{j}[\delta(w-w_0)-\delta(w+w_0)]$
$u(t)$	$\frac{1}{jw}+\pi\delta(w)$
$\sum_{n=-\infty}^{+\infty}\delta(t-nT)$	$\frac{2\pi}{T}\sum_{k=-\infty}^{+\infty}\delta(w-\frac{2\pi k}{T})$
$\prod(\frac{t}{2T})=\begin{cases}1 & \lvert t\rvert<T\\ 0 & \lvert t\rvert>T\end{cases}$	$2\frac{\sin wT}{w}$
$\frac{\sin Wt}{\pi t}$	$\prod(\frac{t}{2W})=\begin{cases}1 & \lvert w\rvert<W\\ 0 & \lvert w\rvert>W\end{cases}$
$e^{-at}u(t),\ \Re\{a\}>0$	$\frac{1}{a+jw}$
$te^{-at}u(t),\ \Re\{a\}>0$	$\frac{1}{(a+jw)^2}$
$\frac{t^{n-1}}{(n-1)!}e^{-at}u(t),\ \Re\{a\}>0$	$\frac{1}{(a+jw)^n}$
$e^{-a\lvert t\rvert},\ \Re\{a\}>0$	$\frac{2a}{a^2+w^2}$

6.7 The Fourier Transform and LTI Systems

Consider an LTI system with the impulse response $h(t)$. Let the input be an aperiodic signal $x(t)$. The Fourier transform of the output signal $y(t)$, and eventually $y(t)$ itself, can be obtained based on the Fourier transform of $x(t)$ and the frequency response of the system. That is, if

$$y(t)=h(t)*x(t)\quad\overset{\mathcal{F.T.}}{\longleftrightarrow}\quad Y(jw)=H(jw)X(jw) \tag{6.16}$$

then

$$H(jw)=\frac{Y(jw)}{X(jw)}, \tag{6.17}$$

where $H(jw)$ is the frequency response of the system which is given by

$$H(jw) = \int_{-\infty}^{+\infty} h(t)e^{-jwt}dt = |H(jw)|\, e^{\angle H(jw)}. \tag{6.18}$$

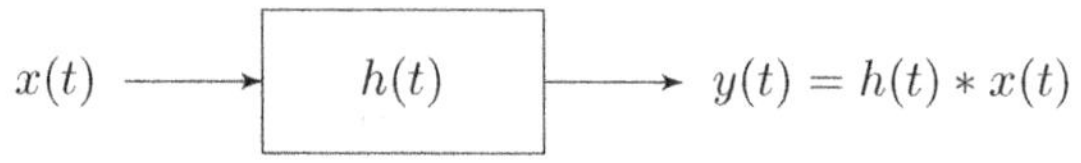

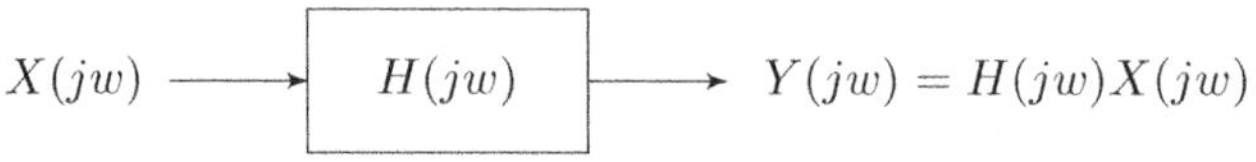

We also have:

$$|Y(jw)| = |H(jw)|\, |X(jw)|, \tag{6.19}$$
$$\angle Y(jw) = \angle H(jw) + \angle X(jw). \tag{6.20}$$

6.8 LTI Systems Characterized by Constant-Coefficient Differential Equations

Consider a continuous-time LTI system which is described by a linear constant-coefficient differential equation of the form

$$\sum_{k=0}^{N} a_k \frac{d^k y(t)}{dt^k} = \sum_{k=0}^{M} b_k \frac{d^k x(t)}{dt^k}, \quad M \leq N. \tag{6.21}$$

Applying the Fourier transform to both sides of (6.21) and using the linearity property (6.6), we have

$$F\left\{\sum_{k=0}^{N} a_k \frac{d^k y(t)}{dt^k}\right\} = F\left\{\sum_{k=0}^{M} b_k \frac{d^k x(t)}{dt^k}\right\}, \tag{6.22}$$

$$\sum_{k=0}^{N} a_k\, F\left\{\frac{d^k y(t)}{dt^k}\right\} = \sum_{k=0}^{M} b_k\, F\left\{\frac{d^k x(t)}{dt^k}\right\}, \tag{6.23}$$

and from the differentiation property, of Table 6.1:

$$\sum_{k=0}^{N} a_k\, (jw)^k Y(jw) = \sum_{k=0}^{M} b_k\, (jw)^k X(jw),$$

$$Y(jw)\left[\sum_{k=0}^{N} a_k\, (jw)^k\right] = X(jw)\left[\sum_{k=0}^{M} b_k\, (jw)^k\right].$$

From (6.17):

$$H(jw) = \frac{Y(jw)}{X(jw)} = \frac{\sum_{k=0}^{M} b_k \, (jw)^k}{\sum_{k=0}^{N} a_k \, (jw)^k}. \tag{6.24}$$

6.9 Filtering

Filtering is an important process in signal processing in which certain frequency components of a give signal are modified or removed. Such a process can very well be analyzed through the Fourier analysis of the signal as well as the system (filter).

An ideal *frequency-selective* filter is defined by

$$H(jw) = \begin{cases} e^{-jwt_0} & |w| < w_c \\ 0 & |w| > w_c \end{cases} \tag{6.25}$$

which is sketched in Fig. 6.1.

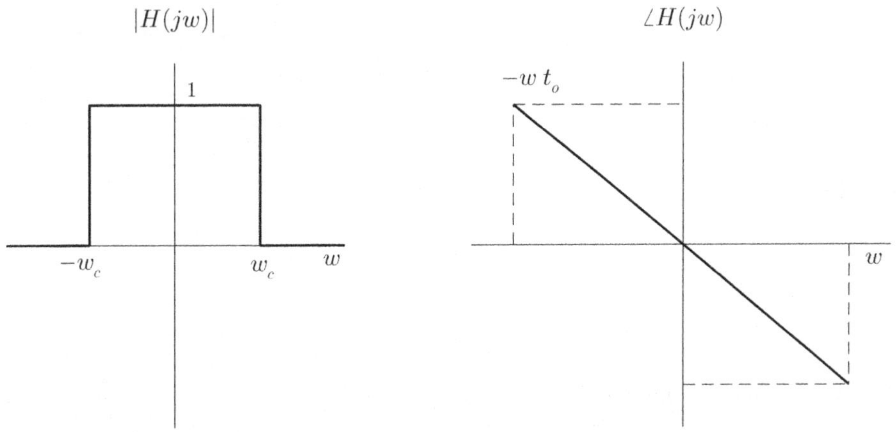

Figure 6.1: An ideal frequency-selective filter

Some important filter types are as follows:

1. **Ideal low-pass Filter**

$$|H(jw)| = \begin{cases} 1 & |w| < w_c \\ 0 & |w| > w_c \end{cases} \tag{6.26}$$

where w_c is the cutoff frequency.

2. **Ideal high-pass Filter**

$$|H(jw)| = \begin{cases} 0 & |w| < w_c \\ 1 & |w| > w_c \end{cases} \tag{6.27}$$

3. **Ideal band-pass Filter**

$$|H(jw)| = \begin{cases} 1 & w_1 < |w| < w_2 \\ 0 & \text{otherwise} \end{cases} \tag{6.28}$$

4. **Ideal band-stop Filter**

$$|H(jw)| = \begin{cases} 0 & w_1 < |w| < w_2 \\ 1 & \text{otherwise} \end{cases} \tag{6.29}$$

5. **Ideal All-pass Filter**

$$|H(jw)| = 1 \qquad \text{for all } w. \tag{6.30}$$

6.10 Problems

Problem 99. Use the Fourier transform definition to obtain the Fourier transform of the following signals:

(a) $x(t) = \delta(t - t_0)$

(b) $x(t) = \prod(\frac{t-2}{2})$

(c) $x(t) = e^t[\mathrm{u}(t) - \mathrm{u}(t-1)]$

(d) $x(t) = t\,\mathrm{u}(t)$

(e) $x(t) = \mathrm{u}(t+2) - 2\,\mathrm{u}(t) + \mathrm{u}(t-2)$

(f) $x(t) = \sum_{m=0}^{\infty} a^m \delta(t-m)$, $m < 1$

(g) $x(t) = e^t\,\mathrm{u}(-t) + e^{-t}\,\mathrm{u}(t)$

(h) $x(t) = t^2 \prod(\frac{t}{4})$

Solution.

(a) $x(t) = \delta(t - t_0)$
By definition,

$$X(jw) = \int_{-\infty}^{\infty} x(t)e^{-jwt}dt.$$

By substituting $x(t)$ in the above definition we get:

$$X(jw) = \int_{-\infty}^{\infty} \delta(t-t_0)e^{-jwt}dt = e^{-jwt_0}.$$

(b) $x(t) = \prod(\frac{t-2}{2})$
The signal is a rectangular pulse where $x(t) = 1$ for $1 \le t \le 3$. Thus we have

$$\begin{aligned} X(jw) &= \int_{-\infty}^{\infty} \prod(\frac{t-2}{2})e^{-jwt}dt \\ &= \int_{1}^{3} e^{-jwt}dt \\ &= \frac{-1}{jw}(e^{-3jw} - e^{-jw}) \\ &= \frac{2e^{-j2w}}{w}\sin(w). \end{aligned}$$

(c) $x(t) = e^t[\mathrm{u}(t) - \mathrm{u}(t-1)]$

The signal is equal to e^t for $0 \le t \le 1$ and 0 elsewhere. Thus we have

$$\begin{aligned} F(jw) &= \int_{-\infty}^{\infty} e^t[\mathrm{u}(t) - \mathrm{u}(t-1)]e^{-jwt}dt \\ &= \int_0^1 e^t e^{-jwt}dt \\ &= \int_0^1 e^{t(1-jw)}dt \\ &= \frac{1}{1-jw}(e^{1-jw} - 1). \end{aligned}$$

(d) $x(t) = t\,\mathrm{u}(t)$

Using the definition of the Fourier transform we write

$$\begin{aligned} F(jw) &= \int_{-\infty}^{\infty} t\,\mathrm{u}(t)e^{-jwt}dt \\ &= \int_0^{\infty} te^{-jwt}dt \\ &= \frac{e^{-jwt}}{-w^2}(-jwt - 1)\Big|_0^{\infty}. \end{aligned}$$

It can be seen that the expression is not bounded as $x(t)$ is not absolutely integrable. As such, Fourier transform is not defined for $x(t)$.

(e) $x(t) = \mathrm{u}(t+2) - 2\,\mathrm{u}(t) + \mathrm{u}(t-2)$

The signal is 1 for $-2 \le t \le 0$ and -1 for $0 \le t \le 2$ and 0 otherwise. By definition of the Fourier transform we write

$$\begin{aligned} F(jw) &= \int_{-\infty}^{\infty} [\mathrm{u}(t+2) - 2\,\mathrm{u}(t) + \mathrm{u}(t-2)]e^{-jwt}dt \\ &= \int_{-2}^{0} e^{-jwt}dt + \int_0^2 (-1)e^{-jwt}dt \\ &= -\frac{1}{jw}(1 - e^{j2w}) + \frac{1}{jw}(-1 + e^{-j2w}) \\ &= -\frac{1}{jw}(2 - e^{j2w} - e^{-j2w}) \\ &= \frac{-4\sin^2(w)}{jw}. \end{aligned}$$

(f) $x(t) = \sum_{m=0}^{\infty} a^m \delta(t-m)$, $a < 1$

By definition of the Fourier transform we have

$$
\begin{aligned}
F(jw) &= \int_{-\infty}^{\infty} \sum_{m=0}^{\infty} a^m \underbrace{\delta(t-m)}_{\text{non-zero when } t=m} e^{-jwt} \\
&= \sum_{m=0}^{\infty} a^m \int_{-\infty}^{\infty} \delta(t-m)e^{-jwt} \\
&= \sum_{m=0}^{\infty} a^m e^{-jwm} \\
&= \sum_{m=0}^{\infty} (ae^{-jw})^m \\
&= \frac{1}{1-a.e^{-jw}},
\end{aligned}
$$

where the last equality results from the Geometric Sequence Sum formulation and the fact that $|a.e^{-jw}| < 1$.

(g) $x(t) = e^t\,\mathrm{u}(-t) + e^{-t}\,\mathrm{u}(t)$

By definition of the Fourier transform and unit step signal we can write

$$
\begin{aligned}
F(jw) &= \int_{-\infty}^{\infty} (e^t\,\mathrm{u}(-t) + e^{-t}\,\mathrm{u}(t))e^{-jwt}dt \\
&= \int_{-\infty}^{\infty} e^t\,\mathrm{u}(-t)e^{-jwt}dt + \int_{-\infty}^{\infty} e^{-t}\,\mathrm{u}(t)e^{-jwt}dt \\
&= \int_{-\infty}^{0} e^t e^{-jwt}dt + \int_{0}^{\infty} e^{-t}e^{-jwt}dt \\
&= \frac{1}{1-jw}(1-0) + \frac{1}{-1-jw}(0-1) \\
&= \frac{1}{1-jw} + \frac{1}{1+jw}.
\end{aligned}
$$

(h) $x(t) = t^2 \prod(\frac{t}{4})$

Fig. 6.2 shows the plot of $x(t)$:

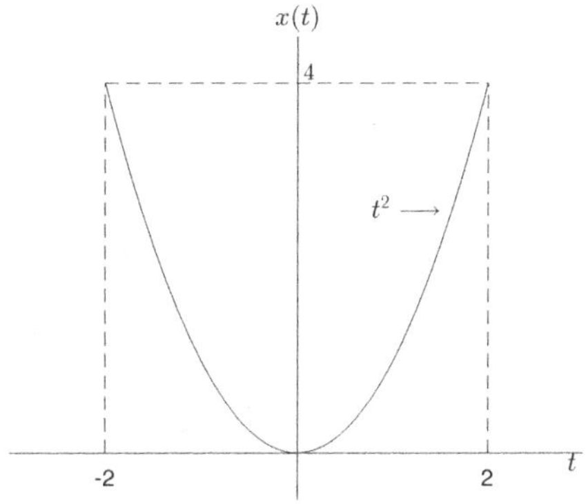

Figure 6.2: Signal $x(t)$, Problem 99 - Part (h).

By definition of the Fourier transform and the above figure,

$$X(jw) = \int_{-2}^{2} t^2 e^{-jwt} dt.$$

The above integral has to be evaluated using the integration by parts method. We know that:

$$\int_{\alpha}^{\beta} t^2 e^{at} dt = \frac{-t^2 e^{at}}{a} - 2\frac{te^{at}}{a^2} - 2\frac{e^{at}}{a^3}\Big|_{\alpha}^{\beta}$$

Using this method we obtain:

$$\begin{aligned} X(jw) &= t^2 \frac{-1}{jw} e^{-jwt} + \frac{2t}{w^2} e^{-jwt} - \frac{2}{jw^3} e^{-jwt}\Big|_{-2}^{2} \\ &= \frac{-4}{jw} e^{-j2w} + \frac{4}{w^2} e^{-j2w} - \frac{2}{jw^3} e^{-j2w} + \frac{4}{jw} e^{j2w} + \frac{4}{w^2} e^{j2w} + \frac{2}{jw^3} e^{j2w} \\ &= 4(\frac{e^{j2w} - e^{-j2w}}{jw}) + 4(\frac{e^{j2w} + e^{-j2w}}{w^2}) + 2(\frac{e^{j2w} - e^{-j2w}}{jw^3}) \\ &= \frac{8\sin(2w)}{w} + \frac{8\cos(2w)}{w^2} + \frac{4\sin(2w)}{w^3}. \end{aligned}$$

Problem 100. For each of the following signals, use Fourier transform properties to determine the corresponding Fourier transforms.

(a) $x(t) = \cos(w_0 t)\,\mathrm{u}(t)$

(b) $x(t) = \frac{1}{\pi t}\cos(4\pi t).\sin(2\pi t)$

(c) $x(t) = \sin(|t|)$

(d) $x(t) = te^{-3|t-2|}$

(e) $x(t) = t^2 e^t\,\mathrm{u}(-t)$

(f) $x(t) = \begin{cases} 1 - t^2 & |t| < 1, \\ 0 & |t| > 1 \end{cases}$

(g) $x(t) = \left[\frac{2\sin(3t)}{\pi t}\right].\left[\frac{\sin(t)}{t}\right]$

Solution.

(a) $x(t) = \cos(w_0 t)\,\mathrm{u}(t)$

Signal $x(t)$ can be rewritten as

$$x(t) = \cos(w_0 t)\,\mathrm{u}(t) = \frac{1}{2}e^{jw_0 t}\,\mathrm{u}(t) + \frac{1}{2}e^{-jw_0 t}\,\mathrm{u}(t).$$

We know that the Fourier transform of the unit step signal is

$$u(t) \leftrightarrow \frac{1}{jw} + \pi\delta(w).$$

Then by the frequency-shifting and the linearity properties, we obtain

$$\frac{1}{2}e^{jw_0 t}\,\mathrm{u}(t) + \frac{1}{2}e^{-jw_0 t}\,\mathrm{u}(t) \xrightarrow{\mathcal{F.T.}} \frac{1}{2}\left[\frac{1}{j(w-w_0)} + \pi\delta(w-w_0)\right] + \frac{1}{2}\left[\frac{1}{j(w+w_0)} + \pi\delta(w+w_0)\right]$$

$$\longrightarrow \quad X(jw) = \frac{\pi}{2}\Big[\delta(w-w_0) + \delta(w+w_0)\Big] + \frac{jw}{w_0^2 - w^2}.$$

(b) $x(t) = \frac{1}{\pi t}\cos(2\pi t).\sin(2\pi t)$

$$\frac{\sin(2\pi t)}{\pi t}.\cos(2\pi t) = \frac{1}{2}e^{j2\pi t}\frac{\sin(2\pi t)}{\pi t} + \frac{1}{2}e^{-j2\pi t}\frac{\sin(2\pi t)}{\pi t}.$$

We know that,

$$\frac{\sin(Wt)}{\pi t} \xrightarrow{\mathcal{F.T.}}$$

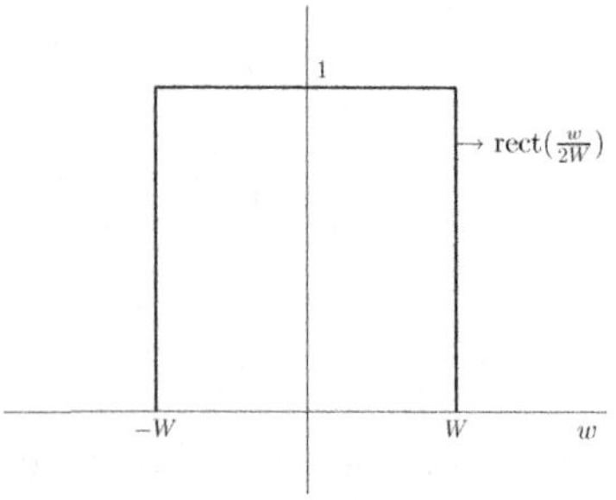

Hence,

$$\frac{sin(2\pi t)}{\pi t} \xrightarrow{\mathcal{F.T.}} \mathrm{rect}(\frac{w}{4\pi})$$

Thus by the frequency-shifting property, we can write

$$X(jw) = \frac{1}{2}\mathrm{rect}(\frac{w-2\pi}{4\pi}) + \frac{1}{2}\mathrm{rect}(\frac{w+2\pi}{4\pi}).$$

(c) $x(t) = \sin(|t|)$

Signal $x(t)$ can be rewritten as

$$\sin(|t|) = \begin{cases} \sin(t) & t \geq 0, \\ -\sin(t) & t < 0 \end{cases}$$

hence we can write

$$\sin(|t|) = \text{sign}(t)\sin(t).$$

Based on Table 6.2

$$\text{sign}(t) \leftrightarrow \frac{2}{jw}.$$

Thus, using the multiplication property we obtain

$$\begin{aligned}
\mathcal{F.T.}\{\sin(|t|)\} &= \frac{1}{2\pi}\mathcal{F.T.}\{\text{sign}(t)\} * \mathcal{F.T.}\{\sin(t)\} \\
&= \frac{1}{2\pi}\frac{2}{jw} * \frac{\pi}{j}\Big(\delta(w-1) - \delta(w+1)\Big) \\
&= \frac{1}{2\pi}\left(\frac{2\pi}{w+1} - \frac{2\pi}{w-1}\right) \\
&= \frac{2}{1-w^2}.
\end{aligned}$$

(d) $x(t) = te^{-3|t-2|}$

We first start by finding the Fourier transform of $e^{-3|t|}$. Then we use time shifting property and frequency differentiation (see Table 6.1) as follows:

$$e^{-a|t|} \xrightarrow{\mathcal{F}} \frac{2a}{a^2+w^2}$$

hence,

$$\Longrightarrow e^{-3|t|} \xrightarrow{\mathcal{F.T.}} \frac{6}{9+w^2}$$

$$\Longrightarrow e^{-3|t-2|} \xrightarrow{\mathcal{F.T.}} \frac{6e^{-2jw}}{9+w^2}$$

$$\Longrightarrow te^{-3|t-2|} \xrightarrow{\mathcal{F.T.}} j\frac{d}{dw}\left(\frac{6e^{-2jw}}{9+w^2}\right)$$

$$X(jw) = j\frac{-12je^{-2jw}(9+w^2) - 12we^{-2jw}}{(9+w^2)^2}.$$

(e) $x(t) = t^2e^t\,\text{u}(-t)$

We know that,

$$e^{-t}\,\text{u}(t) \xrightarrow{\mathcal{F.T.}} \frac{1}{1+jw}.$$

Hence,

$$\Longrightarrow e^t\,\mathrm{u}(-t) \xrightarrow{\mathcal{F.T.}\text{(time reversal))}} \frac{1}{1-jw}$$

$$\Longrightarrow t^2e^t\,\mathrm{u}(-t) \xrightarrow{\mathcal{F.T.}\text{(frequency differentiation twice)}} (j)^2\frac{d^2}{dw^2}(\frac{1}{1-jw})$$

So

$$X(jw) = \frac{2}{(1-jw)^3}.$$

(f) $x(t) = \begin{cases} 1-t^2 & |t| < 1, \\ 0 & |t| > 2 \end{cases}$

Signal $x(t)$ can be rewritten as

$$x(t) = (1-t^2)\big(\mathrm{u}(t+1) - \mathrm{u}(t-1)\big).$$

We differentiate the above expression twice to obtain

$$x'(t) = (-2t)\big(\mathrm{u}(t+1) - \mathrm{u}(t-1)\big) + \underbrace{(1-t^2)\big(\delta(t+1) - \delta(t-1)\big)}_{\to 0}$$

$$\begin{aligned} x''(t) &= (-2)\big(\mathrm{u}(t+1) - \mathrm{u}(t-1)\big) + (-2t)\big(\delta(t+1) - \delta(t-1)\big), \\ &= (-2)\big(\mathrm{u}(t+1) - \mathrm{u}(t-1)\big) + 2\delta(t+1) + 2\delta(t-1). \end{aligned}$$

Taking Fourier transform from both sides we obtain (note that $(jw)(jw) = -w^2$):

$$\begin{aligned} -w^2X(jw) &= -4\frac{\sin(w)}{w} + 2e^{jw} + 2e^{-jw}, \\ \Longrightarrow X(jw) &= \frac{4}{w^2}(\frac{\sin(w)}{w} - \cos(w)). \end{aligned}$$

(g) $x(t) = \left[\frac{2\sin(3t)}{\pi t}\right].\left[\frac{\sin(t)}{t}\right]$

Using the multiplication property we obtain

$$X(jw) = \frac{1}{2\pi}\Big(\mathcal{F.T.}\Big\{\frac{2\sin(3t)}{\pi t}\Big\} * \mathcal{F.T.}\Big\{\frac{\sin(t)}{t}\Big\}\Big),$$

$$\xrightarrow{\frac{\sin(Wt)}{\pi t} \leftrightarrow \prod(\frac{w}{2W})} \quad X(jw) = \frac{1}{2\pi}\Big\{2\prod(\frac{w}{6}) * \pi\prod(\frac{w}{2})\Big\}$$

Now all we have to do is to find the convolution of 2 rectangular pulse signals:

$$\Rightarrow \quad X(jw) = \begin{cases} 4-w & 2<w\leq 4 \\ 2 & -2<w\leq 2 \\ 4+w & -4<w\leq -2 \\ 0 & o.w \end{cases}$$

Problem 101. Use the definition to obtain the Fourier transform of each of the following signals.

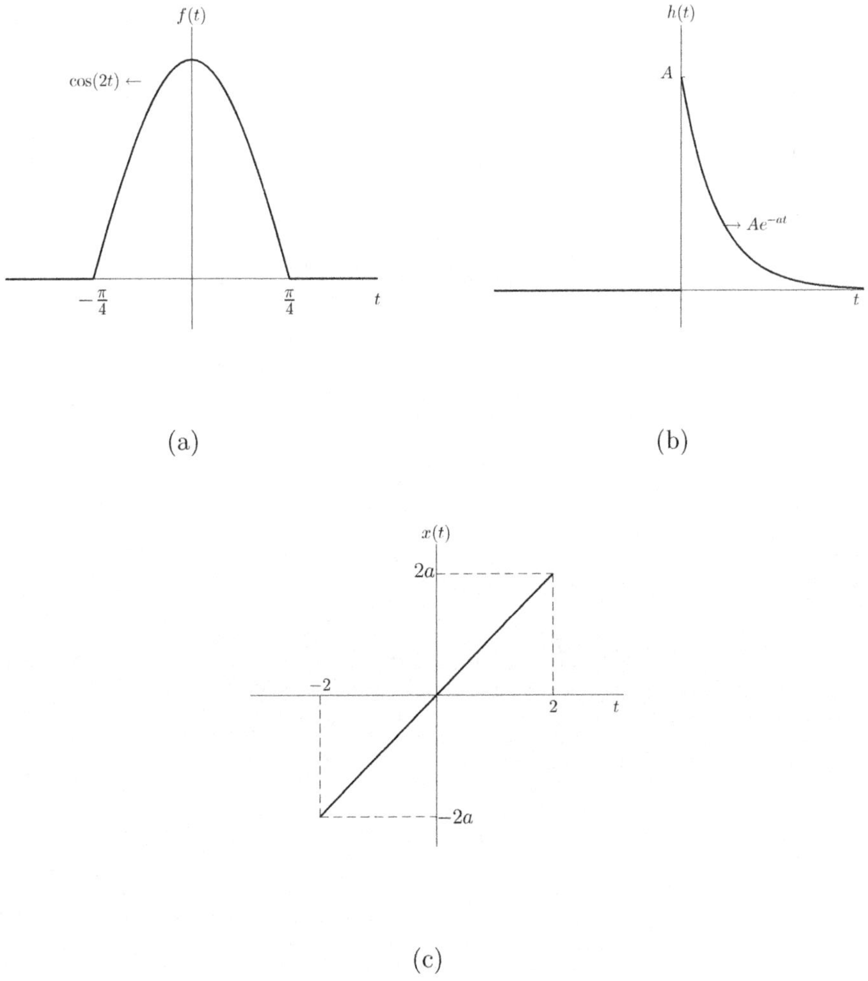

Figure 6.6: Problem 101.

Solution.

(a) By definition,

$$
\begin{aligned}
F(jw) &= \int_{-\frac{\pi}{4}}^{\frac{\pi}{4}} \cos(2t)e^{-jwt}dt \\
&= \frac{1}{2}\int_{-\frac{\pi}{4}}^{\frac{\pi}{4}} (e^{j2t}+e^{-j2t})e^{-jwt}dt \\
&= \frac{1}{2j(2-w)}e^{j(2-w)t}\Big|_{-\frac{\pi}{4}}^{\frac{\pi}{4}} + \frac{1}{-2j(2+w)}e^{-j(2+w)t}\Big|_{-\frac{\pi}{4}}^{\frac{\pi}{4}} \\
&= \frac{1}{2(2-w)}(e^{-jw\frac{\pi}{4}}+e^{jw\frac{\pi}{4}}) - \frac{1}{2(2+w)}(-e^{jw\frac{\pi}{4}}-e^{-jw\frac{\pi}{4}}) \\
&= \frac{1}{2-w}\cos(\frac{w\pi}{4}) + \frac{1}{2+w}\cos(\frac{w\pi}{4}) \\
&= \frac{4}{4-w^2}\cos(\frac{w\pi}{4}).
\end{aligned}
$$

(b) By definition,

$$
\begin{aligned}
H(jw) &= \int_{-\infty}^{\infty} Ae^{-at}\,\mathrm{u}(t)e^{-jwt}dt \\
&= \int_{0}^{\infty} Ae^{-at}e^{-jwt}dt \\
&= \int_{0}^{\infty} Ae^{-(a+jw)t}dt \\
&= \frac{A}{a+jw}.
\end{aligned}
$$

(c) $x(t) = at$ for $-2 \le t < 2$ and 0 otherwise. So

$$
\begin{aligned}
X(jw) &= \int_{-\infty}^{\infty} x(t)e^{-jwt}dt \\
&= \int_{-2}^{2} ate^{-jwt}dt \\
&= ae^{-jwt}(\frac{1}{w^2}-\frac{t}{jw})\Big|_{-2}^{2} \\
&= a\left(e^{-j2w}(\frac{1}{w^2}-\frac{2}{jw}) - e^{j2w}(\frac{1}{w^2}+\frac{2}{jw})\right) \\
&= -j\frac{2a}{w^2}\sin(2w) + j\frac{4a}{w}\cos(2w).
\end{aligned}
$$

Problem 102. For each of the following signals, use Fourier transform properties to determine Fourier transforms.

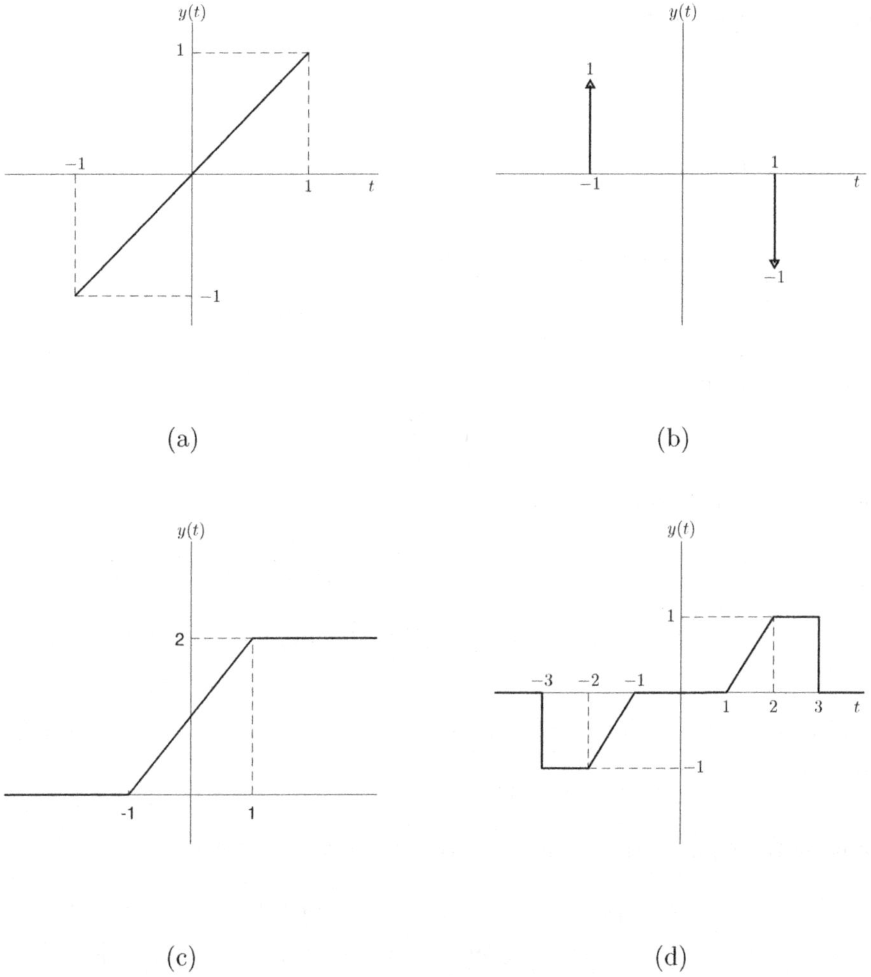

Figure 6.11: Problem 102.

Solution. Let

$$x(t) = \begin{cases} 1 & |t| < 1 \\ 0 & o.w \end{cases}$$

The Fourier transform for signal $x(t)$ is already known as

$$X(jw) = \frac{2\sin(w)}{w}.$$

(a) According to the figure,

$$y(t) = tx(t).$$

Using the frequency differentiation property,

$$Y(jw) = j\frac{d}{dw}X(jw) = j\frac{2w\cos(w) - 2\sin(w)}{w^2}.$$

(b) According to the figure,

$$y(t) = x^{'}(t).$$

Thus using the time differentiation property,

$$Y(jw) = jwX(jw) = j2\sin(w).$$

(c) According to the figure,

$$y(t) = \int_{-\infty}^{t} x(t)dt.$$

Thus by using integration property,

$$\begin{aligned} Y(jw) &= \frac{X(jw)}{jw} + \pi X(j0)\delta(w), \\ &= \frac{-j2\sin(w)}{w^2} + 2\pi\delta(w). \end{aligned}$$

(d) We differentiate $y(t)$ with respect to t to get

$$y^{'}(t) = x(t-1) + x(t+1) - \delta(t-3) + \delta(t+3).$$

Now using time differentiation and linearity properties we have

$$\begin{aligned} \mathcal{F.T}.\{y^{'}(t)\} &= X(jw)e^{-jw} + X(jw)e^{jw} - e^{-3jw} + e^{3jw} \\ &= 2\frac{\sin(w)}{w}(e^{-jw} + e^{jw}) - e^{-3jw} + e^{3jw} \\ &= 4\frac{\sin(w)}{w}\cos(w) + 2j\sin(3w). \\ \text{As such : } Y(jw) &= \frac{\mathcal{F.T}.\{y^{'}(t)\}}{jw} \\ &= 4\frac{\sin(w)}{jw^2}\cos(w) + 2\frac{\sin(3w)}{w}. \end{aligned}$$

Problem 103. Use the definition to determine the inverse Fourier transforms of the following items.

(a) $F(jw) = e^{-|w|}$

(b) $F(jw) = \begin{cases} j\sin(w) & |w| < \pi \\ 0 & o.w \end{cases}$

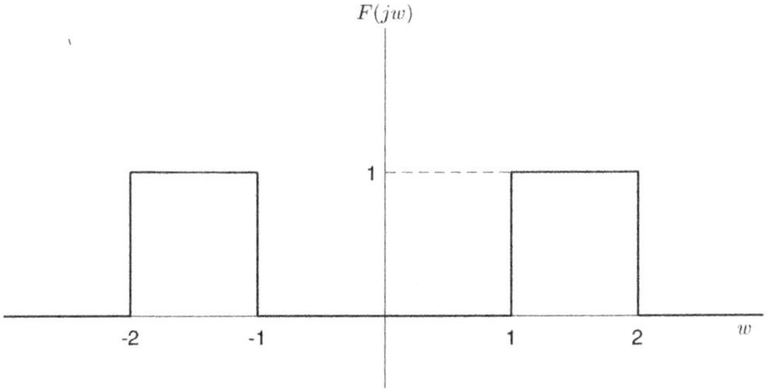

(c)

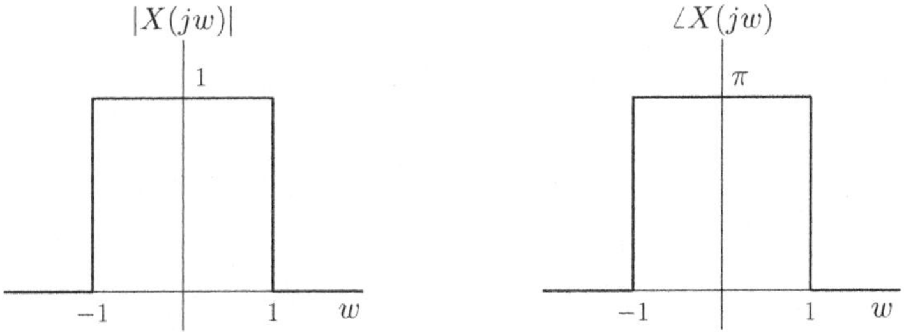

(d)

Figure 6.14: Problem 103 - Parts (c) and (d).

Solution.

(a) $F(jw) = e^{-|w|}$

$$
\begin{aligned}
f(t) &= \frac{1}{2\pi}\int_{-\infty}^{\infty} F(jw)e^{jwt}dw \\
&= \frac{1}{2\pi}\int_{-\infty}^{\infty} e^{-|w|}e^{jwt}dw \\
&= \frac{1}{2\pi}\int_{-\infty}^{0} e^{w}e^{jwt}dw + \frac{1}{2\pi}\int_{0}^{\infty} e^{-w}e^{jwt}dw \\
&= \frac{1}{2\pi}\left(\frac{1}{1+jt}e^{(1+jt)w}\Big|_{-\infty}^{0} + \frac{1}{-1+jt}e^{(-1+jt)w}\Big|_{0}^{\infty}\right) \\
&= \frac{1}{2\pi}\frac{2}{1+t^2}.
\end{aligned}
$$

(b) $F(jw) = \begin{cases} j\sin(w) & |w| < \pi \\ 0 & o.w \end{cases}$

$$\begin{aligned}
f(t) &= \frac{1}{2\pi}\int_{-\infty}^{\infty} F(jw)e^{jwt}dw \\
&= \frac{1}{2\pi}\int_{-\pi}^{\pi} j\sin(w)e^{jwt}dw \\
&= \frac{1}{2\pi}\int_{-\pi}^{\pi} \Big(j\sin(w)\cos(wt) - \sin(w)\sin(wt)\Big)dw \\
&= \frac{1}{2\pi}\int_{-\pi}^{\pi} j\sin(w)\cos(wt)dw - \frac{1}{2\pi}\int_{-\pi}^{\pi} \sin(w)\sin(wt)dw \\
&= \frac{1}{2\pi}\int_{-\pi}^{\pi} \underbrace{j\sin(w)\cos(wt)dw}_{\text{Odd function}} \to 0 - \frac{1}{2\pi}\int_{-\pi}^{\pi} \underbrace{\sin(w)\sin(wt)dw}_{\text{Even function}} \\
&= -\frac{1}{\pi}\int_{0}^{\pi} \sin(w)\sin(wt)dw \\
&= \frac{1}{2\pi}\int_{0}^{\pi} \big(\cos(w+wt) - \cos(w-wt)\big)dw \\
&= \frac{1}{2\pi(1+t)}\sin((1+t)w)\Big|_0^{\pi} - \frac{1}{2\pi(1-t)}\sin((1-t)w)\Big|_0^{\pi} \\
&= \frac{1}{2\pi(1+t)}(-\sin(\pi t)) - \frac{1}{2\pi(1-t)}(\sin(\pi t)) \\
&= \frac{-\sin(\pi t)}{\pi(1-t^2)}.
\end{aligned}$$

(c)

$$\begin{aligned}
f(t) &= \frac{1}{2\pi}\int_{-\infty}^{\infty} F(jw)e^{jwt}dw \\
&= \frac{1}{2\pi}\int_{-2}^{-1} e^{jwt}dw + \frac{1}{2\pi}\int_{1}^{2} e^{jwt}dw \\
&= \frac{1}{2\pi}\left(\frac{1}{jt}(e^{-jt} - e^{-2jt}) + \frac{1}{jt}(e^{2jt} - e^{jt})\right) \\
&= \frac{1}{\pi t}\big(\sin(2t) - \sin(t)\big).
\end{aligned}$$

(d) Based on the given magnitude and phase, $x(t)$ is $e^{j\pi} = -1$ for $-1 \leq t < 1$. Thus

$$\begin{aligned}
x(t) &= \frac{1}{2\pi}\int_{-\infty}^{\infty} |X(jw)|\angle X(jw)e^{jwt}dw \\
&= \frac{1}{2\pi}\int_{-1}^{1} (-1)e^{jwt}dw
\end{aligned}$$

$$
\begin{aligned}
&= \frac{-\sin(t)}{\pi t}. \quad \text{for} \quad t \neq 0, \\
\text{for } t = 0\colon x(0) &= \frac{1}{2\pi} \int_{-\infty}^{\infty} |X(jw)| \angle X(jw) dw \\
&= \frac{1}{2\pi} \int_{-1}^{1} (-1) dw \\
&= \frac{1}{2\pi}(-1-1) = -\frac{1}{\pi}.
\end{aligned}
$$

Problem 104. Consider a real signal $x(t)$ where

$$
X(jw) = \frac{1}{2 + j\frac{w}{3}}.
$$

Determine the Fourier transform of each of the following signals.

(a) $x_a(t) = \frac{dx(t)}{dt}$

(b) $x_b(t) = x(-1-t)$

(c) $x_c(t) = e^{-jt}x(t-2)$

(d) $x_d(t) = x(t)\cos(2\pi t)$

Solution.

(a) $x_a(t) = \frac{dx(t)}{dt}$

Using the time differentiation property we obtain

$$
\begin{aligned}
X_a(jw) &= jwX(jw) \\
&= \frac{jw}{2 + j\frac{w}{3}}.
\end{aligned}
$$

(b) $x_b(t) = x(-1-t)$

Using time reversal and shifting properties we obtain

$$
X_b(jw) = \frac{e^{-jw}}{2 - j\frac{w}{3}}.
$$

(c) $x_c(t) = e^{-jt}x(t-2)$

By time and frequency shifting properties we obtain

$$
X_c(jw) = \frac{1}{2 + j\frac{w+1}{3}} e^{-2j(w+1)}.
$$

(d) $x_d(t) = x(t)\cos(2\pi t)$

$$
x_d(t) = x(t)\cos(2\pi t)
$$

$$= x(t).\left(\frac{1}{2}e^{-2j\pi t} + \frac{1}{2}e^{2j\pi t}\right)$$
$$= \frac{1}{2}x(t)e^{-2j\pi t} + \frac{1}{2}x(t)e^{2j\pi t}.$$

Now using frequency shifting and linearity properties we have

$$X_d(jw) = \frac{0.5}{2 + j\frac{w+2\pi}{3}} + \frac{0.5}{2 + j\frac{w-2\pi}{3}}.$$

Problem 105. For each of the following signals, use Fourier transform properties to determine the inverse Fourier transform.

(a) $X(jw) = \frac{e^{2jw}}{(1+jw)^2}$

(b) $X(jw) = e^{-w}\,\mathrm{u}(w)$

(c) $X(jw) = \frac{d}{dw}\{\frac{\sin(2w) - jcos(2w)}{1+j(\frac{w}{3})}\}$

(d) $X(jw) = 2\frac{w\cos(w) - \sin(w)}{w^2}$

(e) $X(jw) = \frac{4\sin^2(w)}{w^2}e^{-j2w}$

Solution.

(a) $X(jw) = \frac{e^{2jw}}{(1+jw)^2}$

We know that

$$e^{-at}\,\mathrm{u}(t) \xrightarrow{\mathcal{F.T.}} \frac{1}{a+jw},$$

thus

$$e^{-2t}\,\mathrm{u}(t) \xrightarrow{\mathcal{F.T.}} \frac{1}{2+jw}.$$

From the time-differentiation property we have

$$ty(t) \xrightarrow{\mathcal{F.T.}} j\frac{d}{dw}Y(jw),$$

hence

$$te^{-t}\,\mathrm{u}(t) \xrightarrow{\mathcal{F.T.}} j\frac{d}{dw}\frac{1}{1+jw} = \frac{1}{(1+jw)^2}.$$

From time-shifting property,

$$x(t) = (t+2)e^{-(t+2)}\,\mathrm{u}(t+2) \xrightarrow{\mathcal{F.T.}} \frac{e^{2jw}}{(1+jw)^2} = X(jw).$$

(b) $X(jw) = e^{-w}\,\mathrm{u}(w)$

we know that

$$e^{-t}\,\mathrm{u}(t) \xrightarrow{\mathcal{F.T.}} \frac{1}{1+jw}.$$

Now by the duality property we have

$$\frac{1}{1+jt} \xrightarrow{\mathcal{F.T.}} 2\pi e^{-(-w)}\,\mathrm{u}(-w) = 2\pi e^{w}\,\mathrm{u}(-w).$$

By the time reversal property

$$\xrightarrow{t \to (-t)} \frac{1}{1-jt} \xrightarrow{\mathcal{F.T.}} 2\pi e^{-w}\,\mathrm{u}(w),$$

hence,

$$x(t) = \frac{1}{2\pi}\frac{1}{1-jt} \xrightarrow{\mathcal{F.T.}} e^{-w}\,\mathrm{u}(w).$$

(c) $X(jw) = \frac{d}{dw}\{\frac{\sin(2w)-j\cos(2w)}{1+j(\frac{w}{3})}\}$

First, we do some manipulation to rewrite $X(jw)$ in a more familiar form:

$$\begin{aligned} X(jw) &= j\frac{d}{dw}\{\frac{-j\sin(2w)-\cos(2w)}{1+j(\frac{w}{3})}\} \\ &= -3j\frac{d}{dw}\{\frac{\cos(2w)+j\sin(2w)}{3+jw}\} \\ &= -3j\frac{d}{dw}\{\frac{e^{2jw}}{3+jw}\}. \end{aligned}$$

let

$$Y(jw) = \frac{e^{2jw}}{3+jw},$$

then

$$X(jw) = -3j\frac{d}{dw}\{Y(jw)\}.$$

Now by using frequency differentiation property

$$X(jw) = -3j\frac{d}{dw}\{Y(jw)\} \xrightarrow{\mathcal{F.T.}^{-1}} x(t) = -3ty(t).$$

Now since

$$Y(jw) = \frac{e^{2jw}}{3+jw} \xrightarrow{\mathcal{F.T.}^{-1}} e^{-3(t+2)}\,\mathrm{u}(t+2),$$

we get

$$x(t) = -3te^{-3(t+2)}\,\mathrm{u}(t+2).$$

(d) $X(jw) = 2\frac{w\cos(w)-\sin(w)}{w^2}$

First, we rewrite $X(jw)$ as

$$X(jw) = 2\frac{w\cos(w)-\sin(w)}{w^2} = 2\frac{d}{dw}(\frac{\sin(w)}{w}) = \frac{d}{dw}(2\frac{\sin(w)}{w}).$$

We know that

$$\prod(\frac{t}{2}) = \begin{cases} 1 & |t| < 1 \\ 0 & o.w \end{cases} \xrightarrow{\mathcal{F.T.}} \frac{2\sin(w)}{w}.$$

Thus using the frequency differentiation property:

$$X(jw) \xrightarrow{\mathcal{F.T.}^{-1}} x(t) = -jt\prod\left(\frac{t}{2}\right).$$

(e) $X(jw) = \frac{4\sin^2(w)}{w^2}e^{-j2w}$

First, we rewrite $X(jw)$ as

$$X(jw) = \frac{4\sin^2(w)}{w^2}e^{-j2w} = \underbrace{\frac{2\sin(w)}{w}}_{R(jw)} \cdot \underbrace{\frac{2e^{-j2w}\sin(w)}{w}}_{S(jw)}.$$

We know that

$$S(jw) \xrightarrow{\mathcal{F.T.}^{-1}} \prod\left(\frac{t-2}{2}\right)$$

and

$$R(jw) \xrightarrow{\mathcal{F.T.}^{-1}} \prod\left(\frac{t}{2}\right).$$

Now from convolution property we get

$$x(t) = s(t) * r(t).$$

Using the convolution techniques we obtain

$$x(t) = \begin{cases} 0, & t < 0 \\ t, & 0 \le t < 2 \\ 4-t, & 2 \le t < 4 \\ 0, & t > 4 \end{cases}$$

Problem 106. For each of the following signals, use Fourier transform properties to determine whether the corresponding Fourier transform is complex, real and even, imaginary and odd, conjugate symmetric, or neither.

(a) $x(t) = e^{-|t-2|}$

(b) $y(t) = \prod\left(\frac{t}{2}\right)$

(c) $z(t) = \sin(2\pi t)$

(d) $w(t) = \cos(2\pi t - \pi/4)$

(e) $b(t) = e^{j2|t|}$

Solution.

(a) $x(t) = e^{-|t-1|}$

The Fourier transform of the real exponential signal is complex but since $x(t)$ is real, its Fourier transform is conjugate symmetric.

$$X(-jw) = X^*(jw)$$

(b) $y(t) = \prod\left(\frac{t}{2}\right)$

Since $y(t)$ is real and even, its Fourier transform is real and even (and obviously conjugate symmetric).

(c) $z(t) = \sin(2\pi t)$

Since $z(t)$ is real and odd, its Fourier transform is imaginary and odd.

(d) $w(t) = \cos(2\pi t - \pi/4)$

Since $w(t)$ is real, its Fourier transform is conjugate symmetric. As the signal is not even nor odd, its Fourier transform is complex.

(e) $b(t) = e^{j2|t|}$

Since $b(t)$ is complex, its Fourier transform is complex and the fact that the signal is even, does not make it to be real.

Problem 107. $x(t)$ is as depicted in the Fig. 6.15.

(a) Evaluate $a = \int_{-\infty}^{\infty} X(jw)dw$

(b) Evaluate $b = \int_{-\infty}^{\infty} |X(jw)|^2 dw$

(c) Evaluate $c = \int_{-\infty}^{\infty} X(jw)e^{4jw}dw$

(d) Evaluate $d = X(j0)$

(e) Find $\angle X(jw)$

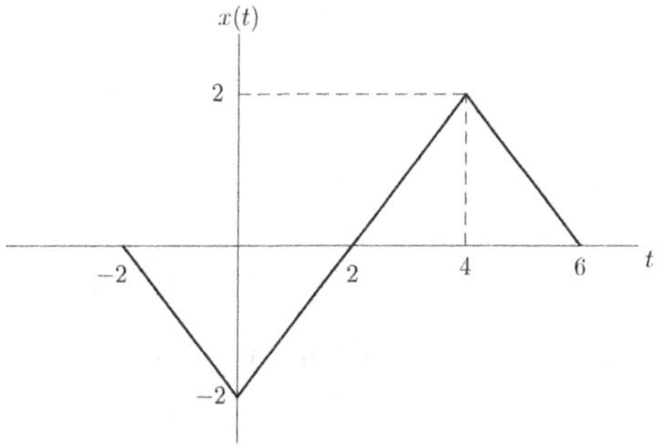

Figure 6.15: Problem 107, Signal $x(t)$.

Solution.

(a) $a = \int_{-\infty}^{\infty} X(jw)dw$

By the inverse Fourier transform definition,

$$x(t) = \frac{1}{2\pi}\int_{-\infty}^{\infty} X(jw)e^{jwt}dw.$$

Setting $t = 0$, we get

$$x(0) = \frac{1}{2\pi}\int_{-\infty}^{\infty} X(jw)dw,$$

$$\Rightarrow \quad a = \int_{-\infty}^{\infty} X(jw)dw = 2\pi x(0) = -4\pi.$$

(b) $b = \int_{-\infty}^{\infty} |X(jw)|^2 dw$

Using Parseval's identity, we have

$$\int_{-\infty}^{\infty} |x(t)|^2 dt = \frac{1}{2\pi}\int_{-\infty}^{\infty} |X(jw)|^2 dw,$$

thus

$$b = \int_{-\infty}^{\infty} |X(jw)|^2 dw = 2\pi \int_{-\infty}^{\infty} |x(t)|^2 dt.$$

As illustrated in the Fig. 6.16, we can write,

$$b = 4 \times 2\pi \int_{-2}^{0} |-t-2|^2 dt = 8\pi \times \frac{8}{3} = \pi \times \frac{64}{3}.$$

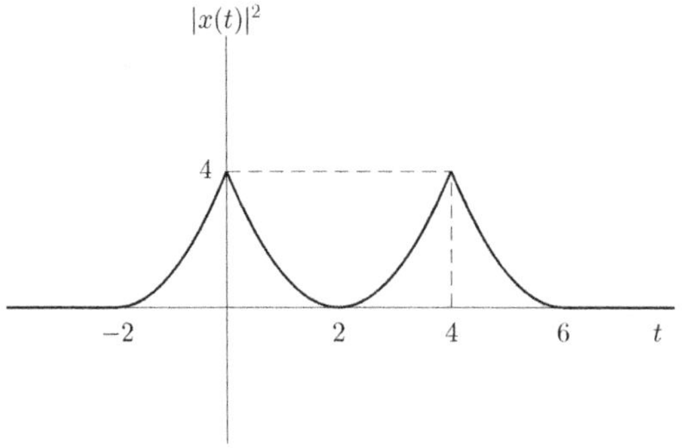

Figure 6.16: Problem 107, Graphical interpretation of $|x(t)|^2$.

(c) $c = \int_{-\infty}^{\infty} X(jw)e^{4jw}dw$

Setting $t = 4$ in the inverse Fourier transform equation we obtain,

$$x(4) = \frac{1}{2\pi} \int_{-\infty}^{\infty} X(jw)e^{4jw}dw,$$

$$\Rightarrow \quad c = \int_{-\infty}^{\infty} X(jw)e^{4jw}dw = 2\pi x(4) = 4\pi.$$

(d) $d = X(j0)$

By definition of the Fourier transform, $X(j0)$ is in fact the average value of $x(t)$:

$$d = X(j0) = \int_{-\infty}^{\infty} x(t)dt = 0.$$

(e) $\angle X(jw)$

Signal $x(t)$ is neither even nor odd, however, $x(t+2)$ is real and odd. Therefore its Fourier transform is imaginary which means that its phase is $\pm\pi/2$:

$$x(t+2) \xrightarrow{\mathcal{F.T.}} e^{j2w}X(jw),$$

$$\angle\big(e^{j2w}X(jw)\big) = \pm\frac{\pi}{2}.$$

So we have

$$\Rightarrow \angle e^{j2w} + \angle X(jw) = \pm\frac{\pi}{2}$$
$$\Rightarrow \angle X(jw) = \pm\frac{\pi}{2} - \angle e^{j2w}$$
$$\Rightarrow \angle X(jw) = \pm\frac{\pi}{2} - 2w.$$

Problem 108. For the following frequency response, determine a differential equation relating the input $x(t)$ and output $y(t)$.

$H(jw) = \frac{2+3jw-4(jw)^2}{1+jw}$

Solution.

We have

$$H(jw) = \frac{Y(jw)}{X(jw)}$$
$$\Longrightarrow \quad (1+jw)Y(jw) = \big(2+3jw-4(jw)^2\big)X(jw).$$

Now by taking the inverse Fourier transform of both sides and assuming zero initial conditions we obtain

$$y(t) + \frac{d}{dt}y(t) = 2x(t) + 3\frac{d}{dt}x(t) - 4\frac{d^2}{dt^2}x(t).$$

Problem 109. Evaluate the following integrals using the Fourier transform properties:

(a) $\int_{-\infty}^{\infty} \frac{\sin(w)\cos(wt)}{w}dw$

(b) $\int_{0}^{\infty} \frac{\sin(w)}{w}dw$

(c) $\int_{0}^{\infty} \frac{2}{(x^2+4)^2}dx$

(d) $\int_{-\infty}^{\infty} \frac{4x^2}{(x^2+4)^2}dx$

Solution.

(a) $\int_{-\infty}^{\infty} \frac{\sin(w)\cos(wt)}{w}dw$

Let $x(t) = \prod\left(\frac{t}{2}\right)$. We know that

$$x(t) = \begin{cases} 1 & |t| < 1 \\ 0 & o.w \end{cases} \xrightarrow{\mathcal{F.T.}} X(jw) = \frac{2\sin(w)}{w}.$$

Thus using the inverse Fourier transform we get

$$x(t) = \frac{1}{2\pi}\int_{-\infty}^{\infty} X(jw)e^{jwt}dw.$$

Thus we can write:

$$\begin{aligned}
\prod\left(\frac{t}{2}\right) &= \frac{1}{2\pi}\int_{-\infty}^{\infty}\frac{2\sin(w)}{w}e^{jwt}dw \\
&= \frac{1}{\pi}\int_{-\infty}^{\infty}\frac{\sin(w)}{w}e^{jwt}dw \\
&= \frac{1}{\pi}\int_{-\infty}^{\infty}\frac{\sin(w)}{w}(\cos(wt)+j\sin(wt))dw \\
&= \frac{1}{\pi}\int_{-\infty}^{\infty}\frac{\sin(w)}{w}\cos(wt)dw + \underbrace{\frac{j}{\pi}\int_{-\infty}^{\infty}\frac{\sin(w)}{w}\sin(wt)dw}_{\text{odd function}\ \to\ 0} \\
&= \frac{1}{\pi}\int_{-\infty}^{\infty}\frac{\sin(w)}{w}\cos(wt)dw
\end{aligned}$$

$$\Rightarrow \int_{-\infty}^{\infty}\frac{\sin(w)}{w}\cos(wt)dw = \pi\prod\left(\frac{t}{2}\right). \tag{6.31}$$

(b) $\int_0^\infty \frac{\sin(w)}{w}dw$
Setting $t=0$ in (6.31) we have

$$\int_{-\infty}^{\infty}\frac{\sin(w)}{w}dw = \pi$$

$$\xrightarrow{\frac{\sin(w)}{w}\text{ is an even func.}} 2\int_{0}^{\infty}\frac{\sin(w)}{w}dw = \pi$$

$$\Rightarrow \int_{0}^{\infty}\frac{\sin(w)}{w}dw = \frac{\pi}{2}.$$

(c) $\int_0^\infty \frac{2}{(x^2+4)^2}dx$

We have

$$e^{-at}\,\mathrm{u}(t) \xrightarrow{\mathcal{F.T.}} \frac{1}{a+jw}$$

thus

$$e^{-2t}\,\mathrm{u}(t) \xrightarrow{\mathcal{F.T.}} \frac{1}{2+jw}.$$

Now using the frequency differentiation property we get

$$te^{-2t}\,\mathrm{u}(t) \xrightarrow{\mathcal{F.T.}} j\frac{d}{dw}\left(\frac{1}{2+jw}\right) = j\frac{-j}{(2+jw)^2} = \frac{1}{(2+jw)^2}. \tag{6.32}$$

Using the Parseval's identity, we have

$$\frac{1}{2\pi}\int_{-\infty}^{\infty}\frac{dw}{|(2+jw)^2|^2} = \int_{-\infty}^{\infty}|te^{-2t}\,\mathrm{u}(t)|^2dt$$

$$\frac{1}{2\pi}\int_{-\infty}^{\infty}\frac{dw}{(4+w^2)^2} = \int_{-\infty}^{\infty} t^2e^{-4t}\,\mathrm{u}(t)dt$$

$$\frac{1}{4\pi}\int_{-\infty}^{\infty}\frac{2dw}{(4+w^2)^2} = \int_{0}^{\infty} t^2e^{-4t}dt$$

$$\frac{1}{4\pi}\int_{-\infty}^{\infty}\frac{2dw}{(4+w^2)^2} = e^{-4t}\Big(-\frac{t^2}{4}-\frac{2t}{16}-\frac{1}{32}\Big)\Big|_0^{\infty} = \frac{1}{32}$$

$$\Rightarrow \int_{-\infty}^{\infty}\frac{2dw}{(4+w^2)^2} = \frac{\pi}{8}.$$

$$\Rightarrow \int_{0}^{\infty}\frac{2dw}{(4+w^2)^2} = \frac{\pi}{16}.$$

(d) $\int_{-\infty}^{\infty}\frac{4x^2}{(x^2+4)^2}dx$.

From (6.32) we have

$$te^{-2t}\,\mathrm{u}(t) \xrightarrow{\mathcal{F.T.}} \frac{1}{(2+jw)^2}.$$

Now using the time differentiation property we obtain

$$\frac{jw}{(2+jw)^2} \xrightarrow{\mathcal{F.T.}^{-1}} \frac{d}{dt}\big(te^{-2t}\,\mathrm{u}(t)\big) = e^{-2t}(1-2t)\,\mathrm{u}(t).$$

Using the Parseval's identity we have

$$\frac{1}{2\pi}\int_{-\infty}^{\infty}\frac{w^2}{|(2+jw)^2|^2}dw = \int_{-\infty}^{\infty}|e^{-2t}(1-2t)\,\mathrm{u}(t)|^2dt$$

$$\begin{aligned}\frac{1}{2\pi}\int_{-\infty}^{\infty}\frac{w^2}{(4+w^2)^2}dw &= \int_{-\infty}^{\infty} e^{-4t}(1-2t)^2\,\mathrm{u}(t)dt\\ &= \int_{0}^{\infty} e^{-4t}(1-2t)^2dt\\ &= -\frac{1}{4}e^{-4t}+e^{-4t}(-t^2-\frac{t}{2}-\frac{1}{8})-e^{-4t}(-t-\frac{1}{8})\Big|_0^{\infty}\\ &= \frac{1}{8}.\end{aligned}$$

$$\Rightarrow \int_{-\infty}^{\infty}\frac{4x^2}{(4+x^2)^2}dx = \pi.$$

Problem 110. Consider an LTI system with impulse response

$$h(t) = 2\frac{\sin(\frac{\pi t}{2})}{\pi t}\cos(5\pi t).$$

Determine the output of the system for each of the following inputs:

(a) $x(t) = \cos(2\pi t) + \sin(5\pi t)$

(b) $x(t) = \sum_{m=-\infty}^{\infty}(-1)^m \delta(t-m)$

(c)

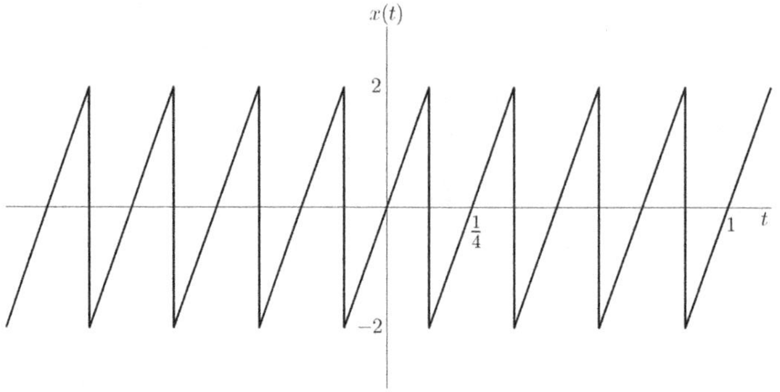

Figure 6.17: Problem 110 - Part (c).

Solution. We first have to obtain the frequency response of the system. The system impulse response is a pulse multiplied by the cosine function. As such

$$\frac{\sin(\pi t/2)}{\pi t} \xrightarrow{\mathcal{F.T.}} \prod(\frac{w}{\pi})$$

$$\underbrace{\frac{2\sin(\pi t/2)}{\pi t}(\frac{e^{5\pi t}+e^{-5\pi t}}{2})}_{h(t)} \xrightarrow{\mathcal{F.T.}} \underbrace{\prod(\frac{w+5\pi}{\pi})+\prod(\frac{w-5\pi}{\pi})}_{H(jw)}$$

(a) $x(t) = \cos(2\pi t) + \sin(5\pi t)$

We take the Fourier transform of $x(t)$:

$$X(jw) = \pi\delta(w-2\pi) + \pi\delta(w+2\pi) + \frac{\pi}{j}\delta(w-5\pi) - \frac{\pi}{j}\delta(w+5\pi).$$

Using $H(jw)$ and $X(jw)$ we obtain

$$\begin{aligned} Y(jw) &= X(jw).H(jw) \\ &= \left(\frac{\pi}{j}\delta(w-5\pi) - \frac{\pi}{j}\delta(w+5\pi)\right). \end{aligned}$$

Now by taking the inverse Fourier transform of $Y(jw)$ we obtain

$$y(t) = \sin(5\pi t).$$

(b) $x(t) = \sum_{m=-\infty}^{\infty}(-1)^m \delta(t-m)$

First we sketch $x(t)$ (see Fig. 6.18). To obtain the Fourier transform of this signal, we first have to obtain the Fourier series coefficients a_k.

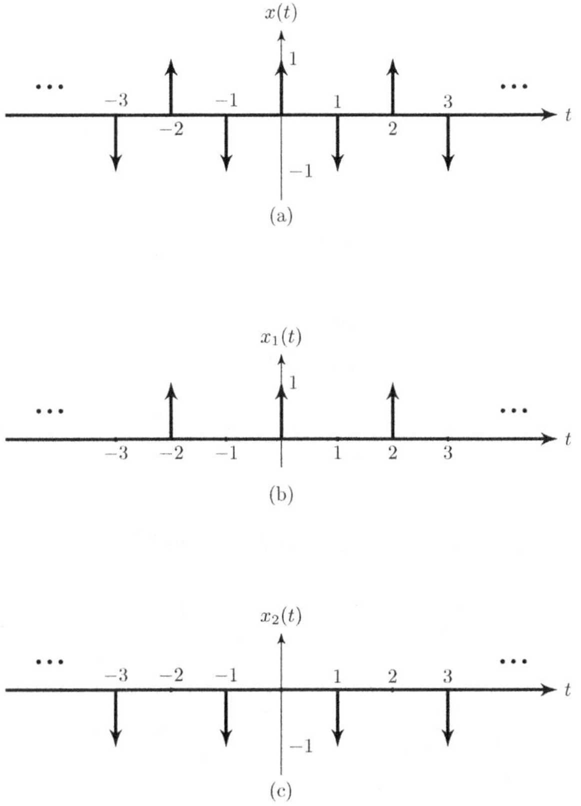

Figure 6.18: Problem 110 - Part (b): (a) Original signal $x(t)$; (b) Signal $x_1(t)$; (c) Signal $x_2(t)$.

We see that signal $x(t)$ is composed of two parts $x_1(t)$ and $x_2(t)$:

$$x(t) = x_1(t) + x_2(t) = \sum_{m=-\infty}^{\infty} \delta(t - 2m) - \sum_{m=-\infty}^{\infty} \delta(t - 1 - 2m). \tag{6.33}$$

The fundamental period T of $x_1(t)$ and $x_2(t)$ is 2. Also the fundamental angular frequency w_0 of $x_1(t)$ and $x_2(t)$ is $w_0 = 2\pi/T = \pi$. Then we have

$$x_1(t) \overset{\mathcal{F.S.}}{\longleftrightarrow} a_{k1} = \frac{1}{T} = \frac{1}{2},$$

$$\begin{aligned} x_2(t) \overset{\mathcal{F.S.}}{\longleftrightarrow} a_{k2} &= (-1)\frac{1}{T}e^{-jkw_0} = \frac{-1}{2}e^{-jk\pi} \\ &= \frac{-1}{2}(-1)^k = \frac{1}{2}(-1)^{k+1}. \end{aligned}$$

Then we have

$$x(t) = x_1(t) + x_2(t) \overset{\mathcal{F.S.}}{\longleftrightarrow} a_k = a_{k1} + a_{k2}$$
$$= \frac{1}{2} + \frac{1}{2}(-1)^{k+1}.$$

Now we use the fact that the Fourier transform of a periodic signal with Fourier series coefficients a_k is an impulse train with a_k's as the magnitude of the impulses, i.e.,

$$X(jw) = 2\pi \sum_{k=-\infty}^{\infty} a_k \delta(w - 2k\pi/T).$$

Thus

$$X(jw) = \pi \sum_{k=-\infty}^{\infty} (1 + (-1)^{k+1})\delta(w - k\pi).$$

The LTI system $h(t)$ only passes the 5th harmonics (k=±5). As such

$$Y(jw) = 2\pi\delta(w - 5\pi) + 2\pi\delta(w + 5\pi).$$

By taking the inverse Fourier transform of $Y(jw)$ we obtain,

$$y(t) = 4\cos(5\pi t).$$

(c) $x(t)$ is periodic with period $T = \frac{1}{4}$ and Fourier series coefficient

$$a_k = 4\int_{-\frac{1}{8}}^{-\frac{1}{8}} 16te^{-jk(8\pi)t}dt = \frac{2j\cos(\pi k)}{\pi k}.$$

Thus

$$X(jw) = 2\pi \sum_{k=-\infty}^{\infty} a_k \delta(w - 8k\pi).$$

As can be seen, $X(jw)$ contains frequencies 8π and higher which can not pass from the filter. As such, $Y(jw) = 0$ and consequently, $y(t) = 0$.

Problem 111. The output $y(t)$ of a system is related to the input $x(t)$ by the equation

$$y(t) = \int_{t}^{t+a} x(\lambda)d\lambda - \int_{t-a}^{t} x(\lambda)d\lambda, \tag{6.34}$$

where a is a positive constant.

(a) Find the frequency response of this system.

(b) Find the response $y(t)$ to the input

$$x(t) = \sum_{k=-\infty}^{\infty} (-1)^k \delta(t - \frac{ak}{2}).$$

Solution.

(a) We can solve this part by two ways.

1. **The first approach:**
   Taking the derivative of both sides of (6.34) with respect to t, we have

$$y'(t) = x(t+a) - x(t) - x(t) + x(t-a) = x(t+a) - 2x(t) + x(t-a). \tag{6.35}$$

   Then, applying the Fourier transform to both sides of (6.35) and using the Fourier transform properties, we get

$$(jw)Y(jw) = e^{jwa}X(jw) - 2X(jw) + e^{-jwa}X(jw) = X(jw)(e^{jwa} - 2 + e^{-jwa}). \tag{6.36}$$

   From (6.17), we obtain

$$\begin{aligned} H(jw) &= \frac{Y(jw)}{X(jw)} = \frac{e^{jwa} + e^{-jwa} - 2}{jw} \\ &= \frac{e^{jwa} - e^{-jwa} + e^{-jwa} + e^{-jwa} - 2}{jw} \\ &= \frac{e^{jwa} - e^{-jwa}}{jw} + \frac{2e^{-jwa} - 2}{jw} \\ &= \frac{2\sin(wa)}{w} + \frac{2(e^{-jwa} - 1)}{jw}. \end{aligned}$$

   We can rewrite the second term of the above equation as

$$\frac{2(e^{-jwa} - 1)}{jw} = \frac{2e^{-jwa/2}(e^{-jwa/2} - e^{jwa/2})}{jw} = \frac{-2e^{-jwa/2}\sin(wa/2)}{w}.$$

   Thus,

$$H(jw) = \frac{2\sin(wa)}{w} - \frac{2e^{-jwa/2}\sin(wa/2)}{w}. \tag{6.37}$$

2. **The second approach:**
   First we try to rewrite the relationship between $y(t)$ and $x(t)$ as the convolution integral. To do so, the integral limits have to be $\pm\infty$. We know that $\mathrm{u}(t+a-\lambda)\,\mathrm{u}(\lambda-t)$ is 1 for $t < \lambda < t+a$ and 0 otherwise. Moreover, $\mathrm{u}(\lambda-t+a)\,\mathrm{u}(t-\lambda)$ is 1 for $t-a < \lambda < t$ and 0 otherwise. Thus we can write

$$\begin{aligned} y(t) &= \int_{-\infty}^{\infty} \mathrm{u}(t+a-\lambda)\,\mathrm{u}(\lambda-t)x(\lambda)d\lambda - \int_{-\infty}^{\infty} \mathrm{u}(\lambda-t+a)\,\mathrm{u}(t-\lambda)x(\lambda)d\lambda \\ &= \int_{-\infty}^{\infty} h(t-\lambda)x(\lambda)d\lambda. \end{aligned}$$

   Hence we obtain $h(t)$ as

$$h(t-\lambda) = \mathrm{u}(t+a-\lambda)\,\mathrm{u}(\lambda-t) - \mathrm{u}(\lambda-t+a)\,\mathrm{u}(t-\lambda)$$

$$\xrightarrow{t\to t+\lambda} h(t) = \mathrm{u}(t+a)\,\mathrm{u}(-t) - \mathrm{u}(-t+a)\,\mathrm{u}(t).$$

In other words,

$$h(t) = \prod\left(\frac{t}{2a}\right) - 2\prod\left(\frac{t}{a} - \frac{1}{2}\right).$$

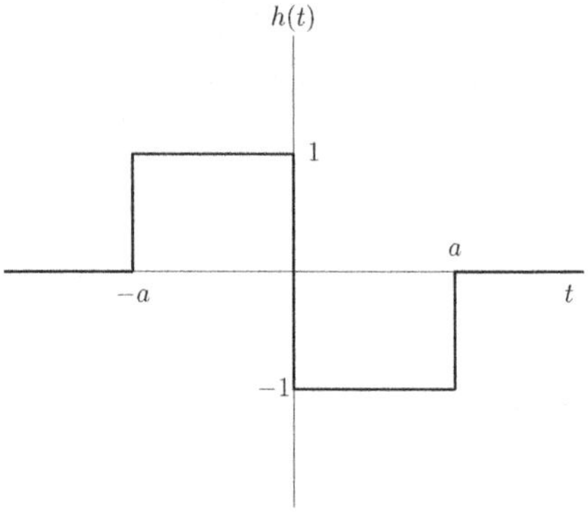

Figure 6.19: Problem 111, Impulse response of the system.

Also the frequency response is

$$H(jw) = \frac{2\sin(aw)}{w} - 2e^{-jw\frac{a}{2}}\frac{\sin(aw/2)}{w}. \tag{6.38}$$

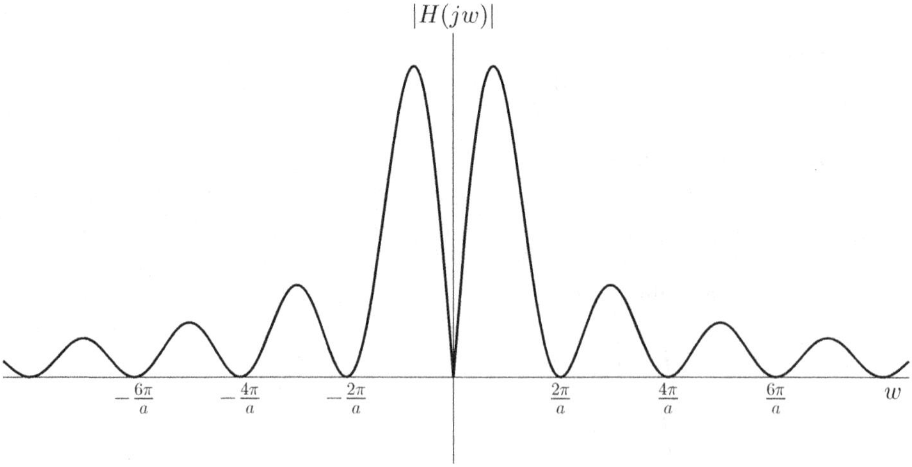

Figure 6.20: Problem 111, Magnitude of the frequency response of the system.

(b) Since $x(t)$ is periodic with period $T = a$, $X(jw)$ consists of impulses in frequency, spaced $\frac{2\pi}{a}$ from each other, that is

$$X(jw) = 2\pi \sum_{k=-\infty}^{\infty} a_k \delta(w - k\frac{2\pi}{a})$$

where the coefficients a_k are the Fourier series coefficients of $x(t)$. It is easy to see that the frequency response $|H(jw)|$ is zero in all harmonics of $x(t)$, thus we do not need to obtain a_k's and already have

$$y(t) = 0.$$

Problem 112. For each of the following signals express the Fourier transforms of the signals in terms of the Fourier transform of $x(t)$.

(a) $y(t) = x(2t - 4)$

(b) $y(t) = x^*(t - 1)$

(c) $y(t) = x^*(1 - \frac{t}{3})$

(d) $y(t) = x(2 - t) + x(-2 - t)$

(e) $y(t) = \int_{-\infty}^{\infty} x(\tau)x(\tau - t)dt$, x(t) is real

Solution.

(a) $y(t) = x(2t - 4) = x(2(t - 2))$

The following block diagram shows how $x(t)$ is related to $y(t)$:

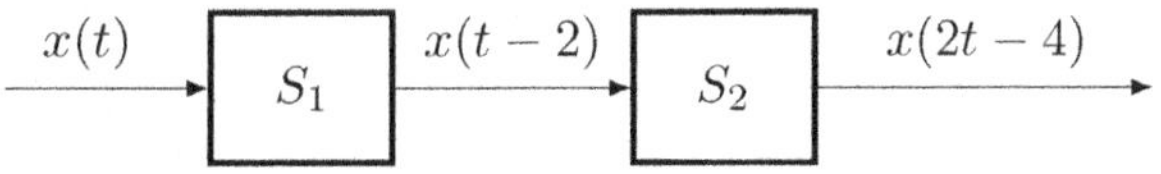

$$\begin{cases} S_1: & \text{input } \mathcal{F.T.}: Z(jw) \to \text{output } \mathcal{F.T.}: \ Z(jw)e^{-2jw} \\ S_2: & \text{input } \mathcal{F.T.}: Z(jw) \to \text{output } \mathcal{F.T.}: \ \frac{1}{2}Z(j\frac{w}{2}) \end{cases}$$

Hence

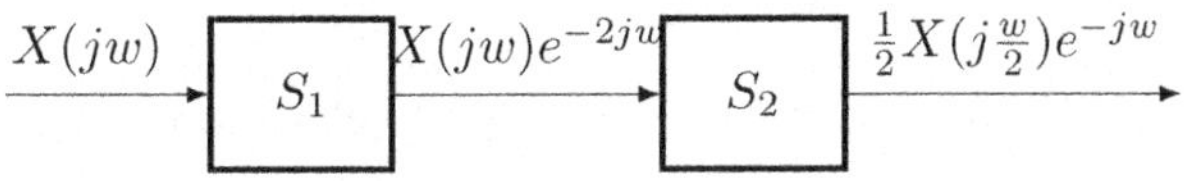

thus

$$Y(jw) = \frac{1}{2}X(j\frac{w}{2})e^{-jw}.$$

(b) $y(t) = x^*(t - 1)$

The following block diagram shows how $y(t)$ is related to $x(t)$:

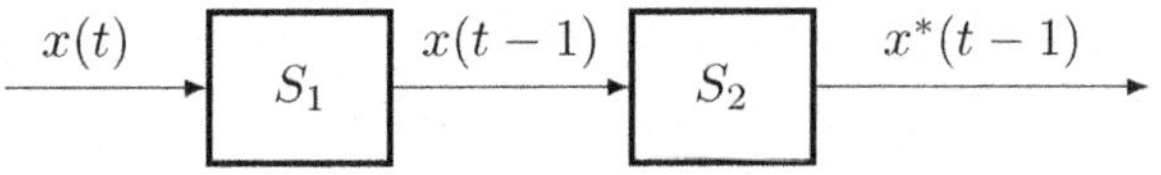

$$\begin{cases} S_1: & \text{input } \mathcal{F.T.}: Z(jw) \to \text{output } \mathcal{F.T.}: Z(jw)e^{-jw} \\ S_2: & \text{input } \mathcal{F.T.}: Z(jw) \to \text{output } \mathcal{F.T.}: Z^*(-jw) \end{cases}.$$

Hence

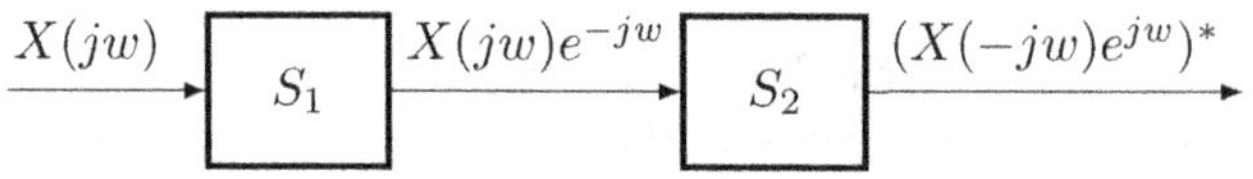

Thus

$$Y(jw) = X^*(-jw)e^{-jw}.$$

(c) $y(t) = x^*(1 - \frac{t}{3})$

The following block diagram shows how $y(t)$ is related to $x(t)$:

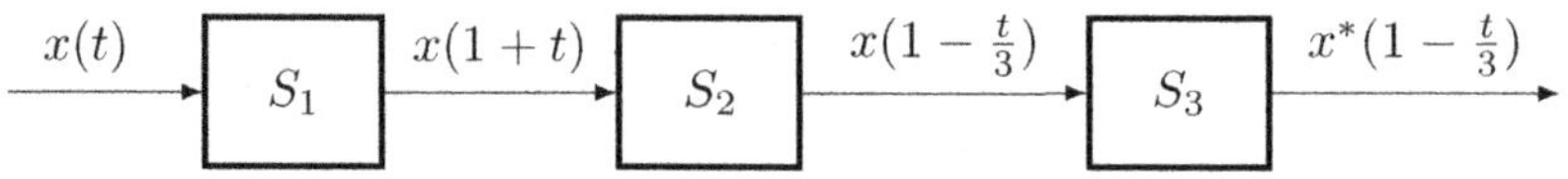

$$\begin{cases} S_1: & \text{input } \mathcal{F.T.}: Z(jw) \to \text{output } \mathcal{F.T.}: Z(jw)e^{jw} \\ S_2: & \text{input } \mathcal{F.T.}: Z(jw) \to \text{output } \mathcal{F.T.}: 3Z(-3jw) \\ S_3: & \text{input } \mathcal{F.T.}: Z(jw) \to \text{output } \mathcal{F.T.}: Z^*(-jw) \end{cases}.$$

Hence

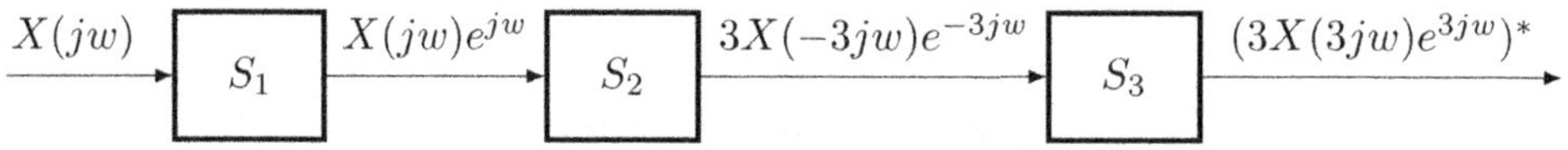

Thus

$$Y(jw) = 3X^*(3jw)e^{-3jw}.$$

(d) $y(t) = x(2 - t) + x(-2 - t)$

Let $x_1(t) = x(2 - t)$ and $x_2(t) = x(-2 - t)$. The following block diagram shows how $x_1(t)$ is related to $x(t)$:

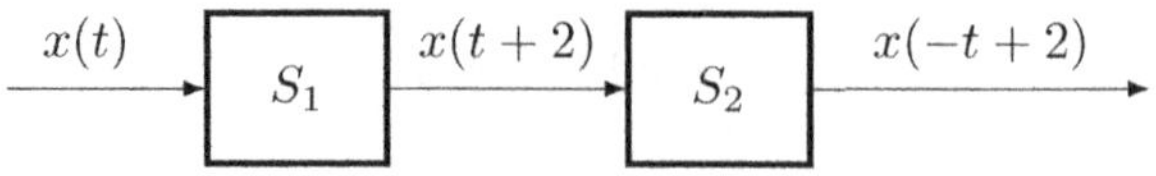

$$\begin{cases} S_1: & \text{input } \mathcal{F.T.}: Z(jw) \to \text{output } \mathcal{F.T.}: Z(jw)e^{2jw} \\ S_2: & \text{input } \mathcal{F.T.}: Z(jw) \to \text{output } \mathcal{F.T.}: Z(-jw) \end{cases}.$$

Hence

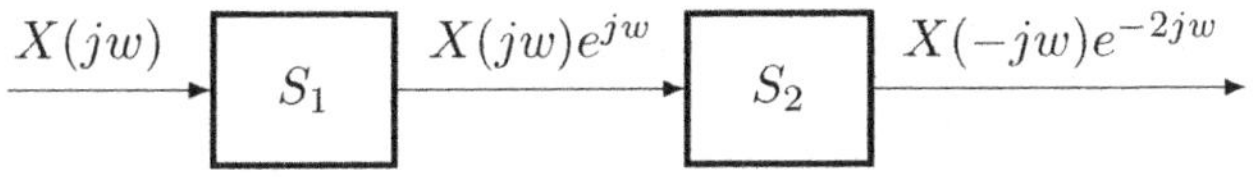

Thus

$$X_1(jw) = X(-jw)e^{-2jw}.$$

The following block diagram shows how $x_2(t)$ is related to $x(t)$,

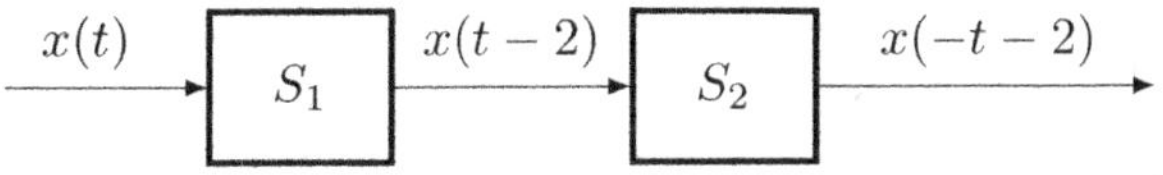

$$\begin{cases} S_1 : & \text{input } \mathcal{F.T.} : Z(jw) \rightarrow \text{output } \mathcal{F.T.} : Z(jw)e^{-2jw} \\ S_2 : & \text{input } \mathcal{F.T.} : Z(jw) \rightarrow \text{output } \mathcal{F.T.} : Z(-jw) \end{cases}.$$

Hence

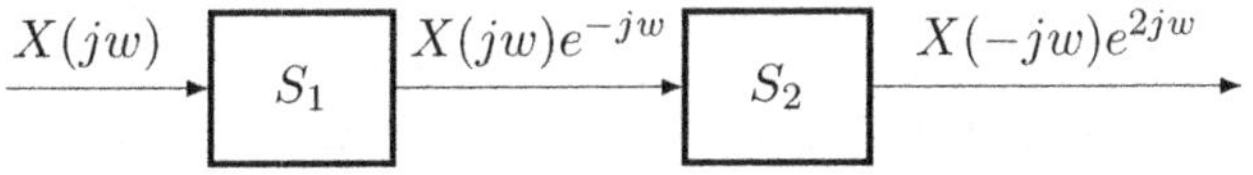

Thus

$$X_2(jw) = X(-jw)e^{2jw}.$$

Finally

$$\begin{aligned} \Rightarrow \quad Y(jw) &= X_1(jw) + X_2(jw) \\ &= X(-jw)e^{-2jw} + X(-jw)e^{2jw} \\ &= 2X(-jw)\cos(2w). \end{aligned}$$

(e) $y(t) = \int_{-\infty}^{\infty} x(\tau)x(\tau - t)dt$

Using the convolution integral definition we can rewrite $y(t)$ as

$$y(t) = x(t) * x(-t),$$

Taking Fourier transforms of both sides we get

$$Y(jw) = X(jw).X(-jw).$$

Also, since $x(t)$ is real valued, $X(jw)$ is conjugate symmetric. That is,

$$X(-jw) = X^*(jw).$$

Thus

$$Y(jw) = X(jw).X^*(jw) = |X(jw)|^2.$$

Problem 113. Consider an LTI system whose response to the input

$$x_1(t) = je^{j100\pi t}$$

is

$$y_1(t) = 2e^{j(100\pi t + \frac{\pi}{4})}.$$

The frequency response of system $|H(jw)|$ is zero except for $0 \leq w \leq 200\pi$ where it has the form $|k|e^{j\theta}$ where k and θ are constant. Determine the output of the system to each of the following inputs:

(a) $x_2(t) = (1 + j)e^{j50\pi t}$

(b) $x_3(t) = \sqrt{2}e^{j300\pi t}$

Solution. The system is LTI, so we have

$$Y(jw) = X(jw)\, H(jw).$$

For $0 \leq w \leq 200\pi$, the magnitude of $X(w)$ is multiplied by $|k|$ and the phase of θ is added to $\angle X(w)$ to obtain $\angle Y(w)$.

To find k and θ we have

$$x_1(t) = je^{100j\pi t} = e^{j(100\pi t + \frac{\pi}{2})} \quad \Longrightarrow \quad y_1(t) = 2e^{j(100\pi t + \frac{\pi}{4})}$$

Thus

$$|k| = 2 \quad , \quad \theta = \frac{\pi}{4} - \frac{\pi}{2} = -\frac{\pi}{4}.$$

(a) $x_2(t) = (1 + j)e^{j50\pi t}$

$$x_2(t) = (1 + j)e^{j50\pi t} = \sqrt{2}e^{j(50\pi t + \frac{\pi}{4})}$$

$$\xrightarrow{w=50\pi \text{ is within the passing band of the system}} y_2(t) = 2\sqrt{2}e^{j(50\pi t + \frac{\pi}{4} - \frac{\pi}{4})} = 2\sqrt{2}e^{j(50\pi t)}.$$

(b) As can be seen, $w = 300\pi$ is not within the passing band of the system. As such, $y(t) = 0$.

Problem 114. A signal $x(t)$ has a Fourier transform $X(jw)$ as illustrated in Fig. 6.21. Evaluate the following without obtaining $x(t)$ directly.

(a) Evaluate $x(0)$

(b) Evaluate $\int_{-\infty}^{\infty} |x(t)|^2 dt$

(c) Evaluate $\int_{-\infty}^{\infty} x(t)\, dt$

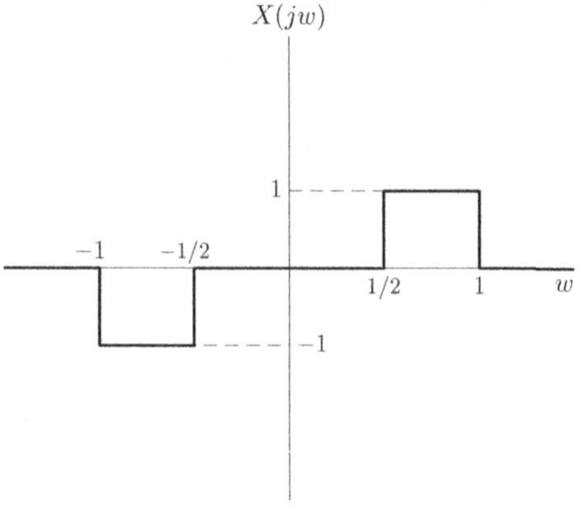

Figure 6.21: The Fourier transforms considered in Problem 114.

Solution.

(a) $x(0)$
By definition,

$$x(t) = \frac{1}{2\pi}\int_{-\infty}^{\infty} X(jw)e^{jwt}dw$$

thus

$$x(0) = \frac{1}{2\pi}\int_{-\infty}^{\infty} X(jw)e^{jw0}dw = \frac{1}{2\pi}\int_{-\infty}^{\infty} X(jw)dw = 0.$$

(b) $\int_{-\infty}^{\infty}|x(t)|^2dt$
by using Parseval's identity, we have

$$\begin{aligned}\int_{-\infty}^{\infty}|x(t)|^2\,dt &= \frac{1}{2\pi}\int_{-\infty}^{\infty}|X(jw)|^2\,dw\\ &= 2\times\frac{1}{2\pi}\int_{\frac{1}{2}}^{1} 1dw\\ &= \frac{1}{2\pi}.\end{aligned}$$

(c) $\int_{-\infty}^{\infty} x(t)\,dt$
By definition,

$$X(jw) = \int_{-\infty}^{\infty} x(t)e^{-jwt}dt.$$

Thus

$$X(0) = \int_{-\infty}^{\infty} x(t)e^{-jw0}dt = \int_{-\infty}^{\infty} x(t)dt$$

$$\Rightarrow \quad \int_{-\infty}^{\infty} x(t)dt = X(0) = 0.$$

Problem 115. Evaluate the following integrals using Fourier transform properties.

(a) $A = \int_{-\infty}^{\infty}\frac{(1-\cos^2(t))^2}{\pi^4t^4}dt$

(b) $B = \int_{-\infty}^{\infty}\frac{\text{sinc}(f)}{1+j2\pi f}df$

Solution.

(a) $A = \int_{-\infty}^{\infty}\frac{(1-\cos^2(t))^2}{\pi^4t^4}dt$
We may be able to use Parseval's identity here. So we rewrite the integral as

$$A = \int_{-\infty}^{\infty}\frac{(1-\cos^2(t))^2}{\pi^4t^4}dt = \int_{-\infty}^{\infty}(\frac{1-\cos^2(t)}{\pi^2t^2})^2dt.$$

Now let $x(t) = \frac{1-\cos^2(t)}{\pi^2t^2}$. We know that

$$1-\cos^2(t) = \sin^2 t,$$

then

$$x(t) = \frac{\sin^2 t}{\pi^2 t^2} = \frac{\sin t}{\pi t} . \frac{\sin t}{\pi t} = z(t).z(t)$$

From the multiplication property of the Fourier transform we have

$$z(t).z(t) \overset{\mathcal{F.T}}{\longleftrightarrow} \frac{1}{2\pi} Z(jw) * Z(jw).$$

where

$$z(t) = \frac{\sin t}{\pi t} \overset{\mathcal{F.T}}{\longleftrightarrow} Z(jw) = \begin{cases} 1 & |w| < 1 \\ 0 & |w| > 1, \end{cases}$$

As such we obtain

$$x(t) = \frac{1 - \cos^2(t)}{\pi^2 t^2} \quad \xrightarrow{\mathcal{F.T.}}$$

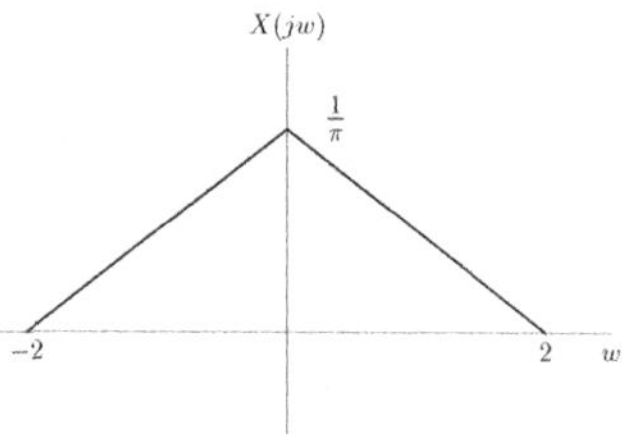

Now using the Parseval's identity we have

$$\begin{aligned} A = \int_{-\infty}^{\infty} |x(t)|^2 \, dt &= \frac{1}{2\pi} \int_{-\infty}^{\infty} |X(jw)|^2 \, dw \\ \Longrightarrow \int_{-\infty}^{\infty} |\frac{1 - \cos^2(t)}{\pi^2 t^2}|^2 \, dt &= \frac{2}{2\pi} \int_0^2 (\frac{1}{2\pi}(2 - w))^2 \, dw \\ &= \frac{1}{4\pi^3} \int_0^2 (2 - w)^2 \, dw = \frac{2}{3\pi^3} \\ \Longrightarrow \quad A &= \frac{2}{3\pi^3}. \end{aligned}$$

(b) $B = \int_{-\infty}^{\infty} \frac{\text{sinc}(f)}{1+j2\pi f} df$

Using the general Parseval's identity (6.12), we have

$$\int_{-\infty}^{\infty} x(t) y^*(t) \, dt = \int_{-\infty}^{\infty} X(f) Y^*(f) \, df. \tag{6.39}$$

We have

$$B = \int_{-\infty}^{\infty} \frac{\text{sinc}(f)}{1 + j2\pi f} \, df$$

$$\xrightarrow{\text{sinc}(f)=\frac{\sin(\pi f)}{\pi f}} = \int_{-\infty}^{\infty} \underbrace{\frac{1}{1+j2\pi f}}_{X(f)} \cdot \underbrace{\frac{\sin(\pi f)}{\pi f}}_{Y^*(f)} \, df = \int_{-\infty}^{\infty} X(f)Y^*(f) \, df.$$

Now we must obtain the inverse transform of $X(f)$ and $Y(f)$ and substitute $x(t)$ and $y(t)$ into the general Parseval's identity. We have

$$X(f) = \frac{1}{1+j2\pi f} \xrightarrow{w=2\pi f} X(jw) = \frac{1}{1+jw} \xrightarrow{\mathcal{F}.\mathcal{T}.^{-1}} x(t) = e^{-t}\,\mathrm{u}(t),$$

$$Y(f) = \frac{\sin(\pi f)}{\pi f} \xrightarrow{w=2\pi f} \frac{2\sin(\frac{w}{2})}{w} \xrightarrow{\mathcal{F}.\mathcal{T}.^{-1}} y(t) = \prod(t).$$

Substituting $x(t)$ and $y(t)$ into (6.39) we obtain

$$\begin{aligned} B &= \int_{-\infty}^{\infty} X(f)Y^*(f) \, df \\ &= \int_{-\infty}^{\infty} x(t)y^*(t) \, dt \\ &= \int_{-\infty}^{\infty} e^{-t}\,\mathrm{u}(t) \prod(t) \, dt. \end{aligned}$$

Since $\prod(t) = 1$ for $-0.5 < t < 0.5$ and 0 otherwise, we continue as

$$\begin{aligned} B &= \int_{0}^{0.5} e^{-t}\,\mathrm{u}(t) \, dt \\ &= \int_{0}^{0.5} e^{-t} \, dt \\ &= 1 - e^{-0.5}. \end{aligned}$$

Problem 116. Consider the block diagram of a continues-time system in Fig. 6.22.

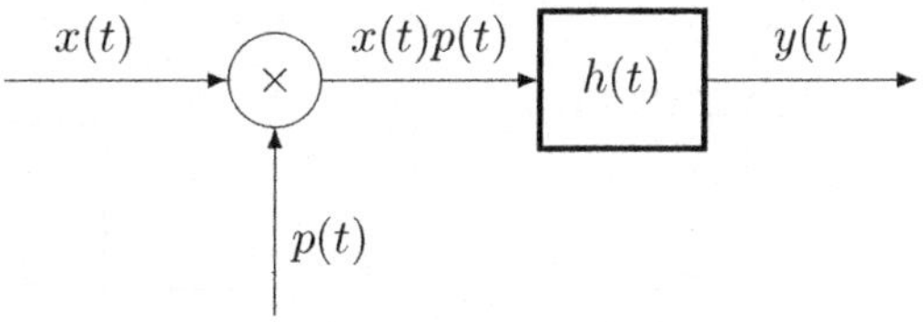

Figure 6.22: Block diagram of a continues-time system in Problem 116.

In this system assume that

$$x(t) = 2\frac{\sin(6\pi t)}{\pi t}e^{j2\pi t}$$

and

$$p(t) = \cos(4\pi t).$$

Also assume $h(t)$ has a Fourier transform $H(jw)$ whose magnitude and phase are as illustrated in Fig. 6.23.

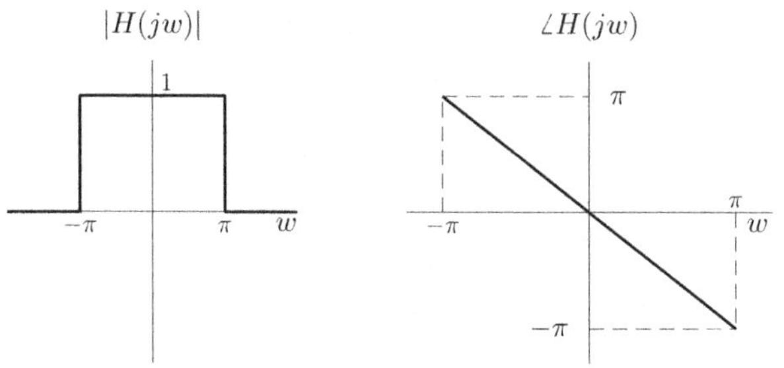

Figure 6.23: Magnitude and phase of the Fourier transform $H(jw)$ for the filter $h(t)$, Problem 116.

(a) Find $Y(jw)$.

(b) Find $y(t)$.

Solution.

(a)

$$2\frac{\sin(6\pi t)}{\pi t} \xrightarrow{\mathcal{F.T.}}$$

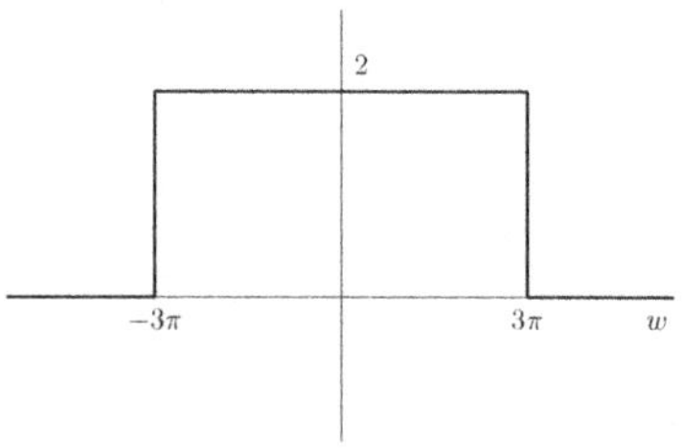

$$2\frac{\sin(6\pi t)}{\pi t}e^{j2\pi t} \xrightarrow{\mathcal{F.T.}}$$

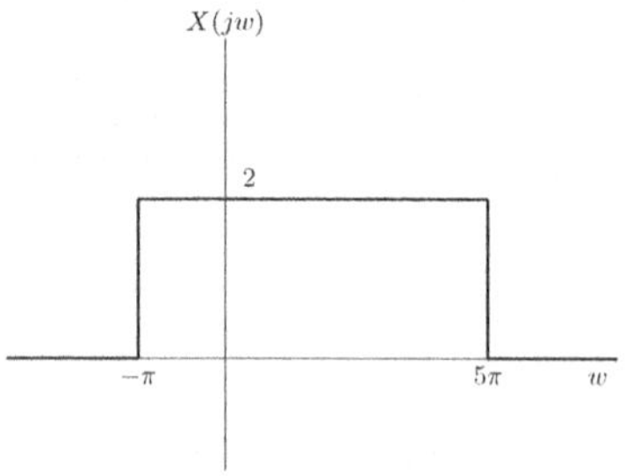

$$p(t) = \cos(4\pi t) \xrightarrow{\mathcal{F.T.}} \pi(\delta(w-4\pi)+\delta(w+4\pi)),$$

From the multiplication property,

$$x_p(t) = x(t).p(t) \quad \xrightarrow{\mathcal{F.T.}} \quad X_P(jw) = \frac{1}{2\pi}(X(jw) * P(jw)),$$

thus $X_P(jw)$ is as illustrated in Fig. 6.24.

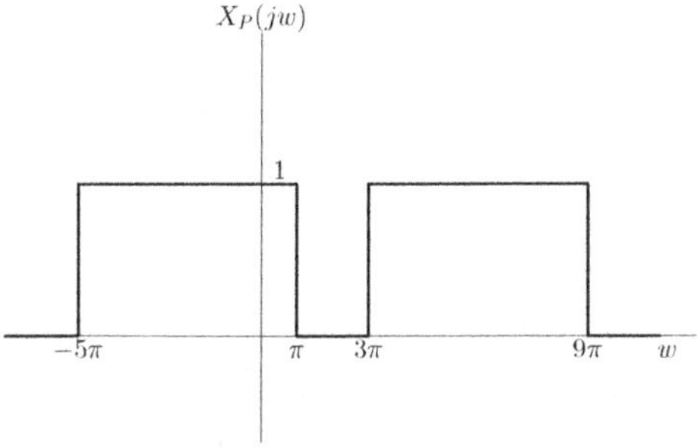

Figure 6.24: Fourier transform of the signal $x(t).p(t)$, Problem 116.

Now we use $XP(jw)$ to find the Fourier transform of the output:

$$|Y(jw)| = |X_P(jw)|.|H(jw)|$$

$$\angle Y(jw) = \angle X_P(jw) + \angle H(jw)$$

$$Y(jw) = \prod(\frac{w}{2\pi})e^{-jw}$$

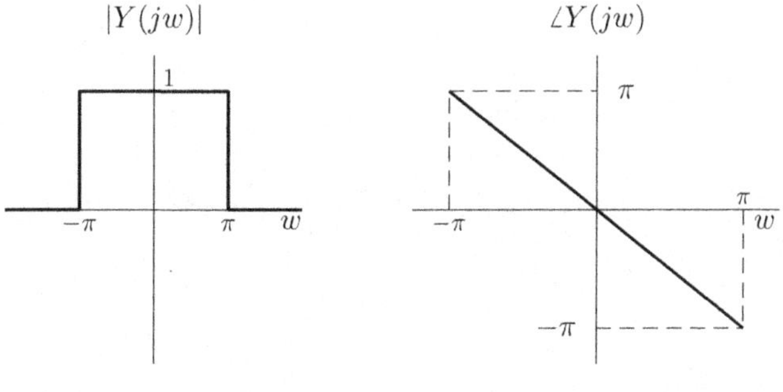

Figure 6.25: Magnitude and phase of the Fourier transform of the output $y(t)$, Problem 116.

(b)

$$\prod(\frac{w}{2\pi}) \quad \xrightarrow{\mathcal{F.T.}^{-1}} \quad \frac{\sin(\pi t)}{\pi t}$$

$$\prod(\frac{w}{2\pi})e^{-jw} \quad \xrightarrow{\mathcal{F.T.}^{-1}} \quad \frac{\sin(\pi(t-1))}{\pi(t-1)}$$

$$\Rightarrow \quad y(t) = \frac{\sin(\pi(t-1))}{\pi(t-1)}.$$

Problem 117. Determine if each of the following systems is time-invariant. In each case, y(t) denotes the system output and x(t) is the system input. Also $Y(jw)$ and $X(jw)$ denote the corresponding Fourier transforms, respectively.

(a) $y(t) = x(0) + X(t-5)$

(b) $y(t) = X(0) + x(t-5)$

(c) $Y(jw) = X(jw) + X(0)$

Solution. A system is time-invariant if its properties or characteristics do not change with time. A mathematical statement corresponding to this definition is that given $y(t) = T\{x(t)\}$ and for any time shift t_0, the time-shifted input $x_{t_0} = x(t-t_0)$ must generate a system output as $y_{t_0} = y(t-t_0)$.

(a) $y(t) = x(0) + X(t-5)$

First, we rewrite the relationship between $y(t)$ and $x(t)$ as

$$\begin{aligned} y(t) &= x(0) + X(jw)\Big|_{jw=t-5} \\ &= x(0) + \int_{-\infty}^{\infty} x(\tau)e^{-jw\tau}\, d\tau\Big|_{jw=t-5} \\ &= x(0) + \int_{-\infty}^{\infty} x(\tau)e^{-j(t-5)\tau}\, d\tau \end{aligned}$$

Now if the input is $x(t-t_0)$, we have:

$$y_{t_0}(t) = x(0) + \int_{-\infty}^{\infty} x(\tau-5)e^{-j(t-5)\tau}\, d\tau$$

Setting $\tau - t_0 = \tau'$:

$$y_{t_0}(t) = x(0) + \int_{-\infty}^{\infty} x(\tau')e^{-j(t-5)(\tau'+t_0)}\, d\tau' = y(t+t_0) \neq y(t-t_0).$$

As such, the system under consideration is not time-invariant.

(b) $y(t) = X(0) + x(t-5)$

We can rewrite the relationship between $y(t)$ and $x(t)$ as

$$y(t) = \int_{-\infty}^{\infty} x(\tau)\, d\tau + x(t-5).$$

Now we check if $T\{x(t-t_0)\}$ and $y(t-t_0)$ are the same:

$$\begin{cases} T\{x(t-t_0)\} = \int_{-\infty}^{\infty} x(\tau - t_0)\, d\tau + x(t-5-t_0) = \int_{-\infty}^{\infty} x(\tau)\, d\tau + x(t-t_0-5), \\ y(t-t_0) = \int_{-\infty}^{\infty} x(\tau)\, d\tau + x(t-t_0-5), \end{cases}$$

$\xrightarrow{T\{x(t-t_0)\}=y(t-t_0)}$ thus system is time-invariant.

(c) $Y(jw) = X(jw) + X(0)$
We take the inverse Fourier transform form both sides:

$$Y(jw) = X(jw) + X(0) \quad \Rightarrow \quad y(t) = x(t) + \left(\int_{-\infty}^{\infty} x(\tau)d\tau\right)\delta(t)$$

$$\begin{cases} T\{x(t-t_0)\} = x(t-t_0) + \left(\int_{-\infty}^{\infty} x(\tau - t_0)d\tau\right)\delta(t), \\ y(t-t_0) = x(t-t_0) + \left(\int_{-\infty}^{\infty} x(\tau)d\tau\right)\delta(t-t_0), \end{cases}$$

$\xrightarrow{T\{x(t-t_0)\}\neq y(t-t_0)}$ thus the system is time-variant.

Problem 118. Consider a continuous-time system whose input $x(t)$ and output $y(t)$ are related through the following relationship:

$$y(t) = \int_{-\infty}^{\infty} \frac{x(\alpha)}{\pi(t-\alpha)}\, d\alpha.$$

If $y(t)$ is the response of this system to $x(t) = e^{-t}u(t)$, find $Y(jw)$.

Solution.

We can write the above relationship as

$$y(t) = \int_{-\infty}^{\infty} \frac{x(\alpha)}{\pi(t-\alpha)}\, d\alpha = x(t) * \frac{1}{\pi t}.$$

As can be seen, $h(t) = \frac{1}{\pi t}$. We now have to find the Fourier transform of $h(t)$. We know from basic Fourier transform pair table that

$$\text{sign}(t) \quad \xrightarrow{\mathcal{F.T.}} \quad (\frac{-2j}{w}).$$

Now using the duality property we can write

$$(\frac{-2j}{t}) \quad \xrightarrow{\mathcal{F.T.}} \quad 2\pi\,\text{sign}(-w) = -2\pi\,\text{sign}(w).$$

Thus,

$$h(t) = \frac{1}{\pi t} \quad \xrightarrow{\mathcal{F.T.}} \quad -j\,\text{sign}(w).$$

On the other hand,

$$x(t) = e^{-t}u(t) \quad \xrightarrow{\mathcal{F.T.}} \quad \frac{1}{1+jw}.$$

Thus

$$\begin{aligned} Y(jw) &= X(jw).H(jw) \\ &= \frac{1}{1+jw}(-j\,\text{sign}(w)) \\ &= \begin{cases} \frac{-j}{1+jw}, & w > 0 \\ \frac{j}{1+jw}, & w \le 0 \end{cases} \end{aligned}$$

Problem 119. A continuous-time signal

$$x(t) = \sum_{k=-\infty}^{\infty} \delta(t - \frac{k\pi}{2})$$

is the input to a filter with frequency response $H(jw)$ whose magnitude and phase are as illustrated in Fig. 6.26.

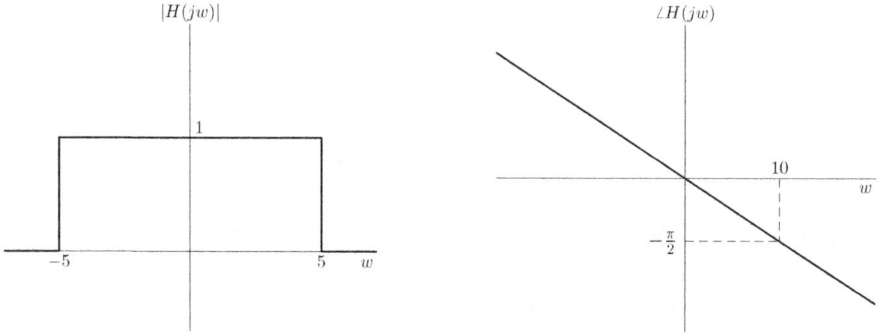

Figure 6.26: Problem 119, magnitude and phase of the Fourier transform for the filter $h(t)$.

Determine the filtered output signal $y(t)$.
Solution.

$$x(t) = \sum_{k=-\infty}^{\infty} \delta(t - \frac{k\pi}{2})$$

The input signal is periodic with period $T = \frac{\pi}{2}$ and the fundamental frequency is $w_0 = 4$. We must evaluate its Fourier series coefficients. The Fourier series coefficients of the impulse train were obtained in Chapter 4 as

$$a_k = \frac{1}{T} = \frac{2}{\pi}.$$

Thus the input signal is expressed as

$$x(t) = \sum_{k=-\infty}^{\infty} a_k e^{jkw_0t} = \sum_{k=-\infty}^{\infty} \frac{2}{\pi} e^{jk4t}.$$

Now we can evaluate the output signal as follows:

$$y(t) = \sum_{k=-\infty}^{\infty} a_k H(kw_0) e^{jkw_0t} = \sum_{k=-\infty}^{\infty} \frac{2}{\pi} H(4k) e^{j4kt}.$$

Based on the magnitude of the system frequency response, all elements except those corresponding to $k = -1, 0, 1$ are filtered. The unfiltered elements correspond to $w = 0$ and $w = \pm 4$ where $\angle H(w = 4) = \frac{4}{10} \times -\frac{\pi}{2} = -\frac{2\pi}{10}$. Similarity we have $\angle H(w = -4) = \frac{2\pi}{10}$.

For these values of k we have:

$$H(k4) \xrightarrow{k=\ -1,\ 0,\ 1} \begin{cases} H(-4) = |H(-4)| e^{j\angle H(-4)} = e^{j(\frac{2\pi}{10})} \\ H(0) = |H(0)| e^{j\angle H(0)} = e^{j(0)} = 1 \\ H(4) = |H(4)| e^{j\angle H(4)} = e^{-j(\frac{2\pi}{10})} \end{cases}.$$

Thus

$$\begin{aligned} y(t) &= \sum_{k=-\infty}^{\infty} a_k H(kw_0) e^{jkw_0t} = \sum_{k=-\infty}^{\infty} \frac{2}{\pi} H(k4) e^{jk4t} \\ &= \frac{2}{\pi} H(-4) e^{-j4t} + \frac{2}{\pi} H(0) + \frac{2}{\pi} H(4) e^{j4t} \\ &= \frac{2}{\pi} e^{j(\frac{2\pi}{10})} e^{-j4t} + \frac{2}{\pi} + \frac{2}{\pi} e^{-j(\frac{2\pi}{10})} e^{j4t} \\ &= \frac{2}{\pi} e^{j(\frac{2\pi}{10} - 4t)} + \frac{2}{\pi} + \frac{2}{\pi} e^{-j(\frac{2\pi}{10} - 4t)} \\ &= \frac{4}{\pi} \cos(4t - \frac{2\pi}{10}) + \frac{2}{\pi}. \end{aligned}$$

Problem 120. Consider an LTI system whose response to the input

$$x(t) = \cos(w_0 t)$$

is

$$y(t) = 4e^{-2|w_0|} \cos(w_0 t + \frac{\pi}{2}),$$

where $w_0 \in \mathcal{R}$. Determine the system's impulse response.

Solution. In general, for an input which only contains frequency w_0, the output has the form

$$y(t) = |H(w_0)| \cos(w_0 t + \angle H(w_0)).$$

As can be seen, the magnitude is multiplied by $4e^{-2|w_0|}$ and the phase is $\frac{\pi}{2}$. As such,

$$H(w) = 4e^{-2|w|}e^{j\frac{\pi}{2}} = 4je^{-2|w|}.$$

To obtain $h(t)$, we see based on the duality property and define $Z(w) = e^{-2|w|}$ that

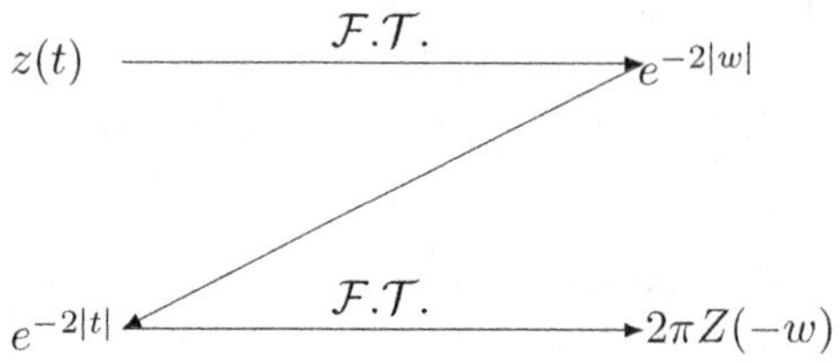

We also know that

$$e^{-2|t|} \xrightarrow{\mathcal{F.T.}} \frac{4}{w^2+4} = 2\pi Z(-w)$$

$$\Rightarrow z(t) = \frac{2}{\pi(t^2+4)}$$

As such

$$h(t) = \frac{8j}{\pi(t^2+4)}.$$

Problem 121. A continuous-time LTI system has a frequency response $H(jw)$ whose magnitude and phase are as illustrated in Figure 6.27. Find the step response $s(t)$ of the system.

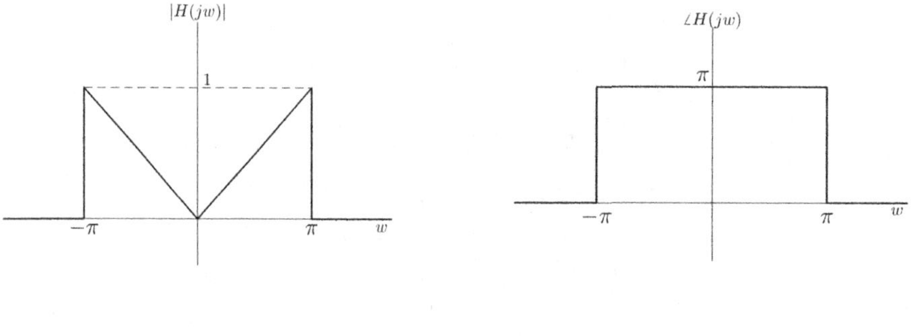

Figure 6.27: Problem 121, Frequency response $H(jw)$.

Solution.
According to the figures, we get

$$H(jw) = |H(jw)|e^{\angle H(jw)} = \begin{cases} \frac{w}{\pi}e^{j\pi} = \frac{-w}{\pi}, & 0 < w < \pi \\ \frac{-w}{\pi}e^{j\pi} = \frac{w}{\pi}, & -\pi < w \le 0\,. \\ 0, & \text{otherwise} \end{cases}$$

We know from Chapter 3 that

$$s(t) = \int_{-\infty}^{t} h(\tau)d\tau.$$

Now using the integration property, we evaluate the Fourier transform of the step response as

$$\begin{aligned} S(w) &= H(w)\left(\frac{1}{jw} + \pi\delta(w)\right) \\ &= \frac{H(w)}{jw} + \cancelto{0}{H(0)\pi\delta(w)} \quad \text{for} \quad -\pi < w < \pi \\ &= \frac{H(w)}{jw} \quad \text{for} \quad -\pi < w < \pi. \end{aligned}$$

Thus we have

$$S(jw) = \begin{cases} \frac{-1}{j\pi}, & 0 < w < \pi \\ \frac{1}{j\pi}, & -\pi < w \le 0 \\ 0, & \text{otherwise} \end{cases} \xrightarrow{\mathcal{F.T.}^{-1}}$$

$$\begin{aligned} s(t) &= \frac{1}{j\pi}\left(\frac{\sin(\pi t)}{\pi t}e^{-j\pi t} - \frac{\sin(\pi t)}{\pi t}e^{j\pi t}\right) \\ &= \frac{2}{\pi}\text{sinc}(t)\sin(\pi t). \end{aligned}$$

Chapter 7

The Discrete-time Fourier Transform

7.1 Introduction

In Chapter 5, we reviewed the discrete-time Fourier series which is mainly defined for periodic signals. We also observed an important aspect of LTI systems when the input is the complex exponential e^{jwn} and the corresponding Fourier analysis. To be able to extend such an analysis to other signals which are not necessarily periodic, the Discrete-Time Fourier Transform is used which is the focus of this chapter.

7.2 The Discrete-Time Fourier Transform

Let $x[n]$ be an aperiodic signal. The Fourier transform of $x[n]$ is denoted by $X(e^{jw})$. The Fourier transform pair of $x[n]$ and $X(e^{jw})$ are defined as

$$X(e^{jw}) = \sum_{n=-\infty}^{+\infty} x[n]e^{-jwn}, \tag{7.1}$$

$$x[n] = \frac{1}{2\pi}\int_{2\pi} X(e^{jw})e^{jwn}dw. \tag{7.2}$$

According to the definition of $X(e^{jw})$, we can see that the difference between the discrete-time and continuous-time Fourier transforms is that $X(e^{jw})$ is periodic in w with period 2π:

$$X(e^{j(w+2\pi)}) = X(e^{jw}), \tag{7.3}$$

where the equation holds because e^{jwn} is periodic in w.

Generally, function $X(e^{jw})$ is complex and it is represented by the *amplitude spectrum*, $|X(e^{jw})|$, and the *phase spectrum*, $\angle X(e^{jw})$, which are usually plotted versus the angular frequency w:

$$X(e^{jw}) = |X(e^{jw})|\ e^{\angle X(e^{jw})}. \tag{7.4}$$

By setting $w = 0$ in (7.1), we see that $X(e^{j0})$ represents the DC component of the aperiodic signal $x[n]$, i.e.,

$$X(e^{j0}) = \sum_{n=-\infty}^{+\infty} x[n]. \tag{7.5}$$

Similar to the case of continuous-time signals, the sufficient condition for the convergence of discrete-time Fourier Transform is that $x[n]$ must be *absolutely summable*; that is,

$$\sum_{n=-\infty}^{+\infty} |x[n]| < \infty. \tag{7.6}$$

7.3 Properties of the Discrete-Time Fourier Transform

(a) Linearity

Consider two given signals $x[n]$ and $y[n]$. If

$$\begin{aligned} x[n] &\overset{\mathcal{F.T.}}{\longleftrightarrow} X(e^{jw}), \\ y[n] &\overset{\mathcal{F.T.}}{\longleftrightarrow} Y(e^{jw}), \end{aligned}$$

Then we have

$$z[n] = ax[n] + by[n] \overset{\mathcal{F.T.}}{\longleftrightarrow} Z(e^{jw}) = aX(e^{jw}) + bY(e^{jw}). \tag{7.7}$$

(b) Time Shifting

Consider a signal $x[n]$ with Fourier Transform $X(e^{jw})$. The Fourier Transform $x[n - n_0]$ is given by

$$x[n - n_0] \overset{\mathcal{F.T.}}{\longleftrightarrow} e^{-jwn_0} X(e^{jw}). \tag{7.8}$$

Equation (7.8) indicates that when an aperiodic signal is shifted in time domain, the magnitudes of its Fourier Transform remains unchanged but the phase is rotated by $-wn_0$.

(c) Frequency Shifting

Consider a signal $x[n]$ with the Fourier Transform $X(e^{jw})$. The frequency shifting property indicates that the multiplication of $x[n]$ by a complex exponential signal $e^{jw_0 n}$ results in a shift of $X(e^{jw})$ in the frequency domain:

$$e^{jw_0 n} x[n] \overset{\mathcal{F.T.}}{\longleftrightarrow} X(e^{j(w-w_0)}). \tag{7.9}$$

(d) Conjugation

The complex conjugation of a signal $x[n]$ causes the complex conjugation and frequency reversal on the corresponding Fourier Transform. That is,

$$\begin{aligned} x[n] &\overset{\mathcal{F.T.}}{\longleftrightarrow} X(e^{jw}), \\ x^*[n] &\overset{\mathcal{F.T.}}{\longleftrightarrow} X^*(e^{-jw}). \end{aligned}$$

Proof:

$$X^*(e^{jw}) = \left[\sum_{n=-\infty}^{+\infty} x[n]e^{-jwn}\right]^*$$
$$= \sum_{n=-\infty}^{+\infty} x^*[n]e^{jwn}.$$

Changing w to $-w$ in the above equation, we have

$$X^*(e^{-jw}) = \sum_{n=-\infty}^{+\infty} x^*[n]e^{-jwn}.$$

(e) Multiplication

Consider two signals $x[n]$ and $y[n]$ with the Fourier transformss $X(e^{jw})$ and $Y(e^{jw})$, respectively. The Fourier transform of $x[n] \cdot y[n]$ is given by

$$x[n] \cdot y[n] \overset{\mathcal{F.T.}}{\longleftrightarrow} \frac{1}{2\pi}\int_{2\pi} X(e^{j\theta})Y(e^{j(w-\theta)})d\theta. \tag{7.10}$$

The right hand side of (7.10) is known as periodic convolution of $X(e^{jw})$ and $Y(e^{jw})$ and is denoted by $X(e^{jw}) \circledast Y(e^{jw})$.

(f) Time Expansion

Let k be a positive integer, where $x_k[n]$ is defined as below. For the Fourier transform we have:

$$x_k[n] = \begin{cases} x[n/k] & \text{if } n = mk \\ 0 & \text{if } n \neq mk \end{cases} \overset{\mathcal{F.T.}}{\longleftrightarrow} X(e^{jkw}). \tag{7.11}$$

(g) Duality

Consider a signal $x[n]$ with the Fourier transform $X(e^{jw})$, then the Fourier transform of the signal $X(e^{jt})$ will be $x[-k]$:

$$x[n] \overset{\mathcal{F.T.}}{\longleftrightarrow} X(e^{jw}),$$
$$X(e^{jt}) \overset{\mathcal{F.T.}}{\longleftrightarrow} x[-k].$$

(h) Parseval's Relation

Consider a signal $x[n]$ with the Fourier transform $X(e^{jw})$, then Parseval's relation is given by

$$\sum_{n=-\infty}^{+\infty} |x[n]|^2 dt = \frac{1}{2\pi}\int_{2\pi} |X(e^{jw})|^2 dw. \tag{7.12}$$

(i) Conjugate Symmetry of Real Signals

If signal $x[n]$ is real, then its Fourier transform is *conjugate symmetric*, i.e.,

$$x[n] = x^*[n] \quad \Rightarrow \quad X(e^{jw}) = X^*(e^{-jw}).$$

As such we have

$$\begin{aligned}
&|X(e^{jw})| \text{ is even} \quad : \ ||X(e^{jw})| = |X(e^{-jw})| \\
&\angle X(e^{jw}) \text{ is odd} \quad : \ \angle X(e^{jw}) = -\angle X(e^{-jw}) \\
&\Re\{X(e^{jw})\} \text{ is even} \quad : \ \Re\{X(e^{jw})\} = \Re\{X(e^{-jw})\} \\
&\Im\{X(e^{jw})\} \text{ is odd} \quad : \ \Im\{X(e^{jw})\} = -\Im\{X(e^{-jw})\} \\
&x_e[n] \ \stackrel{\mathcal{F.S.}}{\longleftrightarrow} \ \Re\{X(e^{jw})\} \\
&x_o[n] \ \stackrel{\mathcal{F.S.}}{\longleftrightarrow} \ j\Im\{X(e^{jw})\}
\end{aligned}$$

Note that if $x[n]$ is real and even, so is its Fourier transform, and if $x[n]$ is real and odd, then its Fourier transform is imaginary and odd.

7.4 The Fourier Transform for Periodic Signals

Consider a periodic signal $x[n]$ with period N, fundamental frequency w_0 and the Fourier series coefficients a_k. From the definition of the Fourier series coefficients, we have

$$x[n] = \sum_{k<N>} a_k e^{jkw_0 n}. \tag{7.13}$$

If we apply the Fourier transform to both sides of (7.13) we get

$$\begin{aligned}
X(e^{jw}) &= \mathcal{F.T.}\left\{\sum_{k<N>} a_k e^{jkw_0 n}\right\} \\
&= \sum_{k<N>} a_k \ \mathcal{F.T.}\left\{e^{jkw_0 t}\right\}.
\end{aligned} \tag{7.14}$$

From Table 7.2, we have

$$e^{jkw_0 n} \ \stackrel{\mathcal{F.T.}}{\longleftrightarrow} \ 2\pi \sum_{l=-\infty}^{+\infty} \delta(w - w_0 - 2\pi l).$$

Therefore, we obtain

$$X(e^{jw}) = \sum_{k=-\infty}^{+\infty} 2\pi a_k \delta(w - k\frac{2\pi}{N}). \tag{7.15}$$

In other words, the Fourier transform of a periodic signal $x[n]$ includes impulses with magnitude $2\pi a_k$, located at the kth harmonic frequency kw_0.

7.5 Fourier Transform of Important Signals and Table of Fourier Transform Properties

Some of the important properties of the Fourier transform discussed above are given in Table 7.1. In Table 7.2, we have summarized some common Fourier transform pairs.

Table 7.1: Properties of the Fourier Transform

Property	**Aperiodic signal**	**Fourier transform**
	$x[n]$	$X(e^{jw})$
	$y[n]$	$Y(e^{jw})$
Linearity	$ax[n] + by[n]$	$aX(e^{jw}) + bY(e^{jw})$
Time Shifting	$x[n - n_0]$	$e^{-jwn_0}X(e^{jw})$
Frequency Shifting	$e^{jw_0 n}x[n]$	$X(e^{j(w-w_0)})$
Conjugation	$x^*[n]$	$X^*(e^{-jw})$
Time Reversal	$x[-n]$	$X(e^{-jw})$
Time Expansion	$x_k[n] = \begin{cases} x[n/k] & \text{if } n = mk \\ 0 & \text{if } n \neq mk \end{cases}$	$X(e^{jkw})$
Convolution	$x[n] * y[n]$	$X(e^{jw})Y(e^{jw})$
Multiplication	$x[n]y[n]$	$\frac{1}{2\pi}\int_{2\pi} X(e^{j\theta})Y(e^{j(w-\theta)})d\theta$
Differentiation in Time	$x[n] - x[n-1]$	$(1 - e^{-jw})X(e^{jw})$
Accumulation	$\sum_{k=-\infty}^{n} x[k]$	$\frac{1}{(1-e^{-jw})}X(e^{jw})$ $+\pi X(e^{j0})\sum_{k=-\infty}^{+\infty}\delta(w - 2\pi k)$
Differentiation in Frequency	$nx[n]$	$j\frac{d}{dw}X(e^{jw})$
Real signal	$x[n] = x_e[n] + x_o[n]$	$X(e^{-jw}) = X^*(e^{jw})$
Even component	$x_e[n]$	$\Re\{X(e^{jw})\}$
Odd component	$x_o[n]$	$j\Im\{X(e^{jw})\}$

Parseval's relation

$$\sum_{n=-\infty}^{+\infty} |x[n]|^2 = \frac{1}{2\pi}\int_{2\pi} |X(e^{jw})|^2 dw$$

Table 7.2: Common Fourier Transforms Pairs

Signal	**Fourier transform**
$\delta[n]$	1
$\delta[n-n_0]$	e^{-jwn_0}
1	$2\pi\sum_{l=-\infty}^{+\infty}\delta(w-2\pi l)$
e^{jw_0n}	$2\pi\sum_{l=-\infty}^{+\infty}\delta(w-w_0-2\pi l)$
$\cos w_0n$	$\pi\sum_{l=-\infty}^{+\infty}\{\delta(w-w_0-2\pi l)+\delta(w+w_0-2\pi l)\}$
$\sin w_0n$	$\frac{\pi}{j}\sum_{l=-\infty}^{+\infty}\{\delta(w-w_0-2\pi l)-\delta(w+w_0-2\pi l)\}$
$u[n]$	$\frac{1}{1-e^{-jw}}+\sum_{k=-\infty}^{+\infty}\pi\delta(w-2\pi k)$
$\sum_{k=-\infty}^{+\infty}\delta[n-kN]$	$\frac{2\pi}{N}\sum_{k=-\infty}^{+\infty}\delta(w-\frac{2\pi k}{N})$
$x[n]=\begin{cases}1 & \lvert n\rvert\le N_1\\0 & \lvert n\rvert>N_1\end{cases}$	$\frac{\sin[w(N_1+\frac{1}{2})]}{\sin(w/2)}$
$\frac{\sin Wn}{\pi n},\ 0<W<\pi$	$X(e^{jw})=\begin{cases}1 & 0\le\lvert w\rvert\le W\\0 & W<\lvert w\rvert\le\pi\end{cases}$
$a^nu[n],\ \lvert a\rvert<1$	$\frac{1}{1-ae^{-jw}}$
$(n+1)a^nu[n],\ \lvert a\rvert<1$	$\frac{1}{(1-ae^{-jw})^2}$
$\frac{(n+r-1)!}{n!(r-1)!}a^nu[n],\ \lvert a\rvert<1$	$\frac{1}{(1-ae^{-jw})^r}$

7.6 The Fourier transform and LTI Systems

Consider an LTI system with the impulse response $h[n]$. Let the input be an aperiodic signal $x[n]$. The Fourier transform of the output signal $y[n]$, and eventually $y[n]$ itself, can be obtained based on the Fourier transform of $x[n]$ and the frequency response of the system. That is, if

$$y[n]=h[n]*x[n]\ \xleftrightarrow{\mathcal{F.T.}}\ Y(e^{jw})=H(e^{jw})X(e^{jw}) \tag{7.16}$$

then

$$H(e^{jw})=\frac{Y(e^{jw})}{X(e^{jw})}, \tag{7.17}$$

where $H(e^{jw})$ is the frequency response of the system which is given by

$$H(e^{jw}) = \int_{-\infty}^{+\infty} h[n]e^{-jwn}dt = |H(e^{jw})|\ e^{\angle H(e^{jw})}. \tag{7.18}$$

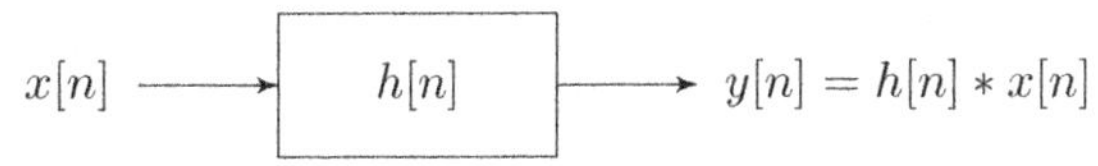

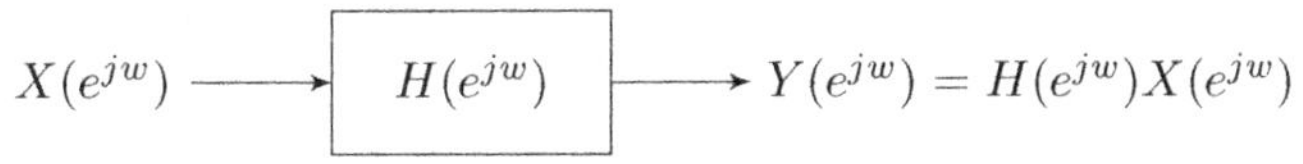

We also have:

$$|Y(e^{jw})| = |H(e^{jw})|\ |X(e^{jw})|, \tag{7.19}$$
$$\angle Y(e^{jw}) = \angle H(e^{jw}) + \angle X(e^{jw}). \tag{7.20}$$

Note from Chapter 5 that if a periodic signal $x[n]$ with the Fourier series coefficients a_k is applied to an LTI system with the impulse response $h[n]$, then the output $y[n]$ is periodic and its Fourier series coefficients, b_k, are given by

$$b_k = H(e^{jk\frac{2\pi}{N}}) \cdot a_k \tag{7.21}$$

Please note that in LTI systems, for input signals $e^{jw_0 n}$, $\cos(w_0 n)$ and $\sin(w_0 n)$, we can easily obtain the output signals as follows:

$$\begin{aligned}
&\text{if } x[n] = e^{jw_0 n} \ \Rightarrow\ y[n] = |H(e^{jw_0})|e^{j\left(w_0 n + \angle H(e^{jw_0})\right)},\\
&\text{if } x[n] = \cos(w_0 n) \ \Rightarrow\ y[n] = |H(e^{jw_0})| \cos\left(w_0 n + \angle H(e^{jw_0})\right),\\
&\text{if } x[n] = \sin(w_0 n) \ \Rightarrow\ y[n] = |H(e^{jw_0})| \sin\left(w_0 n + \angle H(e^{jw_0})\right).
\end{aligned}$$

7.7 LTI Systems Characterized by Constant-Coefficient Difference Equations

Consider a discrete-time LTI system which is described by a linear constant-coefficient Difference equation of the form

$$\sum_{k=0}^{N} a_k y[n-k] = \sum_{k=0}^{M} b_k x[n-k], \quad M \le N. \tag{7.22}$$

Applying the Fourier transform to both sides of (7.22) and using the linearity property (7.7), we have

$$\mathcal{F}\left\{\sum_{k=0}^{N} a_k\ y[n-k]\right\} = \mathcal{F}\left\{\sum_{k=0}^{M} b_k\ x[n-k]\right\}, \tag{7.23}$$

$$\sum_{k=0}^{N} a_k\ \mathcal{F}\{y[n-k]\} = \sum_{k=0}^{M} b_k\ \mathcal{F}\{x[n-k]\}, \tag{7.24}$$

and from the differentiation property in Table 7.1,

$$\sum_{k=0}^{N} a_k\ e^{-jkw} Y(e^{jw}) = \sum_{k=0}^{M} b_k\ e^{-jkw} X(e^{jw}).$$

Thus, from (7.17),

$$H(e^{jw}) = \frac{Y(e^{jw})}{X(e^{jw})} = \frac{\sum_{k=0}^{M} b_k\ e^{-jkw}}{\sum_{k=0}^{N} a_k\ e^{-jkw}}. \tag{7.25}$$

7.8 Filtering

Similar to the case of continuous-time, an ideal *frequency-selective* filter is defined by

$$H(jw) = \begin{cases} k\ e^{-jwt_d} & |w| < w_c \\ 0 & |w| > w_c \end{cases}. \tag{7.26}$$

Note that the difference between discrete-time and continuous-time filters is that the frequency response of discrete-time filters is periodic with period 2π and also their highest frequency is located at $w = \pi$.

In figures below, we have depicted different types of ideal *frequency-selective* filters.

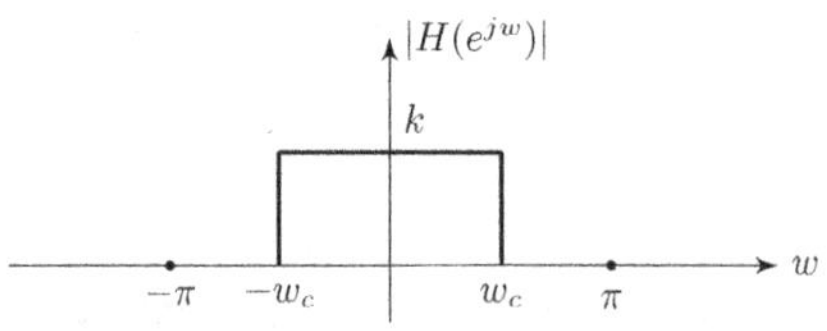

Figure 7.1: Ideal low-pass filter.

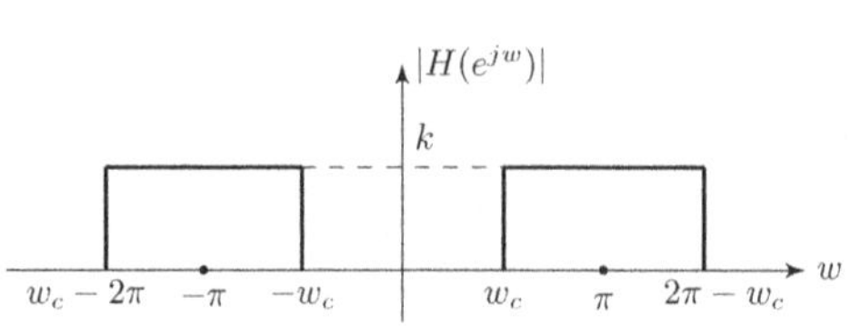

Figure 7.2: Ideal high-pass filter.

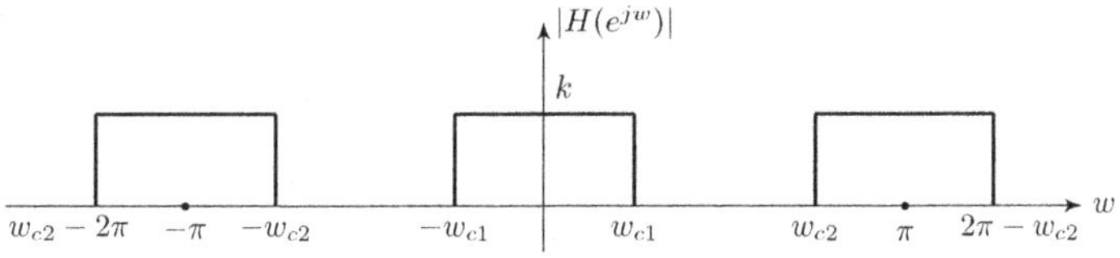

Figure 7.3: Ideal band-stop filter.

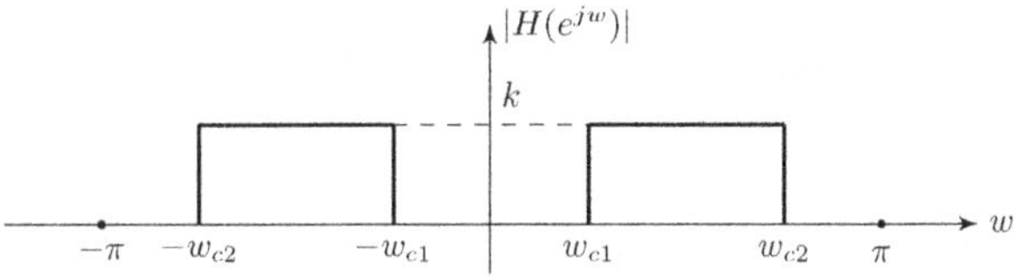

Figure 7.4: Ideal band-pass filter.

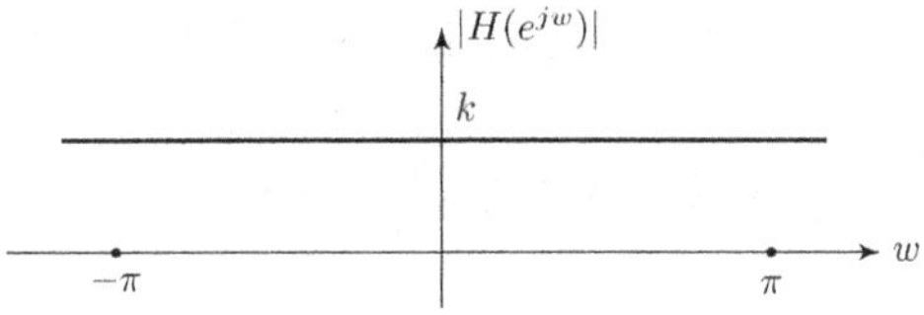

Figure 7.5: Ideal All-pass filter.

7.9 Problems

Problem 122. Use the discrete-time Fourier transform analysis definition to compute the Fourier transforms of each of the following signals:

(a) $x[n] = (\frac{3}{5})^n u[n-3]$

(b) $x[n] = a^{|n|}, \quad |a| < 1$

(c) $x[n] = \delta[4-2n]$

(d) $x[n] = (\frac{1}{3})^{-n} u[-n-1]$

(e) $x[n] = \sin(\frac{5\pi}{4}n) + \cos(\frac{3\pi}{4}n)$

Solution.

(a) $x[n] = (\frac{3}{5})^n u[n-3]$

By definition of the Fourier transform

$$X(e^{jw}) = \sum_{n=-\infty}^{+\infty} x[n]e^{-jwn}$$

$$= \sum_{n=3}^{+\infty} (\frac{3}{5})^n e^{-jwn} = \sum_{n=3}^{+\infty} (\frac{3}{5} e^{-jw})^n,$$

and by using infinite geometric series formula, we have

$$X(e^{jw}) = \sum_{n=3}^{+\infty} (\frac{3}{5} e^{-jw})^n = \frac{(\frac{3}{5}e^{-jw})^3}{1 - \frac{3}{5}e^{-jw}}.$$

(b) $x[n] = a^{|n|}, \quad |a| < 1$

By definition of the Fourier transform we can write

$$X(e^{jw}) = \sum_{n=-\infty}^{+\infty} a^{|n|} e^{-jwn}$$
$$= \sum_{n=-\infty}^{-1} a^{-n} e^{-jwn} + \sum_{n=0}^{+\infty} a^n e^{-jwn}.$$

Substituting $m = -n$ in the first summation, we obtain

$$X(e^{jw}) = \sum_{m=1}^{+\infty} (ae^{jw})^m + \sum_{n=0}^{+\infty} (ae^{-jw})^n.$$

Both of these summations are infinite geometric series so we can write

$$X(e^{jw}) = \frac{ae^{jw}}{1 - ae^{jw}} + \frac{1}{1 - ae^{-jw}}$$
$$= \frac{1 - a^2}{1 + a^2 - 2a\cos(w)}.$$

(c) $x[n] = \delta[4 - 2n]$

The signal only has 1 nonzero sample in $n = 2$, thus we have

$$X(e^{jw}) = e^{-2jw}.$$

(d) $x[n] = (\frac{1}{3})^{-n} u[-n-1]$

By definition of the Fourier transform, we can write

$$X(e^{jw}) = \sum_{n=-\infty}^{+\infty} (\frac{1}{3})^{-n} u[-n-1] e^{-jwn} = \sum_{n=-\infty}^{-1} (\frac{1}{3})^{-n} e^{-jwn}.$$

Substituting $m = -n$ and interchanging the order of summation, we have

$$X(e^{jw}) = \sum_{m=1}^{+\infty} (\frac{1}{3})^m e^{jwm} = \sum_{m=1}^{+\infty} (\frac{1}{3} e^{jw})^m$$
$$= \frac{\frac{1}{3}e^{jw}}{1 - \frac{1}{3}e^{jw}} = \frac{e^{jw}}{3 - e^{jw}}.$$

(e) $x[n] = \sin(\frac{5\pi}{4}n) + \cos(\frac{3\pi}{4}n)$

We have

$$x_1[n] = \sin(w_0 n) \xrightarrow{\mathcal{F.T.}} X_1(e^{jw}) = \frac{\pi}{j}[\delta(w - w_0) - \delta(w + w_0)],$$
$$x_2[n] = \cos(w_0 n) \xrightarrow{\mathcal{F.T.}} X_2(e^{jw}) = \pi[\delta(w - w_0) + \delta(w + w_0)].$$

Using the linearity property we write

$$X(e^{jw}) = X_1(e^{jw}) + X_2(e^{jw})$$
$$= \frac{\pi}{j}[\delta(w - \frac{5\pi}{4}) - \delta(w + \frac{5\pi}{4})] + \pi[\delta(w - \frac{3\pi}{4}) + \delta(w + \frac{3\pi}{4})].$$

Using the fact that the Fourier transform of discrete signals is periodic with period 2π, we can write

$$X(e^{jw}) = \frac{\pi}{j}[\delta(w + \frac{3\pi}{4}) - \delta(w - \frac{3\pi}{4})] + \pi[\delta(w - \frac{3\pi}{4}) + \delta(w + \frac{3\pi}{4})]$$
$$= (\pi + \frac{\pi}{j})\delta(w + \frac{3\pi}{4}) + (\pi - \frac{\pi}{j})\delta(w - \frac{3\pi}{4}).$$

Problem 123. Use the inverse Fourier transform formulation to compute the inverse Fourier transforms of the following signals.

(a) $X(e^{jw}) = \cos^2(w)$

(b) $X(e^{jw}) = \cos(w) + j\sin(w)$

(c) As in Fig. 7.6.

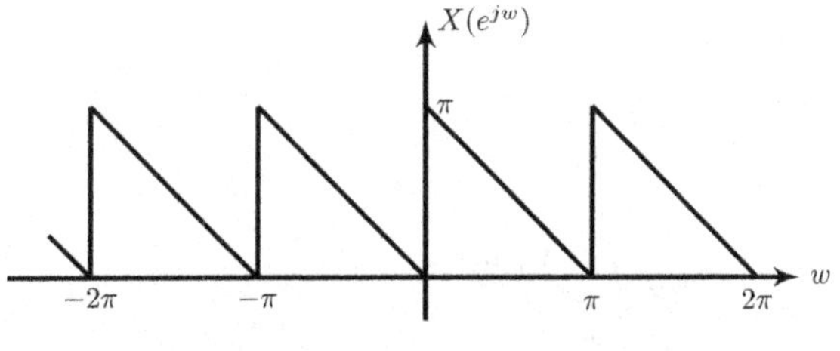

Figure 7.6: Problem 123 - Part (c).

(d) As in Fig. 7.7 where $\angle X(e^{jw}) = 0 \, \forall w$.

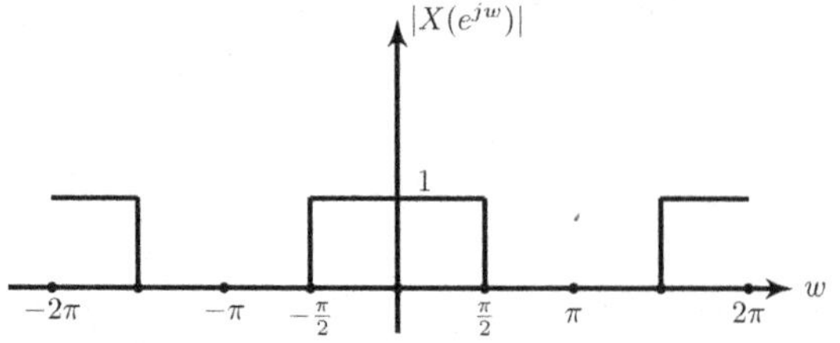

Figure 7.7: Problem 123 - Part (d).

Solution.

(a) $X(e^{jw}) = \cos^2(w)$

Using Euler's formula, we get

$$\begin{aligned} X(e^{jw}) &= \frac{1}{4}(e^{jw} + e^{-jw})^2 \\ &= \frac{1}{2} + \frac{1}{4}e^{2jw} + \frac{1}{4}e^{-2jw}. \end{aligned}$$

Then we have

$$x[n] = \frac{1}{2}\delta[n] + \frac{1}{4}\delta[n+2] + \frac{1}{4}\delta[n-2].$$

(b) $X(e^{jw}) = \cos(w) + j\sin(w)$

Again by Euler's formula we have

$$X(e^{jw}) = \frac{1}{2}e^{jw} + \frac{1}{2}e^{-jw} + \frac{1}{2}e^{jw} - \frac{1}{2}e^{-jw} = e^{jw}.$$

Thus

$$x[n] = \delta[n+1].$$

(c) From Fig. 7.6 we have

$$\begin{aligned} x[n] &= \frac{1}{2\pi}\int_{-\pi}^{\pi} X(e^{jw})e^{jwn}dw \\ &= \frac{1}{2\pi}\int_{-\pi}^{0} -we^{jwn}dw + \frac{1}{2\pi}\int_{0}^{\pi} (\pi - w)e^{jwn}dw \end{aligned}$$

$$
\begin{aligned}
&= \frac{1}{2\pi}\int_{-\pi}^{\pi} -we^{jwn}dw + \frac{1}{2}\int_{0}^{\pi} e^{jwn}dw \\
&= \frac{j(-1)^n}{n} + \frac{1}{j2n}[(-1)^n - 1] \\
&= -\frac{1}{j2n}[(-1)^n + 1] = \frac{j}{2n}[(-1)^n + 1].
\end{aligned}
$$

Finally

$$
x[n] = \begin{cases} \frac{j}{n} & n \text{ even} \\ 0 & n \text{ odd} \end{cases}.
$$

(d) From Fig. 7.7 we have

$$
\begin{aligned}
x[n] &= \frac{1}{2\pi}\int_{-\pi}^{\pi} X(e^{jw})e^{jwn}dw \\
&= \frac{1}{2\pi}\int_{-\frac{\pi}{2}}^{\frac{\pi}{2}} e^{jwn}dw \\
&= \frac{1}{j2n\pi}(e^{jn(\pi/2)} - e^{-jn(\pi/2)}) \\
&= \frac{1}{n\pi}\sin(\frac{n\pi}{2}), \quad n \neq 0.
\end{aligned}
$$

If $n = 0$ then we write:

$$
\begin{aligned}
x[n] &= \frac{1}{2\pi}\int_{-\pi}^{\pi} X(e^{jw})dw \\
&= \frac{1}{2\pi}\int_{-\frac{\pi}{2}}^{\frac{\pi}{2}} dw = \frac{1}{2}.
\end{aligned}
$$

Problem 124. For each of the following signals, use Fourier transform properties to determine the corresponding Fourier transform.

(a) $x[n] = (\frac{1}{3})^n u[n+3]$

(b) $x[n] = \cos(\frac{\pi}{2}n)(\frac{1}{3})^n u[n-1]$

(c) $x[n] = \frac{\sin(\frac{\pi}{2}n)}{n\pi} * \frac{\sin(\frac{\pi}{2}(n-3))}{(n-3)\pi}$

(d) $x[n] = na^n u[n], \qquad |a| < 1$

Solution.

(a) $x[n] = (\frac{1}{3})^n u[n+3]$

Signal $x[n]$ can be rewritten as

$$x[n] = 27(\frac{1}{3})^{n+3}u[n+3].$$

We know that

$$a^n u[n] \xrightarrow{\mathcal{F.T.}} \frac{1}{1 - ae^{-jw}}, \quad |a| < 1$$

hence,

$$(\frac{1}{3})^n u[n] \xrightarrow{\mathcal{F.T.}} \frac{1}{1 - \frac{1}{3}e^{-jw}},$$

and by using the time shifting property, we obtain

$$(\frac{1}{3})^{n+3}u[n+3] \xrightarrow{\mathcal{F.T.}} \frac{e^{j3w}}{1 - \frac{1}{3}e^{-jw}}.$$

Finally, we have

$$x[n] = 27(\frac{1}{3})^{n+3}u[n+3] \xrightarrow{\mathcal{F.T.}} X(e^{jw}) = \frac{27e^{j3w}}{1 - \frac{1}{3}e^{-3jw}}.$$

(b) $x[n] = \cos(\frac{\pi}{2}n)(\frac{1}{3})^n u[n-1]$

We can rewrite signal $x[n]$ as

$$x[n] = \cos(\frac{\pi}{2}n)(\frac{1}{3})(\frac{1}{3})^{n-1}u[n-1].$$

Let's define signal $y[n]$ as follows:

$$y[n] = (\frac{1}{3})^{n-1}u[n-1].$$

Now since

$$(\frac{1}{3})^n u[n] \xrightarrow{\mathcal{F.T.}} \frac{1}{1 - \frac{1}{3}e^{-jw}},$$

using the time shifting property, we obtain

$$y[n] = (\frac{1}{3})^{n-1}u[n-1] \xrightarrow{\mathcal{F.T.}} Y(e^{jw}) = \frac{e^{-jw}}{1 - \frac{1}{3}e^{-jw}}.$$

Using Euler's formula, signal $x[n]$ can be rewritten as

$$\begin{aligned} x[n] &= \frac{1}{3}\cos(\frac{\pi}{2}n)y[n] \\ &= \frac{1}{3}(\frac{e^{jn(\pi/2)} + e^{-jn(\pi/2)}}{2})y[n] \\ &= \frac{1}{6}\big(e^{jn(\pi/2)}y[n] + e^{-jn(\pi/2)}y[n]\big). \end{aligned}$$

Now using the frequency shifting property, we obtain

$$\begin{aligned} X(e^{jw}) &= \frac{1}{6}\big(Y(e^{j(w-\frac{\pi}{2})}) + Y(e^{j(w+\frac{\pi}{2})})\big) \\ &= \frac{1}{6}\Big(\frac{e^{-j(w-\frac{\pi}{2})}}{1-\frac{1}{3}e^{-j(w-\frac{\pi}{2})}} + \frac{e^{-j(w+\frac{\pi}{2})}}{1-\frac{1}{3}e^{-j(w+\frac{\pi}{2})}}\Big) \\ &= \frac{1}{2}e^{-jw}\Big(\frac{e^{j\frac{\pi}{2}}}{3-e^{-j(w-\frac{\pi}{2})}} + \frac{e^{-j\frac{\pi}{2}}}{3-e^{-j(w+\frac{\pi}{2})}}\Big). \end{aligned}$$

(c) $x[n] = \dfrac{\sin(\frac{\pi}{2}n)}{n\pi} * \dfrac{\sin(\frac{\pi}{2}(n-3))}{(n-3)\pi}$

We see that signal $x[n]$ is the convolution of two signals $s[n]$ and $s_d[n]$, where

$$s[n] = \frac{\sin(\frac{\pi}{2}n)}{n\pi}, \qquad s_d[n] = s[n-3].$$

We know that

$$z[n] = \frac{\sin(Wn)}{\pi n} \xrightarrow{\mathcal{F.T.}} Z(w) = \prod(\frac{w}{2W}) \quad 0 < W < \pi,$$

where $Z(w)$ is periodic with period 2π. Hence

$$s[n] = \frac{\sin(\frac{\pi}{2}n)}{n\pi} \xrightarrow{\mathcal{F.T.}} S(e^{jw}) = \prod(\frac{w}{\pi}) \quad -\pi \le w \le \pi.$$

Also using the time shifting property, the Fourier transform for $s_d[n]$ is given by

$$s_d[n] = s[n-3] \xrightarrow{\mathcal{F.T.}} S_d(e^{jw}) = e^{-j3w}\prod(\frac{w}{\pi}) \quad -\pi \le w \le \pi.$$

According to the convolution property, we have

$$X(e^{jw}) = S(e^{jw})\, S_d(e^{jw}) = e^{-j3w}\prod(\frac{w}{\pi}) \quad -\pi \le w \le \pi.$$

(d) $x[n] = na^n u[n], \qquad |a| < 1$

We know that

$$a^n u[n] \xrightarrow{\mathcal{F.T.}} \frac{1}{1-ae^{-jw}}, \quad |a|<1.$$

According to the frequency-domain differentiation property, we have

$$nx[n] \xrightarrow{\mathcal{F.T.}} j\frac{d}{dw}X(e^{jw}),$$

and so the Fourier transform of $x[n]$ is given by

$$\begin{aligned} X(e^{jw}) &= j\frac{d}{dw}(\frac{1}{1-ae^{-jw}}) \\ &= \frac{j(-jae^{-jw})}{(1-ae^{-jw})^2} \\ &= \frac{ae^{-jw}}{(1-ae^{-jw})^2}. \end{aligned}$$

Problem 125. For each of the following Fourier transforms, use Fourier transform properties to determine the inverse Fourier transform.

(a) $X(e^{jw}) = j\sin(2w)$

(b) $X(e^{jw}) = \dfrac{1}{1-\frac{1}{2}e^{-j5w}}$

(c) $X(e^{jw}) = \cos(2w)[\dfrac{\sin(\frac{5w}{2})}{\sin(\frac{w}{2})}]$

(d) $X(e^{jw}) = e^{-j(2w+\pi/2)}\frac{d}{dw}[\dfrac{2}{1-\frac{1}{3}e^{-j(w-\pi/3)}} + \dfrac{2}{1-\frac{1}{3}e^{-j(w+\pi/3)}}]$

(e) $X(e^{jw}) = \dfrac{10}{-e^{-j2w}+3e^{-jw}+10}$

Solution.

(a) $X(e^{jw}) = j\sin(2w)$

Using Euler's formula, we can rewrite $X(e^{jw})$ as follows:

$$X(e^{jw}) = \frac{1}{2}e^{j2w} - \frac{1}{2}e^{-j2w},$$

and from time shifting properties, we have

$$x[n] = \frac{1}{2}\delta[n+2] - \frac{1}{2}\delta[n-2].$$

(b) $X(e^{jw}) = \dfrac{1}{1 - \frac{1}{2}e^{-j5w}}$

Let $b[n]$ and its Fourier transform be defined as

$$b[n] = (\frac{1}{2})^n u[n] \xrightarrow{\mathcal{F.T.}} B(e^{jw}) = \frac{1}{1 - \frac{1}{2}e^{-jw}}.$$

In this case we can write $X(e^{jw}) = B(e^{j5w})$. Using time expansion property, $x[n]$ is given by

$$x[n] = \begin{cases} (\frac{1}{2})^{n/5} & n = 0, \pm 5, \pm 10, \cdots \\ 0 & \text{otherwise} \end{cases}.$$

(c) $X(e^{jw}) = \cos(2w)[\dfrac{\sin(\frac{5w}{2})}{\sin(\frac{w}{2})}]$

Let $b[n]$ and its Fourier transform be defined as

$$B(e^{jw}) = \frac{\sin(\frac{5w}{2})}{\sin(\frac{w}{2})} \xrightarrow{\mathcal{F.T.}^{-1}} b[n] = \prod(\frac{n}{4}).$$

We rewrite $b[n]$ as

$$b[n] = u[n+2] - u[n-2]. \tag{7.27}$$

Using Euler's formula, $X(e^{jw})$ can be rewritten as

$$X(e^{jw}) = [\frac{e^{j2w} + e^{-j2w}}{2}]B(e^{jw}). \tag{7.28}$$

From (7.27) and (7.28) and using time shifting property, we have

$$\begin{aligned} x[n] &= \frac{1}{2}b[n+2] + \frac{1}{2}b[n-2] \\ &= \frac{1}{2}u[n+4] + u[n] - \frac{1}{2}u[n-4]. \end{aligned}$$

(d) $X(e^{jw}) = e^{-j(2w+\pi/2)}\frac{d}{dw}[\dfrac{2}{1 - \frac{1}{3}e^{-j(w-\pi/3)}} + \dfrac{2}{1 - \frac{1}{3}e^{-j(w+\pi/3)}}]$

Let's define $S_1(e^{jw})$ as

$$S_1(e^{jw}) = \frac{2}{1 - \frac{1}{3}e^{-jw}}. \tag{7.29}$$

Then the inverse Fourier transform, $s_1[n]$, can be expressed as

$$s_1[n] = 2(\frac{1}{3})^n u[n]. \quad (7.30)$$

From the frequency shifting property and (7.30), we can write

$$\begin{aligned} S_1(e^{j(w-\pi/3)}) &\xrightarrow{\mathcal{F.T.}} e^{j(\pi/3)n} s_1[n]. \\ S_1(e^{j(w+\pi/3)}) &\xrightarrow{\mathcal{F.T.}} e^{-j(\pi/3)n} s_1[n]. \end{aligned}$$

Now if $S(e^{jw})$ is defined as

$$S(e^{jw}) = S_1(e^{j(w-\pi/3)}) + S_1(e^{j(w+\pi/3)}), \quad (7.31)$$

then we have

$$\begin{aligned} s[n] &= e^{-j(\pi/3)n} s_1[n] + e^{j(\pi/3)n} s_1[n] \\ &= 2\cos(\frac{\pi}{3}n) s_1[n]. \end{aligned} \quad (7.32)$$

We know that

$$e^{-j(2w+\pi/2)} = e^{-j2w} e^{-j(\pi/2)} = -je^{-j2w}. \quad (7.33)$$

Now we define $Y(e^{jw})$ as follows:

$$Y(e^{jw}) = -j\frac{d}{dw} S(e^{jw}),$$

and using (7.31) and (7.33) and differentiation in frequency property, we obtain

$$\begin{aligned} y[n] &= -ns[n] \\ &= -2n\cos\big(\frac{\pi}{3}n\big)(\frac{1}{3})^n u[n]. \end{aligned}$$

Using the time-shifting property, since $X(e^{jw}) = e^{-2jw} Y(e^{jw})$, signal $x[n]$ is obtained as

$$\begin{aligned} x[n] &= -(n-2)s[n-2] \\ &= -2(n-2)\cos\big(\frac{\pi}{3}(n-2)\big)(\frac{1}{3})^{(n-2)} u[n-2]. \end{aligned}$$

(e) $X(e^{jw}) = \dfrac{10}{-e^{-j2w} + 3e^{-jw} + 10}$

Using partial fractions, we write

$$X(e^{jw}) = \frac{1}{(1+\frac{1}{2}e^{-jw})(1-\frac{1}{5}e^{-jw})} = \frac{A}{1+\frac{1}{2}e^{-jw}} + \frac{B}{1-\frac{1}{5}e^{-jw}}, \tag{7.34}$$

where coefficients A and B are given by

$$A + B = 1, \quad -\frac{1}{5}A + \frac{1}{2}B = 0$$

$$\Rightarrow \quad A = \frac{5}{7}, \quad B = \frac{2}{7}.$$

Substituting these values into (7.34), we have

$$X(e^{jw}) = \frac{\frac{5}{7}}{1+\frac{1}{2}e^{-jw}} + \frac{\frac{2}{7}}{1-\frac{1}{5}e^{-jw}}.$$

The inverse Fourier transform of each term can be given by inspection where

$$x[n] = \frac{5}{7}(-\frac{1}{2})^n u[n] + \frac{2}{7}(\frac{1}{5})^n u[n].$$

Problem 126. Consider the discrete-time signal $x[n]$ with Fourier transform $X(e^{jw})$. Determine the Fourier transforms for each of the following cases in terms of $X(e^{jw})$.

(a) $y[n] = x^*[-n]$

(b) $y[n] = x[n] * x^*[-n]$

(c) $y[n] = x[n] - x[n-2]$

(d) $y[n] = x[n] * x[n-1]$

(e) $y[n] = \begin{cases} x[n] & n = \text{multiple of } 3 \\ 0 & n \neq \text{multiple of } 3 \end{cases}$

(f) $y[n] = x[2-n] + x[-2-n]$

(g) $y[n] = x^*[-n+2]$

Solution.

(a) $y[n] = x^*[-n]$

By definition of the Fourier transform

$$
\begin{aligned}
Y(e^{jw}) &= \sum_{n=-\infty}^{+\infty} x^*[-n]e^{-jwn} \\
&= \sum_{n=-\infty}^{+\infty} x^*[n]e^{jwn} \\
&= \Big[\sum_{n=-\infty}^{+\infty} x[n]e^{-jwn}\Big]^* = X^*(e^{jw}).
\end{aligned}
$$

(b) $y[n] = x[n] * x^*[-n]$

Using the convolution property and from the result of Part (a), we have

$$
Y(e^{jw}) = X(e^{jw})X^*(e^{jw}) = |X(e^{jw})|^2.
$$

(c) $y[n] = x[n] - x[n-2]$

Using the linearity and time shifting properties, we obtain

$$
\begin{aligned}
Y(e^{jw}) &= X(e^{jw}) - e^{-j2w}X(e^{jw}) \\
&= (1 - e^{-j2w})X(e^{jw}).
\end{aligned}
$$

(d) $y[n] = x[n] * x[n-1]$

We know that convolution in the time domain corresponds to multiplication in the frequency domain. Thus, using the convolution property, the Fourier transform of $y[n]$ is given by

$$
Y(e^{jw}) = X(e^{jw}) \cdot X(e^{jw})e^{-jw} = X^2(e^{jw})e^{-jw}.
$$

(e) $y[n] = \begin{cases} x[n] & n = \text{multiple of } 3 \\ 0 & n \neq \text{multiple of } 3 \end{cases}$

First we rewrite the relationship between $y[n]$ and $x[n]$ as follows:

$$
y[n] = x[n] \cdot \begin{cases} 1 & n = \text{multiple of } 3 \\ 0 & n \neq \text{multiple of } 3 \end{cases} = x[n] \sum_{k=-\infty}^{+\infty} \delta[n-3k]. \tag{7.35}
$$

As we see, $y[n]$ can be seen as the multiplication of $x[n]$ with an impulse train. Using the Fourier transform of the impulse train (see Table 7.2), we have

$$
\sum_{k=-\infty}^{+\infty} \delta[n-3k] \xrightarrow{\mathcal{F.T.}} \frac{2\pi}{3} \sum_{k=-\infty}^{+\infty} \delta(w - \frac{2k\pi}{3}).
$$

Using the multiplication property of the discrete-time Fourier transform, we obtain

$$Y(e^{jw}) = \frac{1}{2\pi} X(e^{jw}) \circledast \frac{2\pi}{3} \sum_{k=-\infty}^{+\infty} \delta(w - \frac{2k\pi}{3}) = \frac{1}{3} \sum_{k=-\infty}^{+\infty} X(e^{j2k\pi/3}) \delta(w - \frac{2k\pi}{3}).$$

(f) $y[n] = x[2-n] + x[-2-n]$

In order to obtain $x[2-n]$, it is easier to first perform the time shifting and then apply the time reversal.

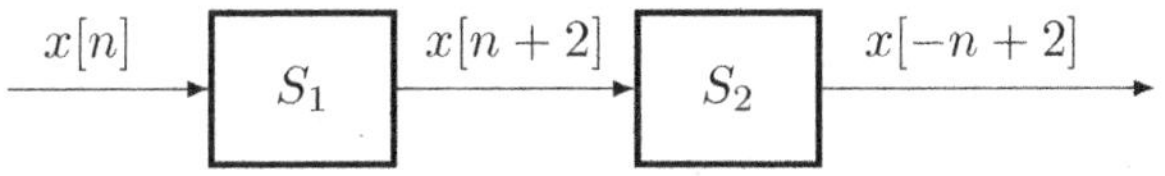

$$\begin{cases} S_1: & \text{input } \mathcal{F.T.}: Z(e^{jw}) \to \text{output } \mathcal{F.T.}: Z(e^{jw})e^{2jw} \\ S_2: & \text{input } \mathcal{F.T.}: Z(e^{jw}) \to \text{output } \mathcal{F.T.}: Z(e^{j(-w)}) \end{cases}.$$

Hence:

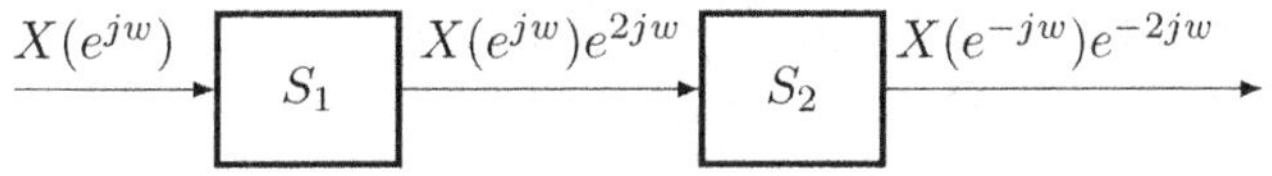

To obtain Fourier transform for $x[-2-n]$ we have

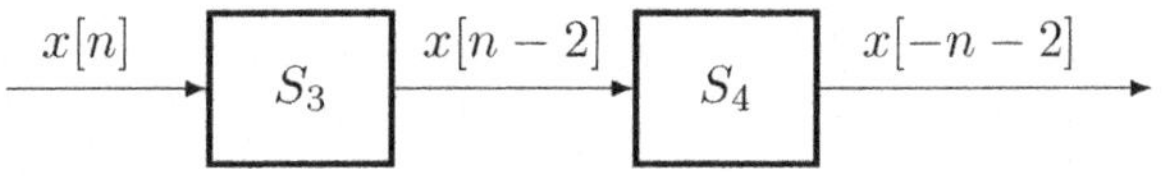

$$\begin{cases} S_3: & \text{input } \mathcal{F.T.}: Z(e^{jw}) \to \text{output } \mathcal{F.T.}: Z(e^{jw})e^{-2jw} \\ S_4: & \text{input } \mathcal{F.T.}: Z(e^{jw}) \to \text{output } \mathcal{F.T.}: Z(e^{j(-w)}) \end{cases}.$$

Hence:

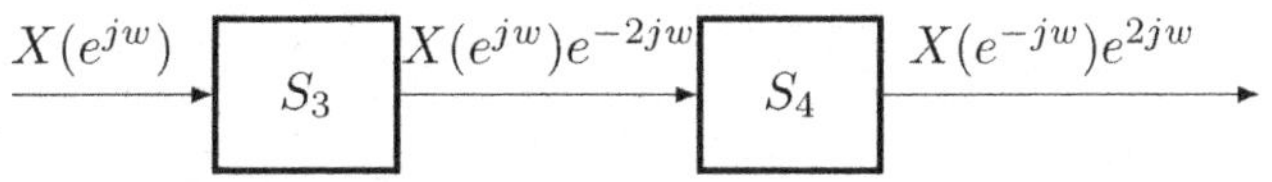

Finally, using the linearity property of the Fourier transform, $Y(e^{jw})$ is given by

$$\begin{aligned} Y(e^{jw}) &= X(e^{-jw})e^{-2jw} + X(e^{-jw})e^{2jw} \\ &= X(e^{-jw})\left[e^{-2jw} + e^{2jw}\right] \\ &= 2\cos(2w)X(e^{-jw}). \end{aligned}$$

(g) $y[n] = x^*[-n+2]$

The following block diagram denotes the steps to produce $y[n]$ from $x[n]$:

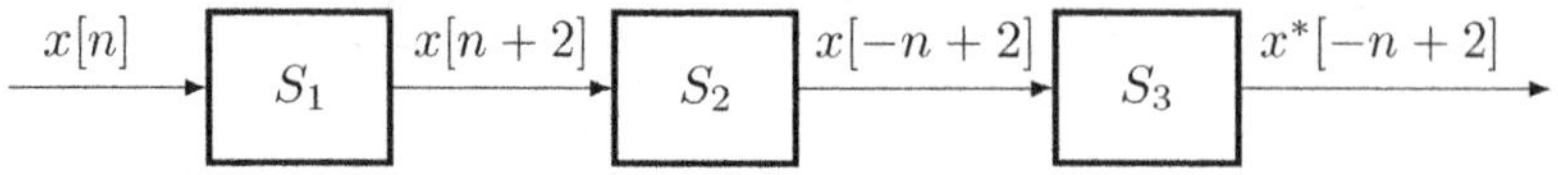

$$\begin{cases} S_1: & \text{input}\,\mathcal{F.T.}: Z(e^{jw}) \to \text{output}\,\mathcal{F.T.}:\ Z(e^{jw})e^{2jw} \\ S_2: & \text{input}\,\mathcal{F.T.}: Z(e^{jw}) \to \text{output}\,\mathcal{F.T.}:\ Z(e^{j(-w)}) \\ S_3: & \text{input}\,\mathcal{F.T.}: Z(e^{jw}) \to \text{output}\,\mathcal{F.T.}:\ Z^*(e^{-jw}) \end{cases}.$$

Hence:

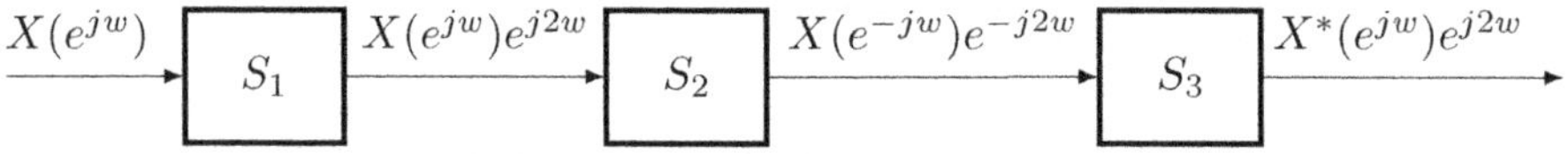

Thus we have

$$Y(e^{jw}) = X^*(e^{jw})e^{j2w}.$$

Problem 127. Consider a real signal $x[n]$ with Fourier transform $X(e^{jw})$. Determine $y[n]$ for each of the following signals based on $x[n]$.

(a) $Y(e^{jw}) = Re\{X(e^{jw})\}$

(b) $Y(e^{jw}) = X(e^{jw}) + X(e^{-jw})$

(c) $Y(e^{jw}) = \frac{d}{dw}\{e^{jw}X(e^{jw})\}$

(d) $Y(e^{jw}) = \frac{d}{dw}\{e^{-jw}[X(e^{j(w+\frac{\pi}{5})}) - X(e^{j(w-\frac{\pi}{5})})]\}$

Solution.

(a) $Y(e^{jw}) = Re\{X(e^{jw})\}$

Note that for real signals, the real part of the Fourier transform corresponds to the even part of the signal. So we have

$$y[n] = \mathcal{E}v\{x[n]\} = \frac{1}{2}x[n] + \frac{1}{2}x[-n].$$

(b) $Y(e^{jw}) = X(e^{jw}) + X(e^{-jw})$

Since $x[n]$ is real, $X(e^{jw}) = X^*(e^{-jw})$. Thus we have

$$Y(e^{jw}) = X(e^{jw}) + X(e^{-jw}) = 2Re\{X(e^{jw})\}.$$

Now based on Part (a) we obtain

$$y[n] = 2\Big(\frac{1}{2}x[n] + \frac{1}{2}x[-n]\Big) = x[n] + x[-n].$$

(c) $Y(e^{jw}) = \frac{d}{dw}\{e^{jw}X(e^{jw})\}$

Using the time shifting property, we have

$$e^{jw}X(e^{jw}) \xleftrightarrow{\mathcal{F.T.}^{-1}} x[n+1],$$

and using differentiation in frequency property, we have

$$j\frac{d}{dw}\{X(e^{jw})\} \xleftrightarrow{\mathcal{F.T.}^{-1}} nx[n].$$

Dividing both sides of the above expression by j, we can write

$$\frac{d}{dw}\{X(e^{jw})\} \xleftrightarrow{\mathcal{F.T.}^{-1}} -jnx[n].$$

Combining these two relationships, we have

$$y[n] = -jnx[n+1].$$

(d) $Y(e^{jw}) = \frac{d}{dw}\{e^{-jw}\big[X(e^{j(w+\frac{\pi}{5})}) - X(e^{j(w-\frac{\pi}{5})})\big]\}$

Let us define $G(e^{jw})$ as

$$G(e^{jw}) = X(e^{j(w+\frac{\pi}{5})}) - X(e^{j(w-\frac{\pi}{5})}). \tag{7.36}$$

Using the frequency shifting property, we have

$$X(e^{j(w+\frac{\pi}{5})}) \xleftrightarrow{\mathcal{F.T.}^{-1}} e^{jn(\pi/5)}x[n],$$
$$X(e^{j(w-\frac{\pi}{5})}) \xleftrightarrow{\mathcal{F.T.}^{-1}} e^{-jn(\pi/5)}x[n].$$

From (7.36) and using the above relationships, we obtain $g[n]$ as

$$g[n] = x[n]\big(e^{jn(\pi/5)} - e^{-jn(\pi/5)}\big). \tag{7.37}$$

Using the time shifting property, we get

$$e^{-jw}G(e^{jw}) \xleftrightarrow{\mathcal{F.T.}^{-1}} g[n-1],$$

and using differentiation in frequency property, we have

$$\frac{d}{dw}\{G(e^{jw})\} \xleftrightarrow{\mathcal{F.T.}^{-1}} -jng[n].$$

Thus from (7.37) and the above relationships, signal $y[n]$ can be written as

$$\begin{aligned} y[n] &= -jng[n-1] \\ &= -jnx[n-1]\big(e^{j(n-1)(\pi/5)} - e^{-j(n-1)(\pi/5)}\big). \end{aligned}$$

Problem 128. A causal discrete-time LTI system is described by

$$y[n] + \frac{1}{2}y[n-1] - \frac{3}{4}y[n-2] = 2x[n-2] + x[n-1] - \frac{5}{2}x[n]. \tag{7.38}$$

where $x[n]$ and $y[n]$ are the input and output of the system, respectively.

(a) Find the frequency response of the system.

(b) Determine the corresponding filter type.

Solution.

(a) Taking the Fourier transform of both sides of (7.38), we have

$$Y(e^{jw}) + \frac{1}{2}e^{-jw}Y(e^{jw}) - \frac{3}{4}e^{-j2w}Y(e^{jw}) = 2e^{-j2w}X(e^{jw}) + e^{-jw}X(e^{jw}) - \frac{5}{2}X(w).$$

So

$$H(e^{jw}) = \frac{Y(e^{jw})}{X(e^{jw})} = \frac{2e^{-j2w} + e^{-jw} - \frac{5}{2}}{1 + \frac{1}{2}e^{-jw} - \frac{3}{4}e^{-j2w}}. \tag{7.39}$$

(b) To determine the filter type, we must evaluate $|H(e^{jw})|$ in low frequency $w = 0$, intermediate frequency $w = \frac{\pi}{2}$ and high frequency $w = \pi$. We have

$$|H(0)| = \frac{\frac{1}{2}}{\frac{3}{4}} = \frac{2}{3}, \quad |H(\frac{\pi}{2})| = |\frac{-2-j-\frac{5}{2}}{1-\frac{1}{2}j+\frac{3}{4}}| = 2\sqrt{(\frac{85}{53})}, \quad |H(\pi)| = \frac{\frac{3}{2}}{\frac{1}{4}} = 6.$$

We see that $|H(w)|$ is bounded away from 0 for $w = 0$, $w = \frac{\pi}{2}$ and $w = \pi$ and thus it can not be associated with a low-pass, band-pass or high pass filter.

Problem 129. Let $x[n]$ have the Fourier transform $X(e^{jw})$. Determine $x[n]$ for each of the following cases:

(a)

$$X(e^{jw}) = \begin{cases} 1 & \frac{\pi}{4} < |w| < \frac{3\pi}{4} \\ 0 & \text{otherwise} \end{cases} \tag{7.40}$$

(b) $x[n]$ is a real and even signal and

$$\mathcal{Re}\{X(e^{jw})\} = \frac{1}{1 - \frac{1}{2}e^{-jw}}. \tag{7.41}$$

Solution.

(a) Let's define $Y(e^{jw})$ as

$$Y(e^{jw}) = \begin{cases} 1 & \frac{-\pi}{4} < |w| < \frac{\pi}{4} \\ 0 & \text{otherwise} \end{cases}. \tag{7.42}$$

$Y(e^{jw})$ represents the pulse $\prod\left(\frac{w}{\frac{\pi}{2}}\right)$ in frequency domain where $X(e^{jw})$ is obtained by shifting $Y(e^{jw})$ to left and right by $\frac{\pi}{2}$ and as such,

$$X(e^{jw}) = Y(e^{j(w-\frac{\pi}{2})}) + Y(e^{j(w+\frac{\pi}{2})}).$$

On the other hand,

$$y[n] = \frac{\sin(\pi n/4)}{\pi n}.$$

Using time and frequency shifting properties

$$\begin{aligned} x[n] &= e^{-j\frac{\pi n}{2}} \frac{\sin(\pi n/4)}{\pi n} + e^{j\frac{\pi n}{2}} \frac{\sin(\pi n/4)}{\pi n} \\ &= 2\cos(\frac{n\pi}{2})\frac{\sin(\pi n/4)}{\pi n}. \end{aligned}$$

(b) We have

$$\mathcal{R}e\{X(e^{jw})\} = \frac{1}{1-\frac{1}{2}e^{-jw}}$$

Since $x[n]$ is real, based on the symmetry property of the Fourier transform we have

$$\mathcal{E}v\{x[n]\} \xrightarrow{\mathcal{F.T.}} \mathcal{R}e\{X(e^{jw})\}.$$

It follows that

$$\mathcal{E}v\{x[n]\} = (\frac{1}{2})^n u[n],$$

and since $x[n]$ is even, we have

$$x[n] = \mathcal{E}v\{x[n]\}.$$

Problem 130. Consider the following system with input $x[n]$ and output $y[n]$ as below. Determine whether each system can be LTI or not and if the system is LTI find the frequency response of the system.

(a) $x[n] = (\frac{1}{4})^n u[n], \quad y[n] = 4(\frac{1}{3})^n u[n].$

(b) $x[n] = \frac{\sin(n\pi/8)}{n\pi}, \quad y[n] = \frac{\sin(n\pi/4)}{n\pi}.$

Solution.

(a) The Fourier transform of $x[n]$ and $y[n]$ are

$$X(e^{jw}) = \frac{1}{1-\frac{1}{4}e^{-jw}}, \qquad Y(e^{jw}) = \frac{4}{1-\frac{1}{3}e^{-jw}}.$$

Since the Fourier transform of $x[n]$ is not zero for all values of w, then the system is LTI and also the frequency response of this system is unique. The frequency response of this system is given by

$$H(e^{jw}) = \frac{Y(e^{jw})}{X(e^{jw})} = \frac{12-3e^{-jw}}{3-e^{-jw}}.$$

(b) We know that

$$\frac{\sin(Wn)}{n\pi} \xrightarrow{\mathcal{F.T.}} \mathrm{rect}(\frac{w}{2W}).$$

Thus the Fourier transforms of $x[n]$ and $y[n]$ are

$$X(e^{jw}) = \prod(\frac{4w}{\pi}), \qquad Y(e^{jw}) = \prod(\frac{2w}{\pi}).$$

Now if the system is LTI, there exists a frequency response $H(e^{jw})$ such that $Y(e^{jw}) = H(e^{jw})X(e^{jw})$ for all values of w. However note that $X(e^{jw})$ is zero for $|w| > \frac{\pi}{8}$ but $Y(e^{jw})$ is not zero in part of this range. As such, the relation $Y(e^{jw}) = H(e^{jw})X(e^{jw})$ does not hold for this range of w and so the system can not be LTI.

Problem 131. Let $X(e^{jw})$ denote the Fourier transform of the signal $x[n]$ depicted in Fig. 7.8.

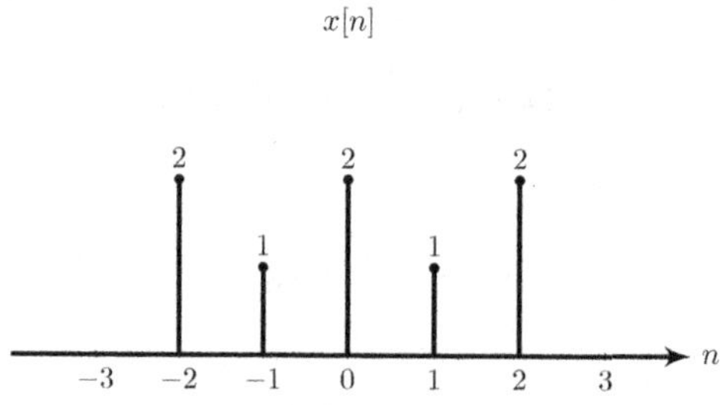

Figure 7.8: Problem 131

(a) Find $X(e^{j0})$.

(b) Find $X(e^{j\pi})$.

(c) Find $\angle X(e^{jw})$.

(d) Find $\int_{-\pi}^{\pi} X(e^{jw})dw$.

(e) Evaluate $\int_{-\pi}^{\pi} |X(e^{jw})|^2 dw$.

(f) Evaluate $\int_{-\pi}^{\pi} |\frac{d}{dw}X(e^{jw})|^2 dw$.

Solution.

(a) By definition of the Fourier transform

$$X(e^{jw}) = \sum_{n=-\infty}^{+\infty} x[n]e^{-jwn}.$$

Setting $w = 0$ in the above equation, and from Fig. 7.8, we have

$$X(e^{j0}) = \sum_{n=-\infty}^{+\infty} x[n] = 2 + 1 + 2 + 1 + 2 = 8.$$

(b) By definition of the Fourier transform and setting $w = \pi$, we have

$$\begin{aligned} X(e^{j\pi}) &= \sum_{n=-\infty}^{+\infty} x[n]e^{-j\pi n} \\ &= \sum_{n=-\infty}^{+\infty} x[n](-1)^n \\ &= 2 - 1 + 2 - 1 + 2 = 4. \end{aligned}$$

(c) From Fig. 7.8, $x[n]$ is real and even. Using the symmetry property for real signals, $X(e^{jw})$ must also be real and even. Thus we have

$$\angle X(e^{jw}) = 0.$$

(d) By the inverse Fourier transform definition we have

$$x[n] = \frac{1}{2\pi}\int_{-\pi}^{\pi} X(e^{jw})e^{jwn}dw.$$

Setting $n = 0$, we get

$$\int_{-\pi}^{\pi} X(e^{jw})dw = 2\pi x[0] = 4\pi.$$

(e) Using the Parseval's identity, we have

$$\sum_{n=-\infty}^{+\infty} |x[n]|^2 = \frac{1}{2\pi}\int_{-\pi}^{\pi} |X(e^{jw})|^2 dw.$$

Thus

$$\int_{-\pi}^{\pi} |X(e^{jw})|^2 dw = 2\pi \sum_{n=-\infty}^{+\infty} |x[n]|^2$$
$$= 2\pi(4+1+4+1+4) = 28\pi.$$

(f) According to the frequency differentiation property, we have

$$nx[n] \xrightarrow{\mathcal{F.T.}} j\frac{d}{dw}X(e^{jw}).$$

Now we can write the Parseval's relation for signal $nx[n]$ as

$$\sum_{n=-\infty}^{+\infty} |nx[n]|^2 = \frac{1}{2\pi}\int_{-\pi}^{\pi} |j\frac{d}{dw}X(e^{jw})|^2 dw$$
$$= \frac{1}{2\pi}\int_{-\pi}^{\pi} |\frac{d}{dw}X(e^{jw})|^2 dw.$$

Thus

$$\int_{-\pi}^{\pi} |\frac{d}{dw}X(e^{jw})|^2 dw = 2\pi \sum_{n=-\infty}^{+\infty} |nx[n]|^2$$
$$= 2\pi((-2\times 2)^2 + (-1\times 1)^2 + (0\times 2)^2 + (2\times 2)^2 + (1\times 1)^2) = 68\pi.$$

Problem 132. Consider an LTI system with input $x[n]$ and output $y[n]$. Find the frequency response and the impulse response of the systems corresponding to each of the following cases.

(a) $x[n] = (\frac{1}{3})^n u[n], \quad y[n] = \frac{1}{2}(\frac{1}{3})^n u[n] + (\frac{1}{4})^n u[n].$

(b) $x[n] = (\frac{1}{3})^n u[n], \quad y[n] = (\frac{1}{3})^n u[n] + (\frac{1}{3})^{n-2} u[n-2].$

Solution.

(a) We know that

$$a^n u[n] \xrightarrow{\mathcal{F.T.}} \frac{1}{1-ae^{-jw}},$$

Hence, the Fourier transforms of $x[n]$ and $y[n]$ are given as

$$X(e^{jw}) = \frac{1}{1-\frac{1}{3}e^{-jw}},$$

$$Y(e^{jw}) = \frac{1}{2}\frac{1}{1-\frac{1}{3}e^{-jw}} + \frac{1}{1-\frac{1}{4}e^{-jw}}.$$

Then, the frequency response $H(e^{jw})$ is obtained as

$$H(e^{jw}) = \frac{Y(e^{jw})}{X(e^{jw})}$$
$$= \frac{1}{2} + \frac{1-\frac{1}{3}e^{-jw}}{1-\frac{1}{4}e^{-jw}}$$
$$= \frac{1}{2} + \frac{1}{1-\frac{1}{4}e^{-jw}} - \frac{\frac{1}{3}e^{-jw}}{1-\frac{1}{4}e^{-jw}},$$

and the impulse response of the system is obtained as

$$h[n] = \frac{1}{2}\delta[n] + (\frac{1}{4})^n u[n] - \frac{1}{3}(\frac{1}{4})^{n-1}u[n-1].$$

(b) The Fourier transforms of $x[n]$ and $y[n]$ are obtained as

$$X(e^{jw}) = \frac{1}{1-\frac{1}{3}e^{-jw}},$$

$$Y(e^{jw}) = \frac{1}{1-\frac{1}{3}e^{-jw}} + \frac{e^{-j2w}}{1-\frac{1}{3}e^{-jw}}.$$

As such, the frequency response $H(e^{jw})$ is

$$H(e^{jw}) = \frac{Y(e^{jw})}{X(e^{jw})} = 1 + e^{-2jw},$$

and the impulse response of the system is

$$h[n] = \delta[n] + \delta[n-2].$$

Problem 133. Consider the causal LTI system described by the following difference equation. Determine the difference equation that characterizes the inverse system, both directly and using the Fourier transform.

$$y[n] + \frac{1}{2}y[n-1] = x[n]. \tag{7.43}$$

Solution.

(1) Direct Method: We can simply swap $x[n]$ and $y[n]$ and obtain

$$x[n] + \frac{1}{2}x[n-1] = y[n].$$

(2) Fourier transform: By taking the Fourier transform of both side of (7.43), we have

$$Y(e^{jw}) + \frac{1}{2}e^{-jw}Y(e^{jw}) = X(e^{jw}),$$

and then

$$H(e^{jw}) = \frac{Y(e^{jw})}{X(e^{jw})} = \frac{1}{1+\frac{1}{2}e^{-jw}}.$$

The frequency response of the inverse system is

$$H_{inv}(e^{jw}) = \frac{1}{H(e^{jw})} = 1 + \frac{1}{2}e^{-jw},$$

and thus

$$Y(e^{jw}) = (1 + \frac{1}{2}e^{-jw})X(e^{jw}). \tag{7.44}$$

Finally by taking the inverse Fourier transform of both side of (7.44), we obtain the difference equation of the inverse system as

$$y[n] = x[n] + \frac{1}{2}x[n-1].$$

Problem 134. For each of the following frequency responses, determine a difference equation that relates the output $y[n]$ to the input $x[n]$.

(a) $H(e^{jw}) = 1 + \frac{7e^{-jw}}{(1-e^{-jw})(3+e^{-jw})}$

(b) $H(e^{jw}) = \frac{e^{j2w}}{1+\cos(w)}$

Solution.

(a) $H(e^{jw}) = 1 + \frac{7e^{-jw}}{(1-e^{-jw})(3+e^{-jw})}$

First we rewrite the frequency response as follows

$$H(e^{jw}) = \frac{3+5e^{-jw}-e^{-j2w}}{3-2e^{-jw}-e^{-j2w}} = \frac{Y(e^{jw})}{X(e^{jw})}.$$

Then we have

$$(3-2e^{-jw}-e^{-j2w})Y(e^{jw}) = (3+5e^{-jw}-e^{-j2w})X(e^{jw}).$$

Now by taking the inverse Fourier transform of both sides of the above equation, we obtain

$$3y[n] - 2y[n-1] - y[n-2] = 3x[n] + 5x[n-1] - x[n-2].$$

(b) $H(e^{jw}) = \frac{e^{2jw}}{1+\cos(w)}$

Using the Euler's formula, $H(e^{jw})$ can be rewritten as

$$H(e^{jw}) = \frac{e^{2jw}}{1 + \frac{1}{2}e^{-jw} + \frac{1}{2}e^{jw}} = \frac{2}{2e^{-2jw} + e^{-3jw} + e^{-jw}} = \frac{Y(e^{jw})}{X(e^{jw})}.$$

Then we have

$$\left(e^{-jw} + 2e^{-2jw} + e^{-j3w}\right)Y(e^{jw}) = 2X(e^{jw}).$$

Now by taking the inverse Fourier transform of both sides of the above equation, we obtain

$$y[n-1] + 2y[n-2] + y[n-3] = 2x[n].$$

Problem 135. Consider a discrete-time system with input $x[n]$ and output $y[n]$. The Fourier transforms of these signals are related through the following equation:

$$Y(e^{jw}) = X(e^{jw}) + 2e^{-2jw}X(e^{jw}) - \frac{d}{dw}X(e^{jw}). \tag{7.45}$$

(a) Is the system linear?

(b) Is the system time-invariant?

(c) Find the output of this system for the input $x[n] = \delta[n]$.

Solution.

(a) First we investigate homogeneity property, i.e., if the input is $\alpha x[n]$, what happens to the output. In this case we have:

$$x_1[n] = \alpha x[n] \quad \Rightarrow \quad X_1(e^{jw}) = \alpha X(e^{jw})$$

From (7.45), we have

$$\begin{aligned}
Y_1(e^{jw}) &= X_1(e^{jw}) + 2e^{-2jw}X_1(e^{jw}) - \frac{d}{dw}X_1(e^{jw}) \\
&= \alpha X(e^{jw}) + 2e^{-2jw}\alpha X(e^{jw}) - \frac{d}{dw}\alpha X(e^{jw}) \\
&= \alpha X(e^{jw}) + 2e^{-2jw}\alpha X(e^{jw}) - \alpha\frac{d}{dw}X(e^{jw}) \\
&= \alpha\left(X(e^{jw}) + 2e^{-2jw}X(e^{jw}) - \frac{d}{dw}X(e^{jw})\right) \\
&= \alpha Y(e^{jw}).
\end{aligned}$$

Thus

$$y_1[n] = \alpha y[n].$$

The system satisfies the homogeneity property. To investigate the additivity property, we let

$$x_3[n] = x_1[n] + x_2[n] \quad \Rightarrow \quad X_3(e^{jw}) = X_1(e^{jw}) + X_2(e^{jw})$$

From (7.45), we have

$$\begin{aligned}
Y_3(e^{jw}) &= X_3(e^{jw}) + 2e^{-2jw}X_3(e^{jw}) - \frac{d}{dw}X_3(e^{jw}) \\
&= \left(X_1(e^{jw}) + X_2(e^{jw})\right) + 2e^{-2jw}\left(X_1(e^{jw}) + X_2(e^{jw})\right) - \frac{d}{dw}\left(X_1(e^{jw}) + X_2(e^{jw})\right) \\
&= X_1(e^{jw}) + 2e^{-2jw}X_1(e^{jw}) - \frac{d}{dw}X_1(e^{jw}) + X_2(e^{jw}) + 2e^{-2jw}X_2(e^{jw}) - \frac{d}{dw}X_2(e^{jw}) \\
&= Y_1(e^{jw}) + Y_2(e^{jw}).
\end{aligned}$$

Thus

$$y_3[n] = y_1[n] + y_2[n].$$

The system has additivity property too. Therefore the system is linear.

(b) From Chapter 2 we know that a system is time-invariant or shift invariant if and only if

$$x[n] \xrightarrow{T} y[n]$$

implies that

$$x[n-k] \xrightarrow{T} y[n-k]$$

for every input signal $x[n]$ and k.

We claim the system is not time-invariant. To prove this, we set the input $x_1[n]$ as follows:

$$x_1[n] = x[n-1] \quad \xrightarrow{\mathcal{F.T.}} \quad X_1(e^{jw}) = e^{-jw}X(e^{jw})$$

and we must have

$$y_1[n] = y[n-1] \quad \xrightarrow{\mathcal{F.T.}} \quad Y_1(e^{jw}) = e^{-jw}Y(e^{jw})$$

From (7.45), the Fourier transforms of the output is obtained as

$$\begin{aligned}
Y_1(e^{jw}) &= X_1(e^{jw}) + 2e^{-2jw}X_1(e^{jw}) - \frac{d}{dw}X_1(e^{jw}) \\
&= e^{-jw}X(e^{jw}) + 2e^{-j3w}X(e^{jw}) - \frac{d}{dw}e^{-jw}X(e^{jw}) \\
&= e^{-jw}X(e^{jw}) + 2e^{-j3w}X(e^{jw}) + je^{-jw}X(e^{jw}) - e^{-jw}\frac{d}{dw}X(e^{jw}) \\
&= e^{-jw}\left(X(e^{jw}) + 2e^{-2jw}X(e^{jw}) - \frac{d}{dw}X(e^{jw})\right) + je^{-jw}X(e^{jw}) \\
&= e^{-jw}Y(e^{jw}) + je^{-jw}X(e^{jw}).
\end{aligned}$$

We see that

$$Y_1(e^{jw}) \neq e^{-jw}Y(e^{jw}),$$

and thus this system is not time-invariant.

(c) We have

$$x[n] = \delta[n], \Rightarrow X(e^{jw}) = 1.$$

Setting $X(e^{jw}) = 1$ in (7.45), we obtain

$$Y(e^{jw}) = 1 + 2e^{-2jw},$$

and by taking the inverse Fourier transform of $Y(e^{jw})$, we have

$$y[n] = \delta[n] + 2\delta[n-2].$$

Problem 136. Consider a discrete-time signal $x[n]$ with Fourier transform as illustrated in Fig. 7.9 and let

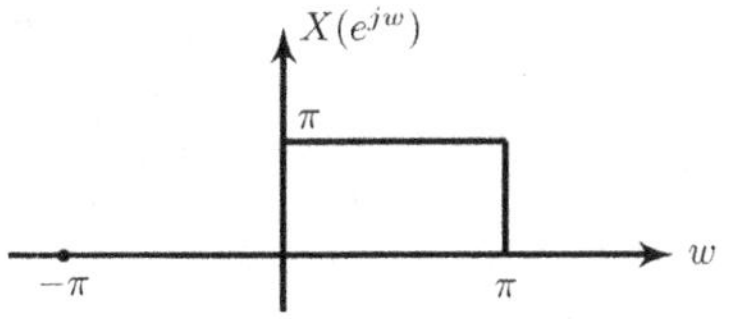

Figure 7.9: Problem 136.

$$A = \sum_{n=-\infty}^{+\infty} jIm\{x[n]\}e^{jn\pi/2}. \tag{7.46}$$

$$h(t) = \sum_{n=-\infty}^{+\infty} 2x[-n]e^{jnt}. \tag{7.47}$$

(a) Evaluate A.

(b) Evaluate $B = \int_0^{\pi/4} h(t)dt$.

Solution.

(a) From (7.46), we have

$$A = \sum_{n=-\infty}^{+\infty} jIm\{x[n]\}e^{jn(\pi/2)}. \tag{7.48}$$

Based on the definition, the discrete-time Fourier transform of signal $jIm\{x[n]\}$ can be written as

$$\mathcal{F.T.}\{jIm\{x[n]\}\} = \sum_{n=-\infty}^{+\infty} jIm\{x[n]\}e^{-jnw}. \tag{7.49}$$

Comparing (7.48) and (7.49), we see that the value of A is equal to the Fourier transform of $jIm\{x[n]\}$ at $w = -\frac{\pi}{2}$:

$$A = \sum_{n=-\infty}^{+\infty} jIm\{x[n]\}e^{jn(\pi/2)} = \mathcal{F.T.}\{jIm\{x[n]\}\}\Big|_{w=-\frac{\pi}{2}}.$$

We know that $jIm\{x[n]\}$ can be written as

$$jIm\{x[n]\} = \frac{x[n] - x^*[n]}{2}.$$

Using the linearity and conjugation properties of the discrete-time Fourier transform, the Fourier transform of $jIm\{x[n]\}$ can be expanded as

$$\begin{aligned} \mathcal{F.T.}\{jIm\{x[n]\}\} &= \mathcal{F.T.}\{\frac{x[n] - x^*[n]}{2}\} \\ &= \mathcal{F.T.}\{\frac{x[n]}{2}\} - \mathcal{F.T.}\{\frac{x^*[n]}{2}\} \\ &= \frac{1}{2}X(e^{jw}) - \frac{1}{2}X^*(e^{-jw}). \end{aligned} \tag{7.50}$$

From Fig. 7.9 and (7.50), we obtain

$$A = \mathcal{F.T.}\{jIm\{x[n]\}\}\Big|_{w=-\frac{\pi}{2}} = \frac{1}{2}X(-\frac{\pi}{2}) - \frac{1}{2}X^*(\frac{\pi}{2}) = -\frac{\pi}{2}.$$

In the above derivation, we used the fact that $X(w)$ is real based on the figure and so $X^*(w) = X(w)$.

(b) $B = \int_0^{\pi/4} h(t)dt$

Based on the definition, the discrete-time Fourier transform of the signal $x[-n]$ is given as

$$\mathcal{F.T.}\{x[-n]\} = \sum_{n=-\infty}^{+\infty} x[-n]e^{-jnw}.$$

Comparing (7.47) and the above equation, we see that $h(t)$ is equal to the Fourier transform of $x[n]$ at $w = -t$:

$$h(t) = \sum_{n=-\infty}^{+\infty} 2x[-n]e^{jnt} = 2 \times \mathcal{F.T.}\{x[-n]\}\big|_{w=-t}.$$

From the time reversal property of the discrete-time Fourier transform, we know that the Fourier transform of $x[-n]$ is equal to $X(e^{-jw})$. Thus, the signal $h(t)$ can be written as

$$h(t) = X(e^{-jw})\big|_{w=-t} = 2X(e^{jt}).$$

Therefore we obtain

$$B = \int_0^{\pi/4} h(t)dt = \int_0^{\pi/4} 2X(e^{jt})dt,$$

By setting the variable of integration to $w = t$, we obtain

$$B = 2\int_0^{\pi/4} X(e^{jw})dw = \frac{\pi^2}{2}.$$

Problem 137. Signal $x[n]$ has Fourier transform $X(e^{jw})$ as illustrated in Fig. 7.10. Obtain the following values:

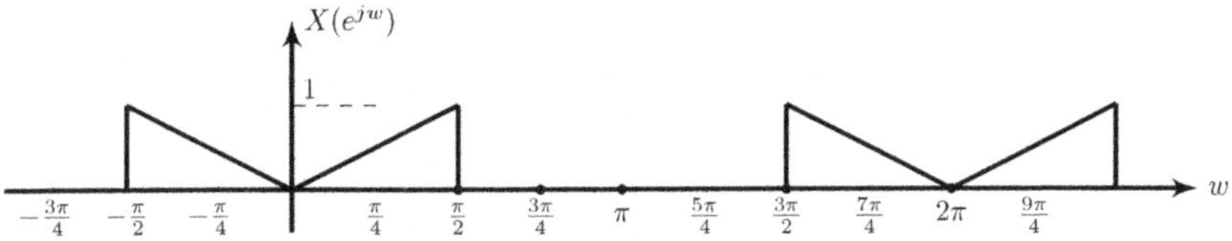

Figure 7.10: Problem 137.

(a) $x[0]$

(b) $\sum_{n=-\infty}^{+\infty} x[n]$

(c) $\sum_{n=-\infty}^{+\infty} |x[n]|^2$

Solution.

(a) $x[0]$

By the inverse Fourier transform definition

$$x[n] = \frac{1}{2\pi}\int_{2\pi} X(e^{jw})e^{jwn}dw.$$

Setting $n = 0$ we get

$$x[0] = \frac{1}{2\pi}\int_{2\pi} X(e^{jw})dw = \frac{1}{2\pi}[2 \times \frac{\pi}{4}] = \frac{1}{4}.$$

(b) $\sum_{n=-\infty}^{+\infty} x[n]$

By the Fourier transform definition,

$$X(e^{jw}) = \sum_{n=-\infty}^{+\infty} x[n]e^{-jwn}.$$

From Fig. 7.10 and by setting $w = 0$ in the above equation, we get

$$\sum_{n=-\infty}^{+\infty} x[n] = X(e^{j0}) = 0.$$

(c) $\sum_{n=-\infty}^{+\infty} |x[n]|^2$

Using Parseval's identity, we have

$$\sum_{n=-\infty}^{+\infty} |x[n]|^2 = \frac{1}{2\pi} \int_{-\pi}^{\pi} |X(e^{jw})|^2 dw.$$

Thus

$$\begin{aligned}\sum_{n=-\infty}^{+\infty} |x[n]|^2 &= \frac{1}{2\pi} \int_{-\pi}^{\pi} |X(e^{jw})|^2 dw \\ &= \frac{1}{2\pi} \times 2 \int_{0}^{\pi/2} (\frac{2}{\pi} w)^2 dw \\ &= \frac{1}{\pi} \times \frac{4}{\pi^2} [\frac{w^3}{3}]_0^{\frac{\pi}{2}} \\ &= \frac{1}{6}.\end{aligned}$$

Problem 138. Consider a system whose input $x[n]$ and output $y[n]$ are related with the following relationship:

$$y[n] = (-1)^n w[n] + w[n],$$
$$w[n] = x[n] * h[n],$$

where the frequency response of the system $H(e^{jw})$ is sketched in Fig. 7.11.

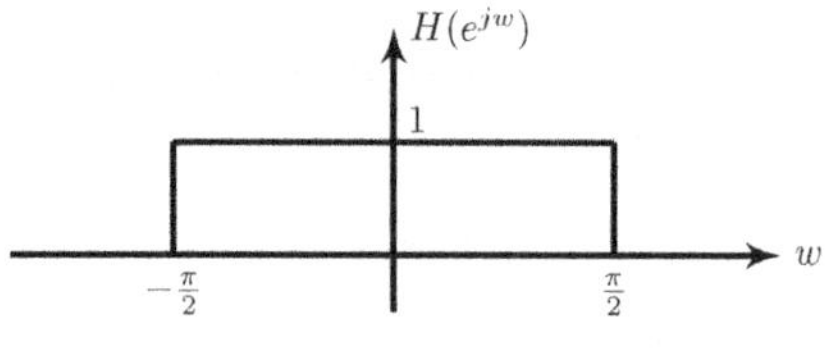

Figure 7.11: Problem 138, The frequency response $H(e^{jw})$.

Find the output of the system to the input

$$x[n] = \delta[n].$$

Solution. We have

$$w[n] = x[n] * h[n]. \tag{7.51}$$

As such, the input-output relation of the system is given by

$$y[n] = (-1)^n w[n] + w[n].$$

Setting $x[n] = \delta[n]$ in (7.51), we have

$$y[n] = (-1)^n h[n] + h[n].$$

Substituting $(-1)^n = e^{jn\pi}$ in the above equation, we obtain

$$y[n] = e^{jn\pi} h[n] + h[n]. \tag{7.52}$$

Taking the Fourier transform of both sides of (7.52) and using the frequency-shifting property, we get

$$Y(e^{jw}) = H(e^{j(w-\pi)}) + H(e^{jw}).$$

Having in mind the fact that the frequency response $H(e^{jw})$ is periodic with period 2π, we comp up with the following figures for $H(e^{jw})$ and its shifted version:

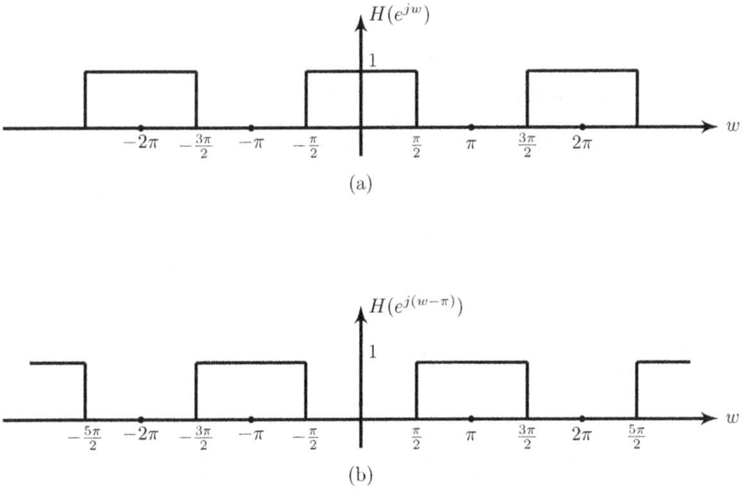

Figure 7.12: Problem 138, (a) The frequency response $H(e^{jw})$, (b) $H(e^{j(w-\pi)})$.

From Fig. 7.12 we see that

$$Y(e^{jw}) = H(e^{j(w-\pi)}) + H(e^{jw}) = 1,$$

and thus

$$y[n] = \delta[n].$$

Problem 139. Consider a discrete-time system with a real impulse response $h[n]$. Let $g[n]$ be another system where $G(e^{jw}) = H(e^{jw})H(e^{-jw})$. Is $g[n]$ real?

Solution.

To show that the system impulse response is real, we must have

$$G(e^{jw}) = G^*(e^{-jw}).$$

As we know $h[n]$ is real and as such

$$H(e^{jw}) = H^*(e^{-jw}),$$

and

$$H(e^{-jw}) = H^*(e^{jw}).$$

Thus, we have

$$\begin{aligned} G^*(e^{jw}) &= \left[H(e^{jw})H(e^{-jw})\right]^* \\ &= H^*(e^{jw})H^*(e^{-jw}) \\ &= H(e^{-jw})H(e^{jw}) \\ &= G(e^{jw}). \end{aligned}$$

On the other hand,

$$G(e^{-jw}) = H(e^{-jw})H(e^{jw}) = G(e^{jw}),$$

i.e., $G(e^{jw})$ is even and thus we have

$$G^*(e^{jw}) = G(e^{jw}) = G(e^{-jw}),$$

or alternatively,

$$G(e^{jw}) = G^*(e^{-jw}).$$

Thus the impulse response $g[n]$ is also real.

Problem 140. Consider a discrete-time system shown in Fig. 7.13,

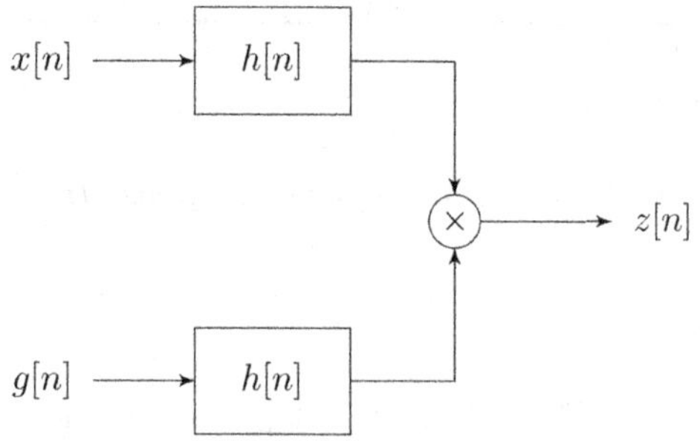

Figure 7.13: Problem 140

where

$$x[n] = 1 + \cos(\frac{\pi}{5}n) + \cos(\frac{3\pi}{5}n). \tag{7.53}$$

$$g[n] = 1 + 2\cos(\frac{2\pi}{5}n), \tag{7.54}$$

$$h[n] = \frac{2\sin(\frac{\pi}{4}n)}{n\pi}\cos(\frac{\pi}{3}n). \tag{7.55}$$

Find and sketch $Z(e^{jw})$ and $z[n]$.

Solution.

The Fourier transforms of $x[n]$ and $g[n]$ are readily obtained as

$$X(e^{jw}) = 2\pi\delta(w) + \pi\delta(w + \frac{\pi}{5}) + \pi\delta(w - \frac{\pi}{5}) + \pi\delta(w + \frac{3\pi}{5}) + \pi\delta(w - \frac{3\pi}{5}),$$

$$G(e^{jw}) = 2\pi\delta(w) + 2\pi\delta(w + \frac{2\pi}{5}) + 2\pi\delta(w - \frac{2\pi}{5}).$$

which are depicted in Fig. 7.14,

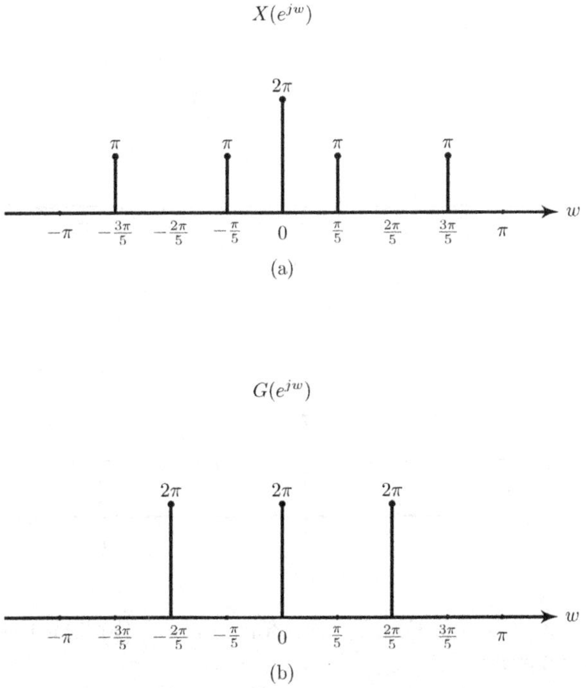

Figure 7.14: Problem 140, (a) the Fourier transform of $x[n]$ (b) the Fourier transform of $g[n]$.

Let us define $v[n]$ and $q[n]$ as the outputs resulting from $x[n]$ and $g[n]$, respectively. Then, we have

$$v[n] = x[n] * h[n], \qquad q[n] = g[n] * h[n].$$

Applying the convolution property of the discrete-time Fourier transform, we obtain

$$V(e^{jw}) = X(e^{jw})H(e^{jw}),$$

$$Q(e^{jw}) = G(e^{jw})H(e^{jw}),$$

where $V(e^{jw})$, $Q(e^{jw})$, and $H(e^{jw})$ are the Fourier transforms of $v[n]$, $q[n]$, and $h[n]$, respectively, depicted in Fig. 7.15.

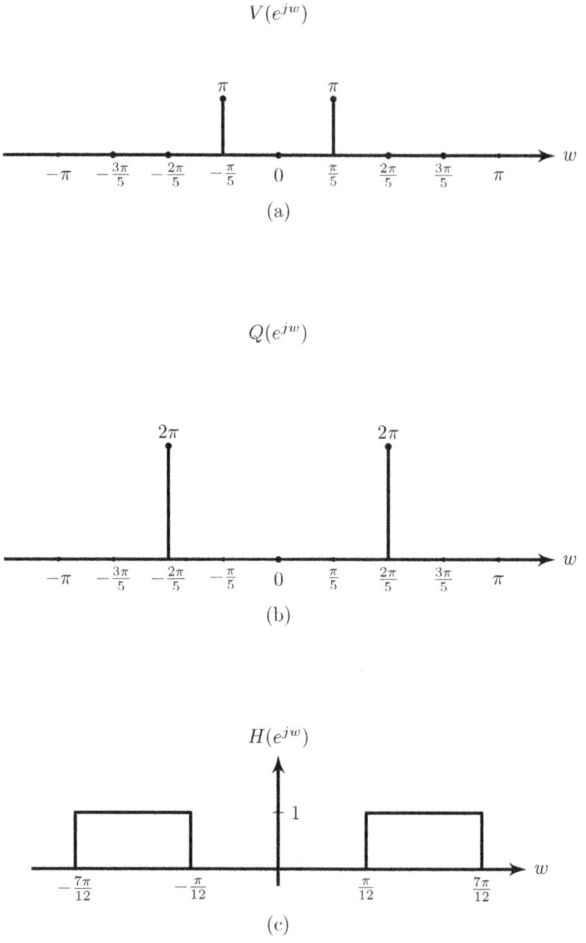

Figure 7.15: Problem 140, (a) The Fourier transform of $v[n]$ (b) The Fourier transform of $q[n]$ (c) The frequency response $H(e^{jw})$.

From Fig. 7.13, we see that

$$z[n] = v[n]q[n],$$

and using the multiplication property of the discrete-time Fourier transform, we obtain

$$Z(e^{jw}) = \frac{1}{2\pi} V(e^{jw}) * Q(e^{jw}).$$

From Fig. 7.15 (b), we see that $Q(e^{jw})$ is composed of two impulses at $w = \frac{2\pi}{5}$ and $w = -\frac{2\pi}{5}$. As we know, convolving a signal with a shifted impulse is the same as shifting the signal. Thus we obtain

$$\begin{aligned} Z(e^{jw}) &= \frac{1}{2\pi} V(e^{jw}) * 2\pi\delta(w - \frac{2\pi}{5}) + \frac{1}{2\pi} V(e^{jw}) * 2\pi\delta(w + \frac{2\pi}{5}) \\ &= V(e^{j(w-2\pi/5)}) + V(e^{j(w+2\pi/5)}). \end{aligned}$$

which is depicted in Fig. 7.16.

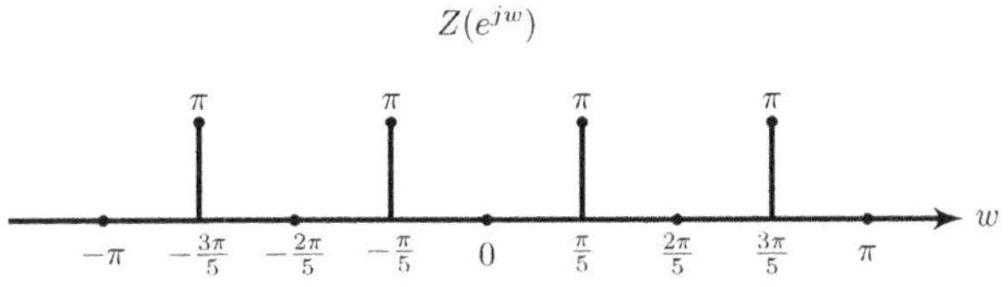

Figure 7.16: Problem 140, the Fourier transform of $z[n]$.

Thus,

$$z[n] = \cos(\frac{\pi}{5}n) + \cos(\frac{3\pi}{5}n).$$

Problem 141. Evaluate each of the following integrals and sums.

(a) $\int_{-\pi}^{\pi} \frac{e^{jw}}{1-0.5e^{-jw}} dw$

(b) $\sum_{n=-\infty}^{+\infty} \frac{\sin(n\frac{\pi}{4})}{n\pi} \cdot \frac{\sin(n\frac{\pi}{6})}{n\pi}$

(c) $\sum_{n=0}^{+\infty} n(\frac{1}{n})^n$

Solution.

(a) $\int_{-\pi}^{\pi} \frac{e^{jw}}{1-0.5e^{-jw}} dw$

We know that

$$(0.5)^n u[n] \xrightarrow{\mathcal{F.T.}} \frac{1}{1-0.5e^{-jw}}.$$

By the inverse Fourier transform definition, we have

$$x[n] = \frac{1}{2\pi}\int_{-\pi}^{\pi} X(e^{jw})e^{jwn}dw.$$

Thus

$$\begin{aligned}(0.5)^n u[n] &= \frac{1}{2\pi}\int_{-\pi}^{\pi} X(e^{jw})e^{jwn}dw \\ &= \frac{1}{2\pi}\int_{-\pi}^{\pi} \frac{1}{1-0.5e^{-jw}}e^{jwn}dw. \qquad (7.56)\end{aligned}$$

Setting $n = 1$ in (7.56), we obtain

$$\int_{-\pi}^{\pi} \frac{1}{1-0.5e^{-jw}}e^{jw}dw = 2\pi(0.5)^1 u[1] = \pi.$$

(b) $\sum_{n=-\infty}^{+\infty} \frac{\sin(n\frac{\pi}{4})}{n\pi} \cdot \frac{\sin(n\frac{\pi}{6})}{n\pi}$

Using Parseval's identity we have

$$\sum_{n=-\infty}^{+\infty} |x[n]|^2 = \frac{1}{2\pi}\int_{-\pi}^{\pi} |X(e^{jw})|^2 dw.$$

Thus

$$\sum_{n=-\infty}^{+\infty} x[n]y^*[n] = \frac{1}{2\pi}\int_{-\pi}^{\pi} X(e^{jw})Y^*(e^{jw})dw. \tag{7.57}$$

Now if we define

$$x[n] = \frac{\sin(n\frac{\pi}{4})}{n\pi}, \qquad y[n] = \frac{\sin(n\frac{\pi}{6})}{n\pi},$$

and knowing that

$$\frac{\sin(Wn)}{n\pi} \xrightarrow{\mathcal{F.T.}} \prod(\frac{w}{2W}),$$

we get

$$X(e^{jw}) = \prod(\frac{2w}{\pi}),$$
$$Y(e^{jw}) = \prod(\frac{3w}{\pi}).$$

which are depicted in Fig. 7.17.

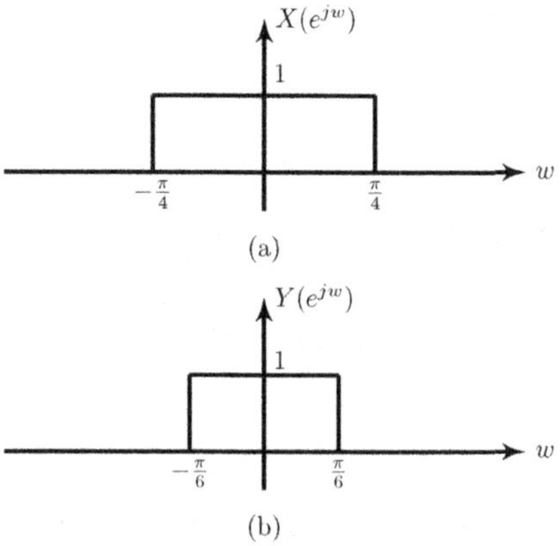

Figure 7.17: Problem 141, (a) The Fourier transform of $x[n]$ (b) The Fourier transform of $y[n]$.

From (7.57) and Fig. 7.17, the sum is evaluated as

$$\sum_{n=-\infty}^{+\infty} \frac{\sin(n\frac{\pi}{4})}{n\pi}\cdot\frac{\sin(n\frac{\pi}{6})}{n\pi} = \sum_{n=-\infty}^{+\infty} x[n]y^*[n]$$
$$= \frac{1}{2\pi}\int_{-\pi}^{\pi} \prod(\frac{2w}{\pi})\prod(\frac{3w}{\pi})dw$$

$$= \frac{1}{2\pi}\int_{-\pi/6}^{\pi/6} dw = \frac{1}{6}.$$

(c) $\sum_{n=0}^{+\infty} n(\frac{1}{n})^n$

We can use the frequency differentiation property, together with the Fourier transform pair

$$(\frac{1}{2})^n u[n] \xrightarrow{\mathcal{F.T.}} \frac{1}{1-\frac{1}{2}e^{-jw}},$$

to conclude that

$$n(\frac{1}{2})^n u[n] \xrightarrow{\mathcal{F.T.}} j\frac{d}{dw}(\frac{1}{1-\frac{1}{2}e^{-jw}}) = \frac{\frac{1}{2}e^{-jw}}{(1-\frac{1}{2}e^{-jw})^2}.$$

Let us define

$$z[n] = n(\frac{1}{2})^n u[n].$$

By the Fourier transform definition, we have

$$Z(e^{jw}) = \sum_{n=-\infty}^{+\infty} z[n]e^{-jwn}.$$

Setting $w = 0$ in the above equation we obtain

$$\sum_{n=-\infty}^{+\infty} z[n]e^{-j0n} = \sum_{n=0}^{+\infty} n(\frac{1}{2})^n = Z(e^{j0}) = \frac{\frac{1}{2}e^{-j0}}{(1-\frac{1}{2}e^{-j0})^2} = 2.$$

Problem 142. Signal $x[n]$ has Fourier transform $X(e^{jw})$ as illustrated in Fig. 7.18. Assume that

$$z[n] = x[n] \cdot y[n], \tag{7.58}$$

where

$$y[n] = \sum_{k=-\infty}^{+\infty} \delta[n-4k]. \tag{7.59}$$

Find $z[n]$ and its Fourier transform.

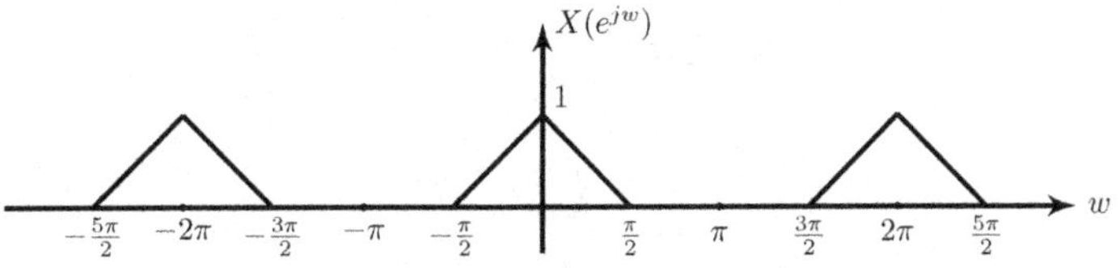

Figure 7.18: Problem 142

Solution.

From (7.58) and (7.59) we have

$$z[n] = x[n] \cdot \sum_{k=-\infty}^{+\infty} \delta[n-4k].$$

Signal $y[n]$ is a periodic train of impulses with period $N = 4$. As we know, the Fourier series coefficients of the impulse train is

$$a_k = \frac{1}{N},$$

and the Fourier transform of a periodic signal can be directly obtained from its Fourier series coefficients, that is

$$Y(e^{jw}) = 2\pi \sum_{k=-\infty}^{+\infty} a_k \delta[w - k\frac{2\pi}{N}] = \frac{2\pi}{N} \sum_{k=-\infty}^{+\infty} \delta[w - k\frac{2\pi}{N}].$$

From the multiplication property we know that $Z(e^{jw})$ is the periodic convolution of $Y(e^{jw})$ and $X(e^{jw})$, where the integral of the convolution can be taken over any interval of length 2π, i.e.,

$$Z(e^{jw}) = \frac{1}{2\pi} X(e^{jw}) \circledast \frac{\pi}{2} \sum_{k=-\infty}^{+\infty} \delta[w - k\frac{2\pi}{N}]. \tag{7.60}$$

To calculate the above convolution, we must convert the equation into an ordinary convolution. So, we define $\hat{X}(e^{jw})$ as

$$\hat{X}(e^{jw}) = \begin{cases} X(e^{jw}) & -\pi < w \leq \pi \\ 0 & \text{otherwise} \end{cases}.$$

Then $Z(e^{jw})$ is given by

$$\begin{aligned} Z(e^{jw}) &= \frac{1}{2\pi} \hat{X}(e^{jw}) * \sum_{k=-\infty}^{+\infty} \frac{\pi}{2} \delta[w - k\frac{\pi}{2}] \\ &= \frac{1}{4} \hat{X}(e^{jw}) * \sum_{k=-\infty}^{+\infty} \delta[w - k\frac{\pi}{2}] \\ &= \frac{1}{4} \sum_{k=-\infty}^{+\infty} \hat{X}(e^{j(w - k\frac{\pi}{2})}), \end{aligned}$$

i.e., $Z(e^{jw})$ is $1/2\pi$ times the (regular) convolution of $\hat{X}(e^{jw})$ and the periodic impulse wave $Y(e^{jw})$, both depicted in Fig. 7.19. The result of this convolution is the Fourier transform $Z(e^{jw})$ which is shown in Fig. 7.20.

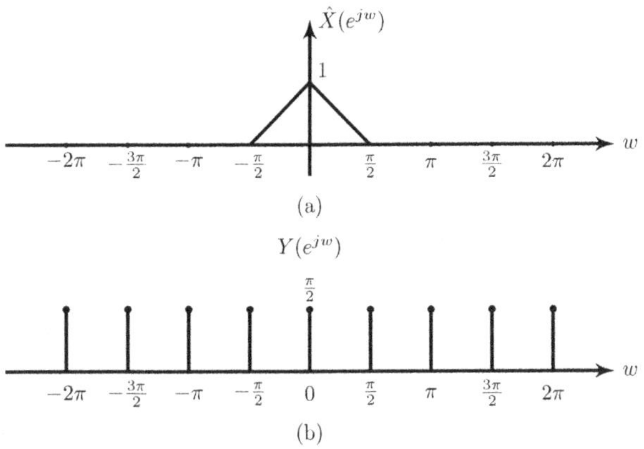

Figure 7.19: Problem 142, (a) The Fourier transform $\hat{X}(e^{jw})$, (b) The Fourier transform of $y[n]$.

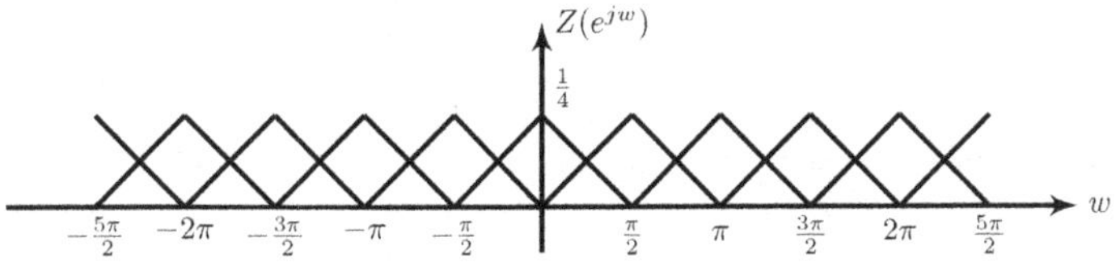

Figure 7.20: Problem 142, the Fourier transform of $z[n]$.

Then, we have

$$Z(e^{jw}) = \frac{1}{4},$$

and $z[n]$ is given by

$$z[n] = \frac{1}{4}\delta[n].$$

Problem 143. Consider an ideal low-pass filter with frequency response

$$H(e^{jw}) = \begin{cases} 1 & |w| < \frac{3\pi}{4} \\ 0 & |w| > \frac{3\pi}{4} \end{cases}.$$

The input to this filter is

$$x[n] = \sum_{k=-\infty}^{+\infty} \left[2(-1)^k\delta[n-2k] + \delta[n-2k]\right].$$

Find the output of this filter.

Solution.

The input $x[n]$ for $-4 \leq n \leq 4$ is shown in Fig. 7.21.

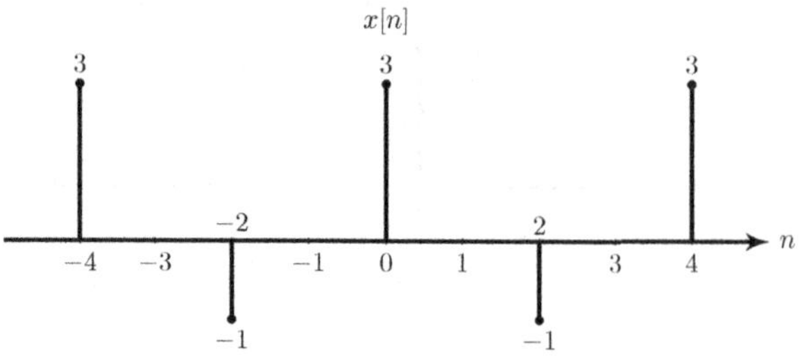

Figure 7.21: Problem 143, the input $x[n]$

From Fig. 7.21, we see that the input $x[n]$ is composed of two periodic trains of impulses with periods $N_1 = 4$ and $N_2 = 4$. From Chapter 5, the Fourier series coefficients of $x[n]$ are

$$a_k = 3 \times \frac{1}{4} - 1 \times \frac{1}{4}(-1)^k = \frac{1}{4}(3 - (-1)^k).$$

The Fourier transform of a periodic signal can be directly obtained from its Fourier series coefficients as

$$\begin{aligned} X(e^{jw}) &= 2\pi \sum_{k=-\infty}^{+\infty} a_k \delta[w - \frac{k\pi}{2}] \\ &= \frac{\pi}{2} \sum_{k=-\infty}^{+\infty} (3 - (-1)^k)\delta[w - \frac{k\pi}{2}], \end{aligned}$$

which is shown in Fig. 7.22.

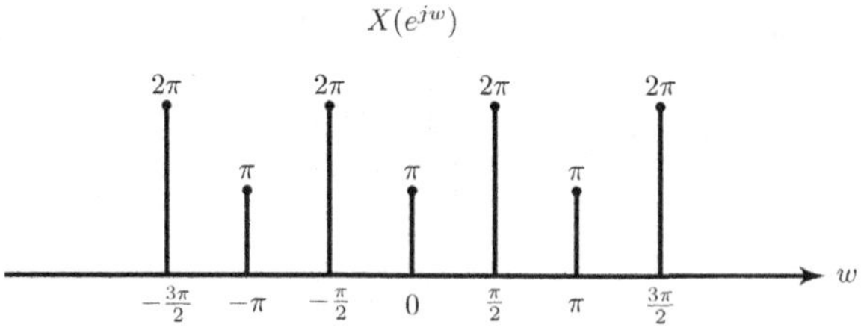

Figure 7.22: Problem 143, The Fourier transform of $x[n]$.

The ideal low-pass filter with the cutoff frequency $w_c = \frac{3\pi}{4}$ is shown in Fig. 7.23

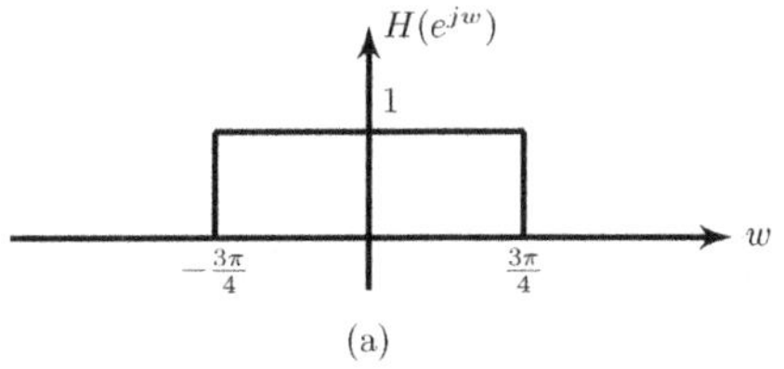

Figure 7.23: Problem 143 - The ideal low-pass filter with the cutoff frequency $w_c = \frac{3\pi}{4}$.

The output of the filter is given by

$$Y(e^{jw}) = H(e^{jw}) \cdot X(e^{jw}).$$

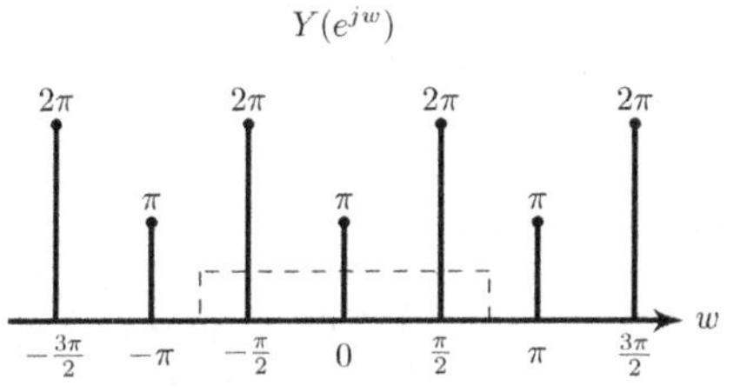

Figure 7.24: Problem 143 - The ideal low-pass filter applied to the input.

From Fig. 7.24 we obtain

$$Y(e^{jw}) = \pi\delta(w) + 2\pi\delta(w - \frac{\pi}{2}) + 2\pi\delta(w + \frac{\pi}{2}),$$

and thus using the Euler's formula,

$$y[n] = \frac{1}{2} + 2\cos[\frac{n\pi}{2}].$$

Problem 144. **(a)** A causal LTI system is described by the difference equation

$$y[n] - ay[n-1] = bx[n] + x[n-1], \tag{7.61}$$

where a is real and with magnitude less than 1.

Find the relation between the values of a and b such that the frequency response of the system satisfies

$$|H(e^{jw})| = 1, \quad \text{for all } w.$$

(b) Find and sketch the output of this system with $a = \frac{1}{3}$ when the input is

$$x[n] = (\frac{1}{3})^n u[n].$$

Use the value of b obtained in Part (a).

Solution.

(a) Taking the Fourier transform of both sides of (7.61), we obtain

$$Y(e^{jw}) - ae^{-jw}Y(e^{jw}) = bX(e^{jw}) + e^{-jw}X(e^{jw}),$$

or

$$(1 - ae^{-jw})Y(e^{jw}) = (b + e^{-jw})X(e^{jw}).$$

Thus, the frequency response of this system is given by

$$H(e^{jw}) = \frac{Y(e^{jw})}{X(e^{jw})} = \frac{b + e^{-jw}}{1 - ae^{-jw}}, \tag{7.62}$$

where the amplitude $|H(e^{jw})|$ is obtained as

$$|H(e^{jw})|^2 = \frac{(b + e^{-jw})(b + e^{jw})}{(1 - ae^{-jw})(1 - ae^{jw})} = \frac{1 + b^2 + 2b\cos(w)}{1 + a^2 - 2a\cos(w)}.$$

Setting the amplitude equal to 1 we have

$$|H(e^{jw})|^2 = \frac{1 + b^2 + 2b\cos(w)}{1 + a^2 - 2a\cos(w)} = 1.$$

Thus we must have

$$a = -b.$$

(b) From the result of Part (a) we have

$$a = \frac{1}{3} \quad \Rightarrow \quad b = -\frac{1}{3}.$$

Substituting these values in (7.62), the frequency response becomes

$$H(e^{jw}) = \frac{-\frac{1}{3} + e^{-jw}}{1 - \frac{1}{3}e^{-jw}}.$$

The Fourier transform of $x[n]$ is

$$X(e^{jw}) = \frac{1}{1 - \frac{1}{3}e^{-jw}},$$

and the Fourier transform of the output $y[n]$ is obtained as

$$\begin{aligned} Y(e^{jw}) &= H(e^{jw})X(e^{jw}) \\ &= \frac{-\frac{1}{3} + e^{-jw}}{(1 - \frac{1}{3}e^{-jw})^2} \\ &= \frac{-\frac{1}{3}}{(1 - \frac{1}{3}e^{-jw})^2} + \frac{e^{-jw}}{(1 - \frac{1}{3}e^{-jw})^2}. \end{aligned}$$

Taking the inverse Fourier transform of $Y(e^{jw})$, the output $y[n]$ is given as

$$y[n] = -\frac{1}{3}(n+1)(\frac{1}{3})^n u[n] + n(\frac{1}{3})^{n-1}u[n-1].$$

Problem 145. Let $x[n]$ be a signal with Fourier transform $X(e^{jw})$.

(a) If

$$X(f) = \prod(\frac{2\pi f}{5})$$

Find the value of

$$A = \sum_{n=-\infty}^{+\infty} n^2 x[n]. \tag{7.63}$$

(b) If

$$x[n] = \cos(\pi n)(u[n+1] - u[n-1]).$$

Find the value of

$$B = \int_{-\pi}^{\pi} |X(e^{jw})\cos(w)|^2 dw. \tag{7.64}$$

Solution.

(a) As can be seen, $X(f)$ is a pulse and as such, $x[n]$ is *sinc* function which makes it difficult to evaluate A directly and thus we try to deploy the Fourier Transform properties.

Applying the frequency-differentiation property, we have

$$nx[n] \xrightarrow{\mathcal{F.T.}} j\frac{d}{dw}X(e^{jw}),$$

noting that

$$w = 2\pi f \Rightarrow f = \frac{w}{2\pi},$$

we can write

$$nx[n] \xrightarrow{\mathcal{F.T.}} \frac{j}{2\pi}\frac{d}{df}X(f).$$

Again by applying the frequency-differentiation property, we obtain

$$n(nx[n]) \xrightarrow{\mathcal{F.T.}} \frac{j}{2\pi}\frac{d}{df}(\frac{j}{2\pi}\frac{d}{df}X(f)) = -\frac{1}{4\pi^2}\frac{d^2}{df^2}X(f).$$

Using the Fourier transform definition, we write

$$-\frac{1}{4\pi^2}\frac{d^2}{df^2}X(f) = \sum_{n=-\infty}^{+\infty} n^2 x[n]e^{-j2\pi fn}. \tag{7.65}$$

Setting $f = 0$ in (7.65), we obtain

$$-\frac{1}{4\pi^2}\frac{d^2}{df^2}X(0) = \sum_{n=-\infty}^{+\infty} n^2 x[n]. \tag{7.66}$$

In order to determine A, we need to evaluate the second derivative of $X(f)$ at $f = 0$. It is easy to see that the 1st and 2nd derivatives of $X(f)$ are 0 at $f = 0$ and thus,

$$A = \sum_{n=-\infty}^{+\infty} n^2 x[n] = -\frac{1}{4\pi^2} X^{''}(0) = 0$$

(b) To evaluate B, we may use Parseval's relation. We first rewrite (7.64) by using Euler's formula as

$$\begin{aligned} B &= \int_{-\pi}^{\pi} |X(e^{jw})\cos(w)|^2 dw \\ &= \int_{-\pi}^{\pi} |X(e^{jw})\frac{e^{jw}+e^{-jw}}{2}|^2 dw \\ &= \frac{1}{4}\int_{-\pi}^{\pi} |X(e^{jw})e^{jw} + X(e^{jw})e^{-jw}|^2 dw. \end{aligned} \tag{7.67}$$

Based on the Parseval's relation, we have

$$\sum_{n=-\infty}^{+\infty} |x[n]|^2 = \frac{1}{2\pi}\int_{-\pi}^{\pi} |X(e^{jw})|^2 dw.$$

Now we have to manipulable the left side of the above equation to be able to form the right side of (7.67) in frequency domain. We act as follows:

$$\sum_{n=-\infty}^{+\infty} \left|x[n+1] + x[n-1]\right|^2 = \frac{1}{2\pi}\int_{-\pi}^{\pi} \left|X(e^{jw})e^{jw} + X(e^{jw})e^{-jw}\right|^2 dw. \tag{7.68}$$

Noting that

$$\begin{aligned} &x[-1] = -1,\ x[0] = 1,\ x[n] = 0,\ n \neq -1, 0, \\ &\Rightarrow x[n] = \delta[n] - \delta[n+1], \end{aligned}$$

we have

$$x[n+1] + x[n-1] = \delta[n+1] - \delta[n+2] + \delta[n-1] - \delta[n].$$

which is shown in Fig. 7.25.

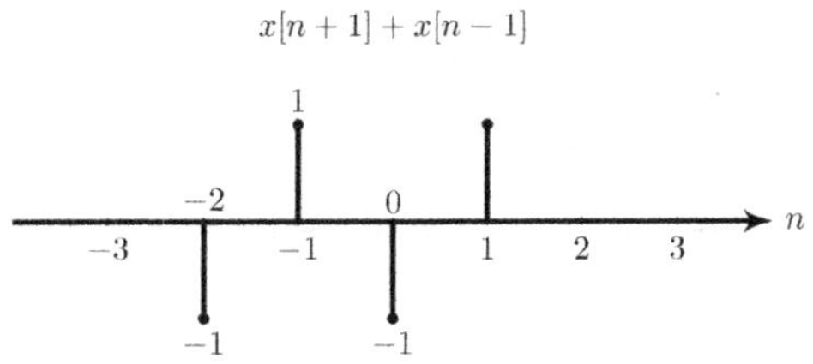

Figure 7.25: Problem 145 - Part (b), signal $x[n+1]+x[n-1]$.

From (7.67) and (7.68), the value of B is given by

$$\begin{aligned} B &= \frac{1}{4}\int_{-\pi}^{\pi} |X(e^{jw})e^{jw} + X(e^{jw})e^{-jw}|^2 dw \\ &= \frac{\pi}{2}\sum_{n=-\infty}^{+\infty} \big|x[n+1]+x[n-1]\big|^2 \\ &= \frac{\pi}{2}\big[(-1)^2 + (1)^2 + (-1)^2 + (1)^2\big] \\ &= 2\pi. \end{aligned}$$

Problem 146. Consider an LTI system with input $x[n]$ and output $y[n]$. Let the input $x[n]$ be

$$x[n] = \cos(\frac{n\pi}{4}) + \cos(\frac{3n\pi}{5}). \tag{7.69}$$

Find the output of this system if the impulse response $h[n]$ is

$$h[n] = \frac{\sin(n\pi/2)}{n} * e^{jn\pi/2}\,\frac{\sin(n\pi/2)}{n}. \tag{7.70}$$

Solution. From (7.70), we see that the impulse response $h[n]$ is the convolution of two filters. Let us define $h_1[n]$ and $h_2[n]$ with the Fourier transform $H_1(e^{jw})$ and $H_2(e^{jw})$, respectively, as

$$h_1[n] = \frac{\sin(n\pi/2)}{n} \quad \Rightarrow \quad H_1(e^{jw}) = \begin{cases} \pi, & |w| < \frac{\pi}{2} \\ 0, & \text{otherwise} \end{cases}.$$

$$h_2[n] = e^{jn\pi/2}\,\frac{\sin(n\pi/2)}{n} \quad \Rightarrow \quad H_2(e^{jw}) = \begin{cases} \pi, & 0 < w < \pi \\ 0, & \text{otherwise} \end{cases}.$$

which are depicted in Fig. 7.26.

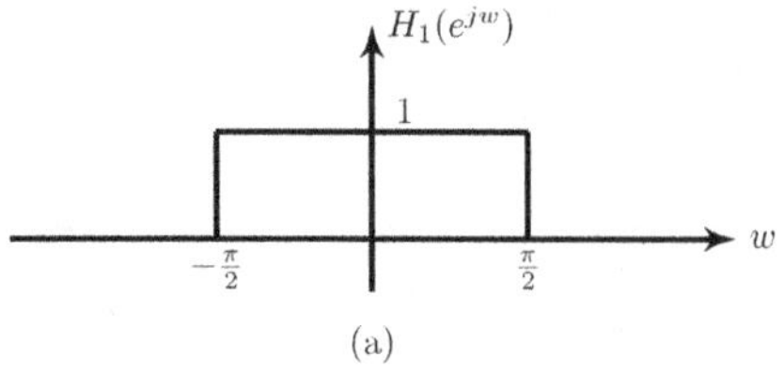

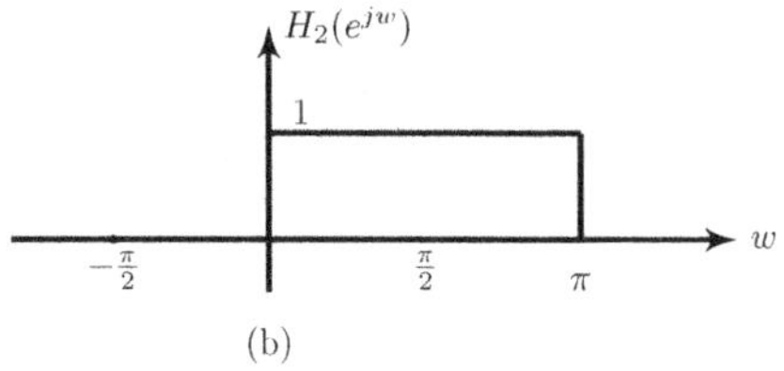

Figure 7.26: Problem 146, (a) The frequency response $H_1(e^{jw})$, (b) The frequency response $H_2(e^{jw})$.

From the convolution property, we know that $H(e^{jw})$ is the multiplication of $H_1(e^{jw})$ and $H_2(e^{jw})$, where $H(e^{jw})$ is the Fourier transform of $h[n]$. Then, the frequency response of the system is given by

$$H(e^{jw}) = H_1(e^{jw})H_2(e^{jw}) = \begin{cases} \pi^2 & 0 < w < \frac{\pi}{2} \\ 0 & \text{otherwise} \end{cases},$$

which is depicted in Fig. 7.27.

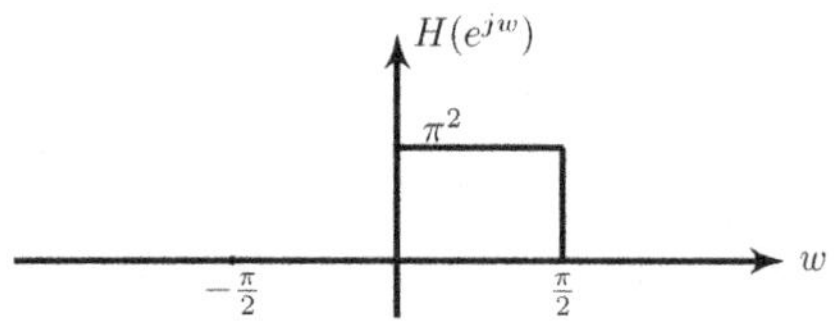

Figure 7.27: Problem 146, the frequency response $H(e^{jw})$.

Now, using Euler's formula, the input $x[n]$ to the system can be rewritten as

$$x[n] = \frac{1}{2}e^{jn\pi/4} + \frac{1}{2}e^{-jn\pi/4} + \frac{1}{2}e^{j3n\pi/5} + \frac{1}{2}e^{-j3n\pi/5}.$$

Thus, the Fourier transform of $x[n]$ over interval $[-\pi\ \pi]$ is given by

$$X(e^{jw}) = \pi\delta(w - \frac{\pi}{4}) + \pi\delta(w + \frac{\pi}{4}) + \pi\delta(w - \frac{3\pi}{5}) + \pi\delta(w + \frac{3\pi}{5}).$$

From Fig. 7.27, we see that only one component of the input can pass from the filter. Thus we obtain

$$Y(e^{jw}) = H(e^{jw})X(e^{jw}) = \pi^3\delta(w - \frac{\pi}{4}),$$

and thus

$$y[n] = \frac{\pi^2}{2} e^{jn\pi/4}.$$

Problem 147. Find the inverse Fourier transform of

$$X(e^{jw}) = \left[e^{-j10w} \frac{\sin(15w/2)}{\sin(w/2)}\right] * \left[\frac{d}{dw}\left(\frac{1}{1-\frac{1}{2}e^{-jw}} + 1\right)\right].$$

Solution. Let us define

$$A(e^{jw}) = \frac{\sin(15w/2)}{\sin(w/2)}, \tag{7.71}$$

and

$$B(e^{jw}) = e^{-j10w} A(e^{jw}). \tag{7.72}$$

As we know,

$$w[n] = \begin{cases} 1, & |n| \le N_1 \\ 0, & |n| > N_1 \end{cases} \quad \Rightarrow \quad W(e^{jw}) = \frac{\sin\left(w(N_1 + \frac{1}{2})\right)}{\sin(w/2)}.$$

Setting $N_1 = 7$ in (7.71), the inverse Fourier transform of $A(e^{jw})$ is

$$a[n] = \begin{cases} 1, & |n| \le 7 \\ 0, & |n| > 5 \end{cases} = u[n+7] - u[n-8]. \tag{7.73}$$

Applying the time-shifting property to (7.72) and from (7.73), we obtain

$$b[n] = a[n-10] = u[n-3] - u[n-18]. \tag{7.74}$$

Now let us define

$$S(e^{jw}) = \frac{1}{1 - \frac{1}{2}e^{-jw}}, \tag{7.75}$$

and

$$Z(e^{jw}) = \frac{d}{dw}\left(S(e^{jw}) + 1\right). \tag{7.76}$$

The inverse Fourier transform of $S(e^{jw})$ is

$$s[n] = (\frac{1}{2})^n u[n]. \tag{7.77}$$

From (7.77) and using the frequency-differentiation property, the inverse Fourier transform of $Z(e^{jw})$ is expressed as

$$z[n] = -jn(\frac{1}{2})^n u[n]. \tag{7.78}$$

Based on the question assumption, $X(e^{iw})$ is defined as the convolution of $B(e^{iw})$ and $Z(e^{iw})$. Then, from the multiplication property, $x[n]$ is the multiplication of $b[n]$ and $z[n]$ times

2π. Finally, from (7.78) and (7.74), $x[n]$ is given by

$$x[n] = 2\pi b[n]z[n] = \begin{cases} -j\pi n(\frac{1}{2})^{n-1}, & 3 \leq n \leq 17 \\ 0, & \text{otherwise} \end{cases}.$$

Problem 148. Consider signals $x_1[n]$ and $x_2[n]$ below with Fourier transforms $X_1(e^{jw})$ and $X_2(e^{jw})$, respectively. Use the Fourier transform properties to determine whether the time-domain signals are complex, real and even, imaginary and odd.

$$X_1(e^{jw}) = e^{-jw}\cos(w)$$
$$X_2(e^{jw}) = j\sin(w)\cos(w).$$

Solution.

First let us focus on $x_1[n]$.

$$X_1(e^{jw}) = e^{-jw}\cos(w). \tag{7.79}$$

From (7.79), we get

$$\angle X_1(e^{jw}) = -w, \qquad |X_1(e^{jw})| = \cos(w).$$

We see that the phase of $X_1(e^{jw})$ is odd and the magnitude of $X_1(e^{jw})$ is even. Then, $X_1(e^{jw})$ is neither odd nor even and thus, $x_1[n]$ is neither odd nor even.

To verify that the signal is real, we must have

$$X(e^{jw}) = X^*(e^{-jw}).$$

We write

$$X_1^*(e^{-jw}) = (e^{-j(-w)})^*\big(\cos(-w)\big)^* = e^{-jw}\cos(w).$$

As can be seen,

$$X_1^*(e^{-jw}) = X_1(e^{jw}),$$

i.e., the Fourier transform of $x_1[n]$ is conjugate symmetric and so $x_1[n]$ is real.
We now focus on $x_2[n]$.

$$X_2(e^{jw}) = j\sin(w)\cos(w). \tag{7.80}$$

From (7.80), we see that the Fourier transform of $x_2[n]$ is a pure imaginary function of w and the product of an even signal and an odd signal. It means that the signal is odd. Thus, the Fourier transform of $x_2[n]$ is odd and imaginary. We know that if $x[n]$ is real and odd, its Fourier transform is imaginary and odd. So signal $x_2[n]$ is real and odd.

Problem 149. Consider the signal depicted in Fig. 7.28. Let the Fourier transform of this signal be written as

$$X(e^{jw}) = A(w) + jB(w).$$

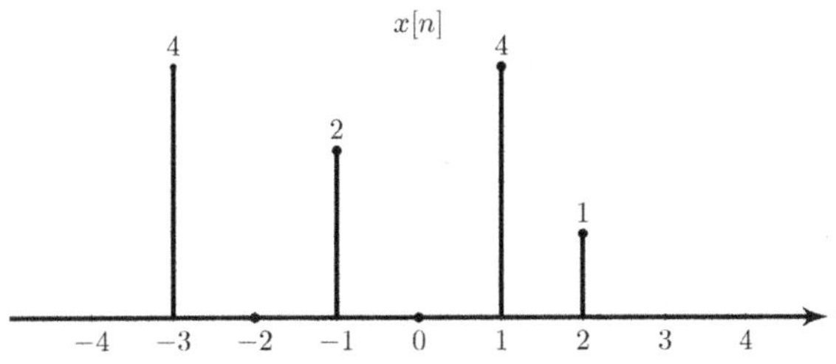

Figure 7.28: Problem 149, the input $x[n]$ to the system.

If the Fourier transform of $y[n]$ is presented as

$$Y(e^{jw}) = 2B(w)e^{jw} + jA(w),$$

find the value of $y[n]$ at $n = 1$.

Solution.

From Fig. 7.28, we see that $x[n]$ is real. Furthermore we know that

$$\mathcal{E}v\{x[n]\} \xrightarrow{\mathcal{F.T.}} \mathcal{R}e\{X(e^{jw})\} = A(w)$$

and

$$\mathcal{O}d\{x[n]\} \xrightarrow{\mathcal{F.T.}} j\mathcal{I}m\{X(e^{jw})\} = jB(w)$$

where $\mathcal{E}v$ and $\mathcal{O}d$ operators denote the even and odd parts of $x[n]$, respectively. The inverse Fourier transform of $A(w)$ and $B(w)$ are $\mathcal{E}v\{x[n]\}$ and $\frac{1}{j}\mathcal{O}d\{x[n]\}$, respectively.

Using the linearity property and the time-shifting property, the inverse Fourier transform of $Y(e^{jw})$ is given by

$$\begin{aligned} y[n] &= \mathcal{F.T.}^{-1}\{2B(w)e^{jw} + jA(w)\} \\ &= \mathcal{F.T.}^{-1}\{2B(w)e^{jw}\} + \mathcal{F.T.}^{-1}\{jA(w)\} \\ &= \frac{2}{j}\mathcal{O}d\{x[n+1]\} + j\mathcal{E}v\{x[n]\}. \end{aligned}$$

Setting $n = 1$ in the above equation, we obtain

$$\begin{aligned} y[1] &= \frac{2}{j}\mathcal{O}d\{x[2]\} + j\mathcal{E}v\{x[1]\} \\ &= \frac{2}{j}\Big(\frac{x[2] - x[-2]}{2}\Big) + j\Big(\frac{x[1] + x[-1]}{2}\Big) = 2j. \end{aligned}$$

Chapter 8

Sampling

8.1 Introduction

In previous sections, we treated continuous-time and discrete-time signals as 2 independent class of signals. However, they are in-fact quite related, that is, a discrete-time signal is usually generated through time-sampling of a continuous-time signal at a certain rate. This is referred to as continuous-to-discrete (C/D) conversion (also called analog to digital conversion). Alternatively, a discrete-time signal can be converted to its continuous-time counterpart through discrete-to-continuous (D/C) conversion. In many cases, we sample a continuous-time signal so that we can perform certain processing (such as filtering) in the discrete domain and then convert the resulting discrete-time signal back to the continuous domain.

8.2 Sampling of Continuous-Time Signals

8.2.1 The Sampling Theorem

The easiest way to model the sampling process is to multiply the continuous-signal $x(t)$ by an impulse train $p(t)$, also referred to as the *sampling function.* The period T_s of the impulse train is called the *sampling period*, and the fundamental frequency of $p(t)$, $f_s = 1/T_s$, is called the *sampling frequency.* Also we define $w_s = 2\pi f_s$.

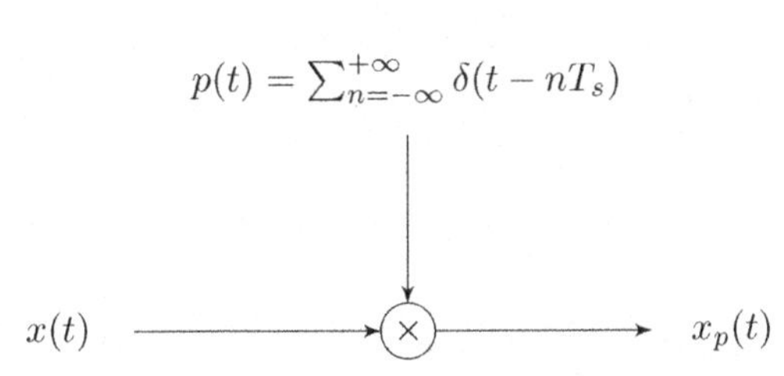

Figure 8.1: Sampling of continuous-time signals.

$$x_p(t) = x(t)\ p(t), \tag{8.1}$$

where

$$p(t) = \sum_{n=-\infty}^{+\infty} \delta(t - nT_s), \tag{8.2}$$

and

$$x_p(t) = \sum_{n=-\infty}^{+\infty} x(nT_s)\delta(t - nT_s). \tag{8.3}$$

The sampling process discussed above is illustrated in Figure 8.2.

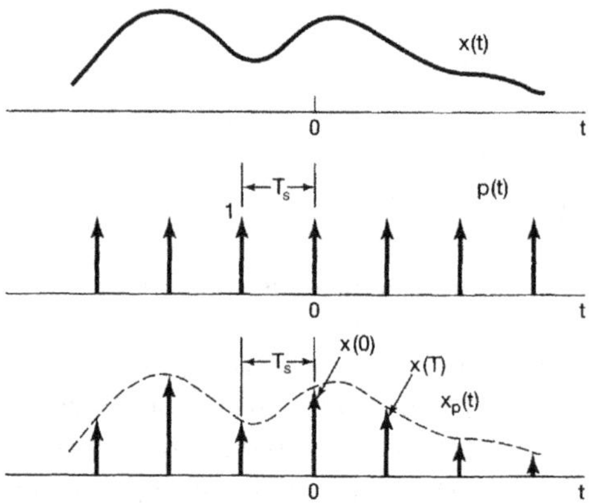

Figure 8.2: Impulse-train sampling.

To see what is happening in the frequency domain through the sampling process, assume that $x(t)$ is a band-limited signal in the frequency domain with the following typical Fourier transform:

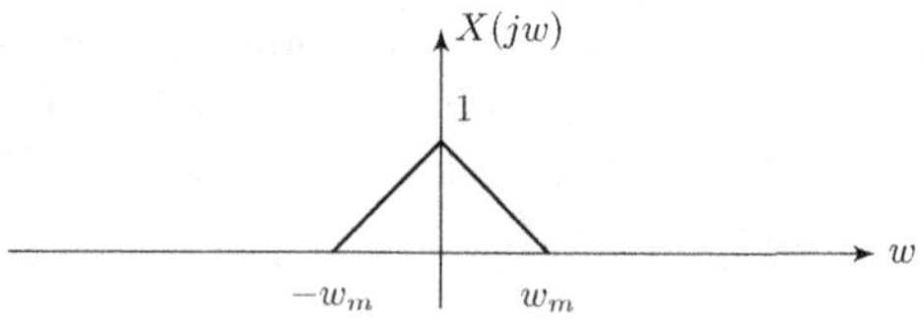

Figure 8.3: Spectrum of band-limited signal $x(t)$.

From the Fourier transform of periodic signals (Section 6.5), we have

$$p(t) = \sum_{n=-\infty}^{+\infty} \delta(t - nT_s) \quad \overset{\mathcal{F.S.}}{\longleftrightarrow} \quad a_k = \frac{1}{T_s}$$

then

$$p(t) \quad \overset{\mathcal{F.T.}}{\longleftrightarrow} \quad P(jw) = \sum_{k=-\infty}^{+\infty} 2\pi a_k \delta(w - kw_s).$$

So we can write:

$$P(jw) = \frac{2\pi}{T_s} \sum_{k=-\infty}^{+\infty} \delta(w - k\frac{2\pi}{T_s}).$$

From the multiplication property (Section 6.4), we know that

$$X_p(jw) = \frac{1}{2\pi} \left\{ X(jw) * P(jw) \right\}. \quad (8.4)$$

Also we know

$$X(jw) * \delta(w - w_0) = X(j(w - w_0)).$$

Therefore,

$$X_p(jw) = \frac{1}{T_s} \sum_{k=-\infty}^{+\infty} X(j(w - kw_s)). \quad (8.5)$$

The process has been depicted in Fig. 8.4.

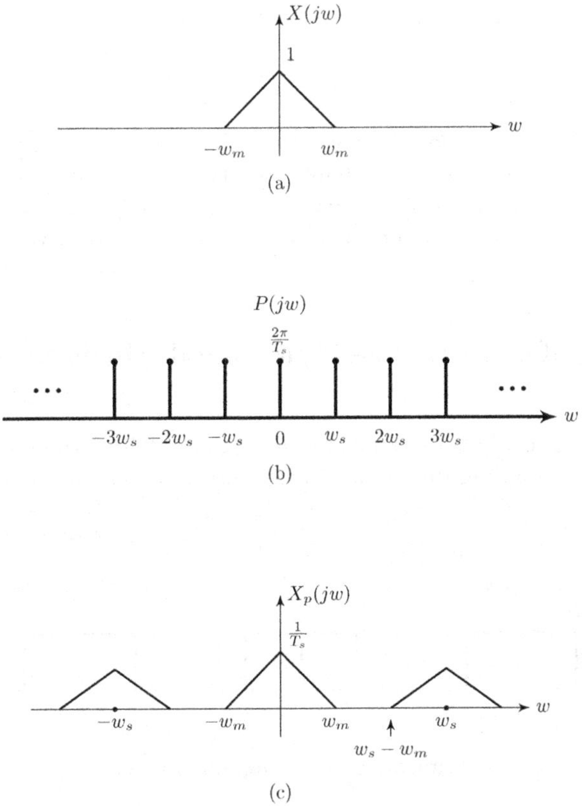

Figure 8.4: Frequency representation of time-domain sampling: (a) Spectrum of the original signal; (b) Spectrum of the sampling function; (c) Spectrum of sampled signal with $w_s > 2w_m$.

To recover $x(t)$ from $x_p(t)$, we use a low-pass filter with gain T_s and a cutoff frequency w_c

which is depicted in Figure 8.5. The cutoff frequency w_c must be greater than w_m and less than $w_s - w_m$, i.e,

$$w_m < w_c < w_s - w_m.$$

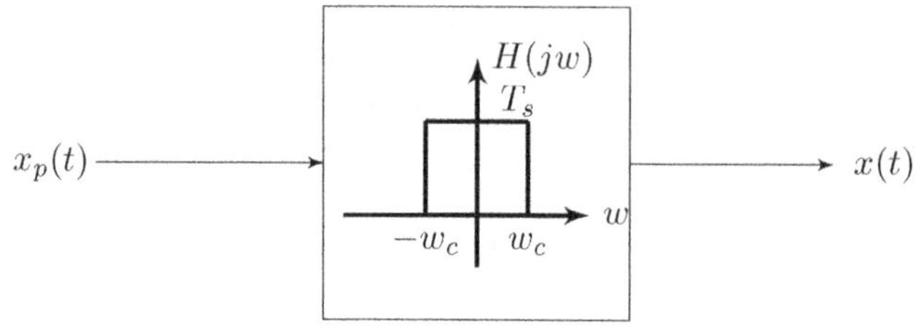

Figure 8.5: Recovering by an ideal low-pass filter.

As can be seen in Figure 8.4, in order not to have an overlap between the shifted replicas of $X(jw)$, we must have

$$w_s - w_m > w_m \qquad \Rightarrow \qquad w_s > 2w_m. \tag{8.6}$$

and equivalently

$$f_s > 2f_m. \tag{8.7}$$

In other words, in order to be able to recover the original signal form its sampled version, the sampling frequency f_s must be greater than twice the signal bandwidth. This statement is known as the *Nyquist Theorem* and frequency $2f_m$ is commonly referred to as the *Nyquist rate.* Note that $w_s = 2\pi f_s$ may also be referred to as the Nyquist rate if we operate in the angular frequency domain.

8.2.2 Sampling of Continuous-Time Signals through Rectangular Pulse Train

In practice, we can not use impulse train to sample continuous-time signals. The next best thing to be used as the sampling function, which is also practical, is the rectangular pulse train shown in Figure 8.6.

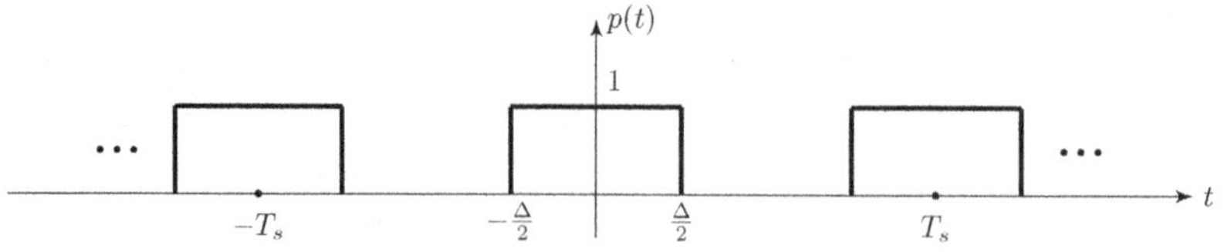

Figure 8.6: Rectangular pulses.

In Figure 8.6, T_s is the fundamental period of $p(t)$, and Δ is the width of each pulse. From Chapter 4, we know that the Fourier series coefficients of this signal are

$$a_k = \frac{\sin(k\pi\Delta/T_s)}{k\pi} = \frac{\sin(kw_s\Delta/2)}{k\pi}.$$

So we have

$$P(jw) = \sum_{k=-\infty}^{+\infty} 2\pi a_k \delta(w - kw_s)$$
$$= \sum_{k=-\infty}^{+\infty} 2\frac{\sin(kw_s\Delta/2)}{k}\delta(w - kw_s). \quad (8.8)$$

Assuming that $w_s > 2w_m$, to recover signal $x(t)$ from $x_p(t)$, we use a low-pass filter $H(jw)$ which satisfies the following conditions:

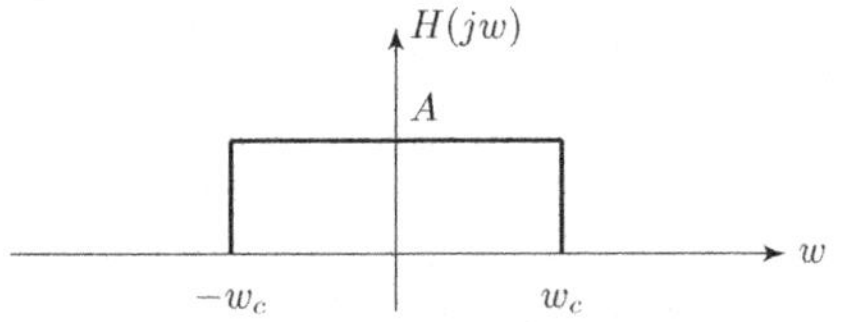

Figure 8.7: Ideal low-pass filter to recover $x(t)$ from $x_p(t)$.

1. Cutoff frequency of the filter: $w_m < w_c < w_s - w_m$.

2. Gain of the filter: $A = \frac{T_s}{\Delta}$.

8.3 Sampling of Discrete-Time Signals

Sampling idea can also be considered for discrete-time signals. In this case, the process is modeled by multiplying signal $x[n]$ by a periodic impulse train $p[n]$:

$$x_p[n] = x[n]\; p[n], \quad (8.9)$$

where

$$p[n] = \sum_{k=-\infty}^{+\infty} \delta[n - kN]. \quad (8.10)$$

We know that

$$p[n] = \sum_{k=-\infty}^{+\infty} \delta[n - kN] \quad \overset{\mathcal{F.S.}}{\longleftrightarrow} \quad a_k = \frac{1}{N}.$$

Thus

$$p[n] \quad \overset{\mathcal{F.T.}}{\longleftrightarrow} \quad P(e^{jw}) = \frac{2\pi}{N} \sum_{k=-\infty}^{+\infty} \delta(w - kw_s), \quad w_s = \frac{2\pi}{N}.$$

From the multiplication property, as also shown in figure below, we have

$$X_p(e^{jw}) = \frac{1}{2\pi}\left\{X(e^{jw}) * P(e^{jw})\right\}$$

$$= \frac{1}{N} \sum_{k=0}^{N-1} X(e^{j(w-kw_s)}). \tag{8.11}$$

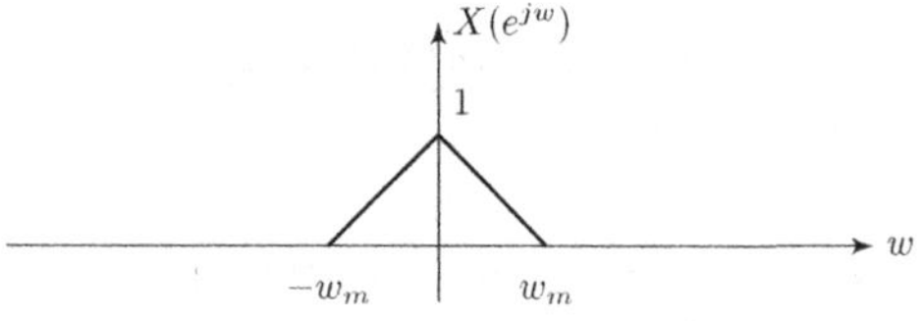

(a)

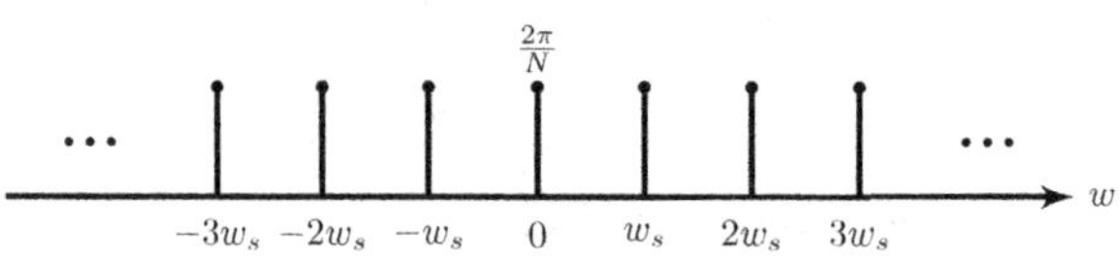

(b)

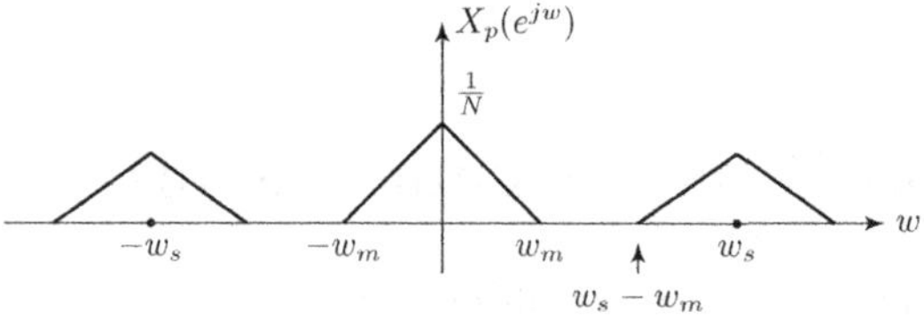

(c)

Figure 8.8: Frequency representation of time-domain sampling: (a) Spectrum of the original signal; (b) Spectrum of the sampling function; (c) Spectrum of sampled signal assuming that $w_s > 2w_m$.

Assuming that $w_s > 2w_m$, to recover signal $x[n]$ from $x_p[n]$, we use a low-pass filter $H(e^{jw})$ which satisfies the following conditions:

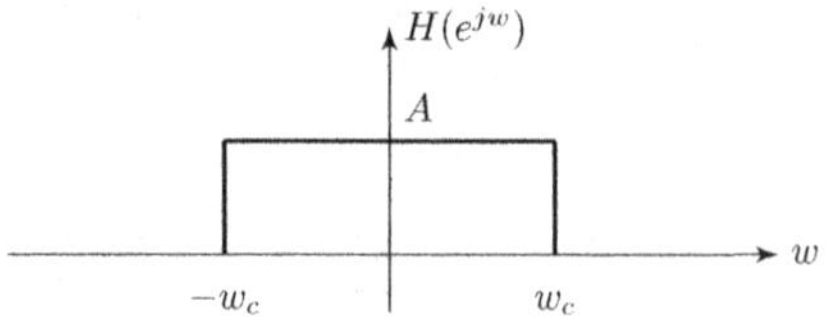

Figure 8.9: Ideal low-pass filter to recover $x[n]$ from $x_p[n]$.

1. Cutoff frequency of filter: $w_m < w_c < w_s - w_m$.

2. Gain of filter: $A = N$.

8.4 Problems

Problem 150. Determine the minimum sampling frequency for the following signals to avoid aliasing.

(a) $x(t) = \left(\frac{\sin(4000\pi t)}{\pi t}\right)^2$ **(b)** $x(t) = \frac{2\sin(4t)\sin(8t)}{\pi t^2}$

Solution.

(a) $x(t) = \left(\frac{\sin(4000\pi t)}{\pi t}\right)^2$

To determine the Nyquist rate, we must obtain the highest frequency of the signal. We have

$$\frac{\sin(Wt)}{\pi t} \xrightarrow{\mathcal{F.T.}}$$

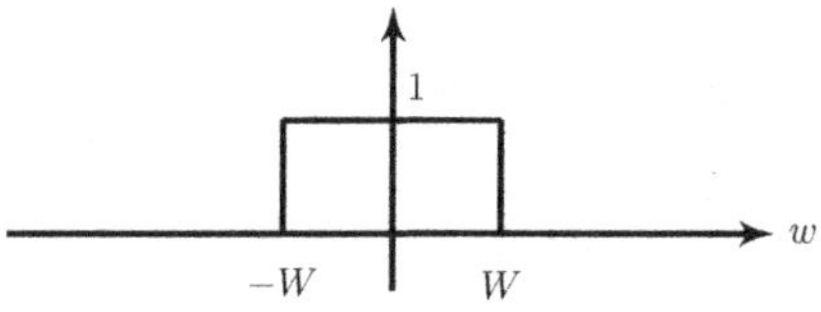

From the multiplication property of the Fourier transform,

$$y(t) = x(t).z(t) \xrightarrow{\mathcal{FT}} Y(jw) = \frac{1}{2\pi}(X(jw) * Z(jw)),$$

$$\left(\frac{\sin(Wt)}{\pi t}\right)^2 \xrightarrow{\mathcal{F.T.}}$$

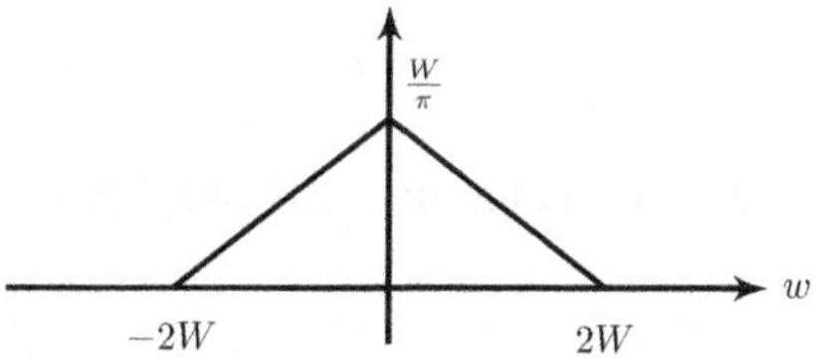

Thus $X(jw)$ is as illustrated in the following figure:

$$x(t) = \left(\frac{\sin(4000\pi t)}{\pi t}\right)^2 \xrightarrow{\mathcal{F.T.}}$$

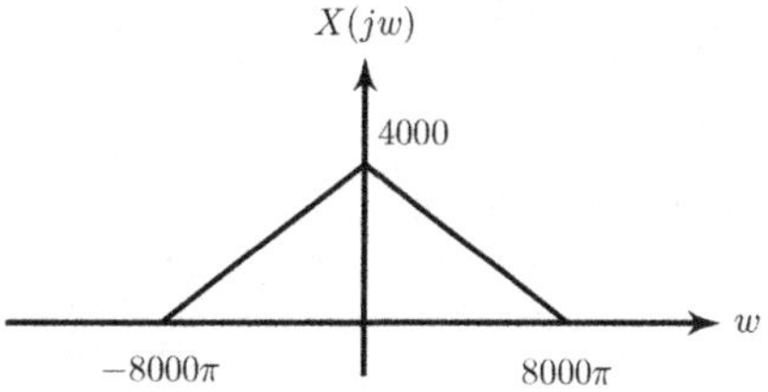

Under the sampling theorem, to avoid aliasing, the sampling frequency must be greater than twice the highest frequency present in the signal. As such

$$w_m = 8000\pi \quad \Rightarrow \quad w_s > 2w_m = 16000\pi,$$

Then, the minimum sampling frequency is obtained by

$$f_s = \frac{w_s}{2\pi} = 8000 \quad Hz.$$

(b) $x(t) = \frac{2\sin(4t)\sin(8t)}{\pi t^2}$

The signal can be decomposed as the product of $x_1(t)$ and $x_2(t)$ signals as follows. From Part (a), we have

$$x_1(t) = \frac{2\sin(4t)}{\pi t} \quad \xrightarrow{\mathcal{F.T.}}$$

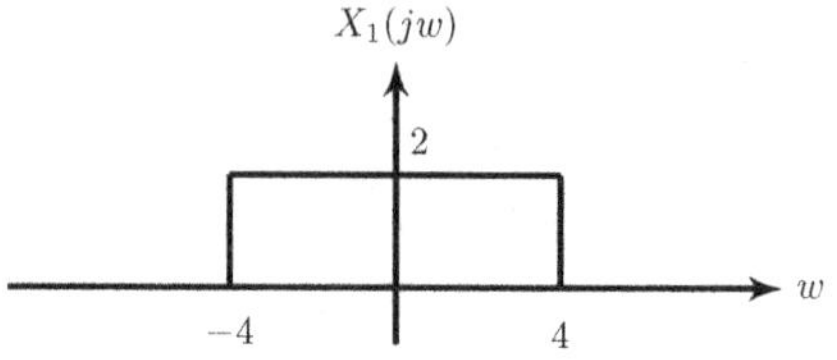

$$x_2(t) = \frac{\sin(8t)}{t} \quad \xrightarrow{\mathcal{F.T.}}$$

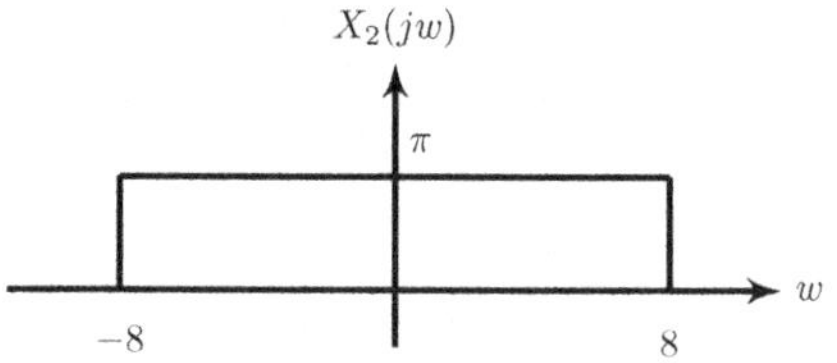

From the multiplication property of the Fourier transform, the Fourier transform of $x(t)$ is given by

$$x(t) = x_1(t).x_2(t) \quad \xrightarrow{\mathcal{F.T.}} \quad X(jw) = \frac{1}{2\pi}\big(X_1(jw) * X_2(jw)\big),$$

where we come up with the figure below for $X(jw)$:

$$x(t) = \frac{2\sin(4t)\sin(8t)}{\pi t^2} \quad \xrightarrow{\mathcal{F.T.}}$$

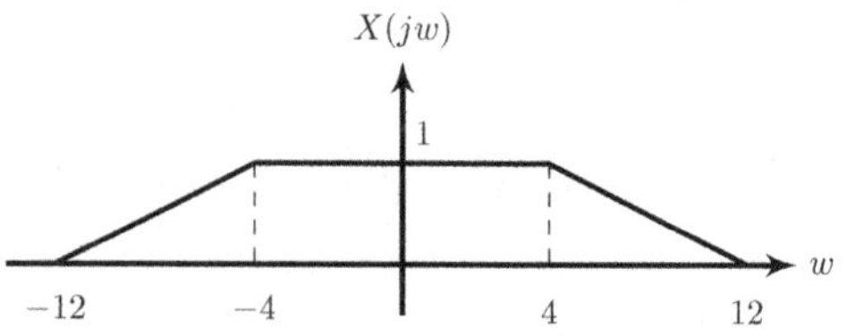

Thus, we obtain

$$w_m = 12 \quad \Rightarrow \quad w_s > 2w_m = 24,$$

and

$$f_s = \frac{w_s}{2\pi} > \frac{12}{\pi} \quad Hz.$$

Problem 151. Let $x(t)$ be a band-limited signal for which the corresponding Nyquist rate is w_0. Determine the Nyquist rate for each of the following signals.

(a) $x(2t)$

(b) $x(t) * x(t)$

(c) $x^2(2t)$

(d) $x(t) \cdot \cos(2w_0 t)$

(e) $\frac{dx(t)}{dt}$

Solution.

(a) $y(t) = x(2t)$

From Chapter 1, we know that signal $y(t)$ is the time-compressed version of the signal $x(t)$. Using the time scaling property of the continuous-time Fourier transform, we obtain

$$Y(jw) = \frac{1}{2} X(\frac{jw}{2}).$$

We see that the largest frequency component of $Y(jw)$ is twice that of $X(jw)$. Thus, the Nyquist rate w_s of $y(t)$ is twice the Nyquist rate w_0 of $x(t)$:

$$w_s = 2w_0.$$

(b) $y(t) = x(t) * x(t)$

From the convolution property of Fourier transform, we have

$$y(t) = x(t) * x(t) \quad \xrightarrow{\mathcal{F.T.}} \quad Y(jw) = X(jw) \cdot X(jw).$$

As such, the highest frequency component of $y(t)$ is equal to the highest frequency present in signal $x(t)$. Thus, the Nyquist rate w_s of $y(t)$ is equal to the Nyquist rate w_0 of $x(t)$:

$$w_s = w_0.$$

(c) $y(t) = x^2(2t)$

From the result of Part (a), the Nyquist rate of $x(2t)$ is twice the Nyquist rate of $x(t)$. Applying the multiplication property of Fourier transform, we obtain

$$x^2(2t) = x(2t).x(2t) \quad \xrightarrow{\mathcal{F.T.}} \quad \frac{1}{2\pi}(X(\frac{jw}{2}) * X(\frac{jw}{2})).$$

Since $X(\frac{jw}{2})$ is convolved with itself in the frequency domain, the highest frequency present in $x^2(2t)$ is twice that of $x(2t)$. So, the highest frequency of $x^2(2t)$ is four times the highest frequency of $x(t)$. As such, the Nyquist rate w_s of $x^2(2t)$ is four times the Nyquist rate w_0 of $x(t)$:

$$w_s = 4w_0.$$

(d) $y(t) = x(t) \cdot \cos(2w_0 t)$

Let the highest frequency in $x(t)$ be denoted by w_m, where $w_0 = 2w_m$. We have:

$$y(t) = x(t) \cdot \cos(2w_0 t) = x(t) \cdot \cos(4w_m t).$$

Now based on the multiplication property of the Fourier transform, we obtain

$$\begin{aligned} Y(jw) &= \frac{1}{2\pi} X(jw) * \left[\delta(w - 4w_m) + \delta(w + 4w_m)\right] \\ &= \frac{1}{2\pi}\left[X\left(j(w - 4w_m)\right) + X\left(j(w + 4w_m)\right)\right]. \end{aligned}$$

It can be seen that the highest frequency component in signal $y(t)$ is $4w_m + w_m = 5w_m$. Therefore, the Nyquist rate w_s of $y(t)$ is five times the Nyquist rate w_0 of $x(t)$:

$$w_s = 5w_0.$$

(e) $y(t) = \frac{dx(t)}{dt}$

By using the differentiation property of the continues-time Fourier transform, we get

$$y(t) = \frac{dx(t)}{dt} \xrightarrow{\mathcal{F.T.}} Y(jw) = jwX(jw).$$

Since $x(t)$ is band-limited, $Y(jw)$ is zero wherever $X(jw)$ is. So the highest frequency component of both signals are the same. As such:

$$w_s = w_0.$$

Problem 152. For each of the following signals, determine the maximum sampling interval T such that the signals are recoverable from the sampled signals through the use of an ideal low-pass filter.

(a) $x(t) = \frac{\sin(3\pi t)}{t} + \cos(2\pi t)$

(b) $x(t) = \frac{\sin(\pi t)}{2t} \cdot \cos(2\pi t)$

(c) $x(t) = e^{-3t}u(t) * \frac{\sin(Wt)}{\pi t}$

Solution.

(a) $x(t) = \frac{\sin(3\pi t)}{t} + \cos(2\pi t)$

To determine the maximum sampling interval T, we must obtain the highest frequency present in the signal. We see that the signal is the summation of signals $x_1(t)$ and $x_2(t)$ where

$$x_1(t) = \frac{\sin(3\pi t)}{t} \xrightarrow{\mathcal{F.T.}}$$

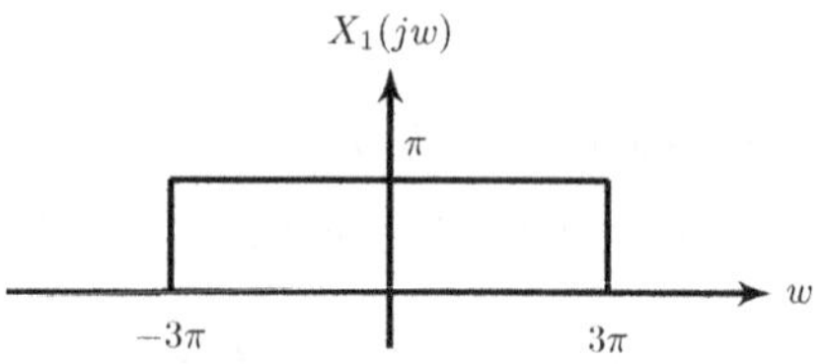

$$x_2(t) = \cos(2\pi t) \quad \xrightarrow{\mathcal{F.T.}}$$

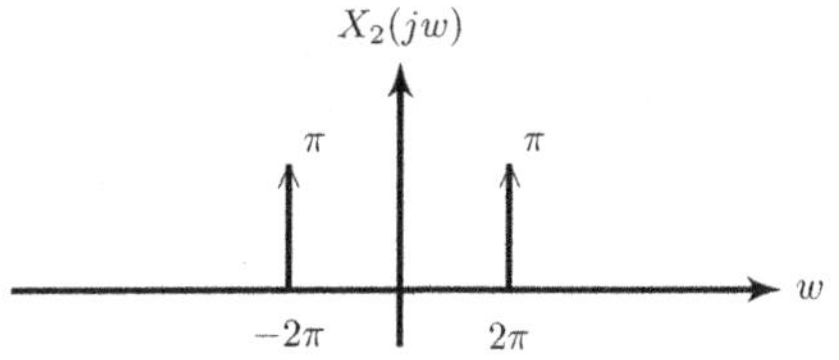

The highest frequency of $X_1(jw)$ is

$$w_m = 3\pi,$$

and the highest frequency of $X_2(jw)$ is

$$w_m = 2\pi.$$

From the linearity property of the Fourier transform, the resulting spectrum is the summation of $X_1(jw)$ and $X_2(jw)$. As such, the highest frequency of $x(t)$ is 3π:

$$w_{max} = 3\pi \quad \Rightarrow \quad w_s > 2w_{max} = 6\pi,$$

then, the maximum sampling interval is given by

$$T_s = \frac{2\pi}{w_s} \quad \Rightarrow \quad T < \frac{1}{3}.$$

(b) $x(t) = \frac{\sin(\pi t)}{2t} \cdot \cos(2\pi t)$
The signal is the product of signals $x_1(t)$ and $x_2(t)$ where

$$x_1(t) = \frac{\sin(\pi t)}{2t} \quad \xrightarrow{\mathcal{F.T.}}$$

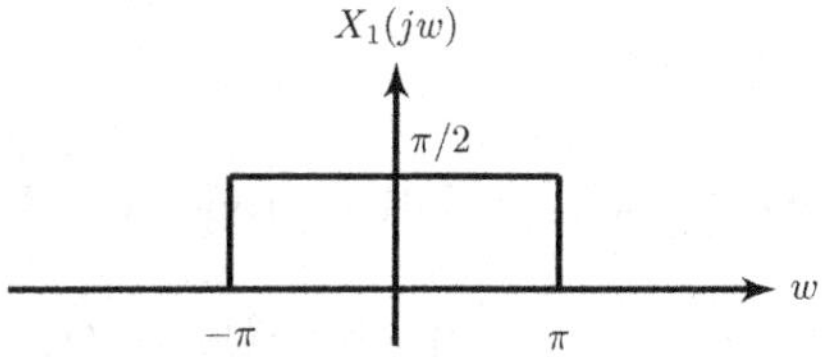

and

$$x_2(t) = \cos(2\pi t) \quad \xrightarrow{\mathcal{F.T.}}$$

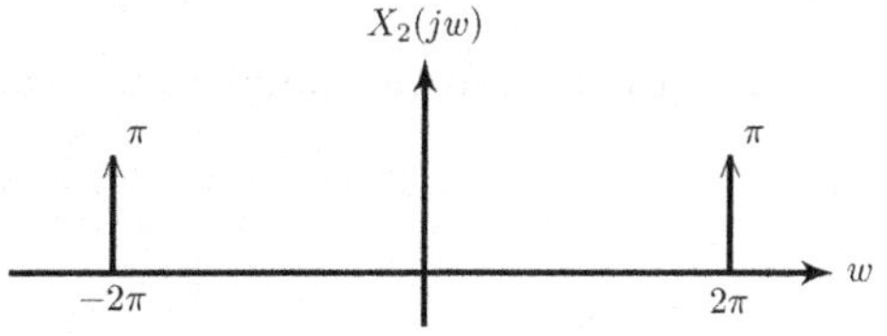

From the multiplication property of the Fourier transform, the Fourier transform of $x(t)$ is given by

$$x(t) = x_1(t).x_2(t) \quad \xrightarrow{\mathcal{F.T.}} \quad X(jw) = \frac{1}{2\pi}\big(X_1(jw) * X_2(jw)\big),$$

and from the convolution properties, the highest frequency present in signal $x(t)$ will be the sum of the highest frequency of $x_1(t)$ and $x_2(2)$. So we have

$$w_{max} = 2\pi + \pi = 3\pi \Rightarrow w_s > w_{max} = 6\pi$$

and

$$T_s = \frac{2\pi}{w_s} \quad \Rightarrow \quad T_s < \frac{2\pi}{w_{max}} = \frac{1}{3}.$$

(c) $x(t) = e^{-3t}u(t) * \frac{\sin(Wt)}{\pi t}$

We know that,

$$x_1(t) = e^{-3t}u(t) \quad \xrightarrow{\mathcal{F.T.}} \quad X_1(jw) = \frac{1}{3+jw} \quad \text{for all } w,$$

and

$$x_2(t) = \frac{\sin(Wt)}{\pi t} \quad \xrightarrow{\mathcal{F.T.}}$$

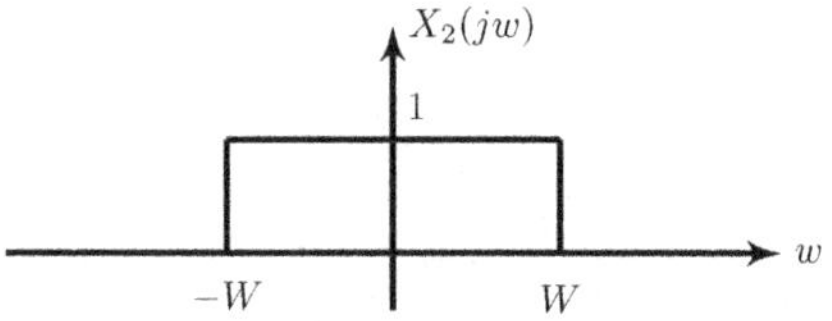

From the convolution property of Fourier transform, we have

$$x(t) = x_1(t) * x_2(t) \quad \xrightarrow{\mathcal{F.T.}} \quad X(jw) = X_1(jw) \cdot X_2(jw).$$

The resulting product does not have any non-zero component outside interval $[-W\ W]$. As such

$$w_{max} = W \Rightarrow w_s > w_{max} = 2W.$$

Then, the maximum sampling interval is given by

$$T_s = \frac{2\pi}{w_s} \quad \Rightarrow \quad T_s < \frac{\pi}{W}.$$

Problem 153. Shown in Fig. 8.10, is a system in which the input signal is multiplied by a periodic square wave. The period of $s(t)$ is 2π. The input signal is band-limited with $|X(jw)| = 0$ for $|w| \geq w_m$. The sampler is followed by an ideal low-pass filter with gain 1 and cutoff frequency w_c, for reconstruction of $x(t)$ from its samples $x_s(t)$. Determine the maximum value of w_m.

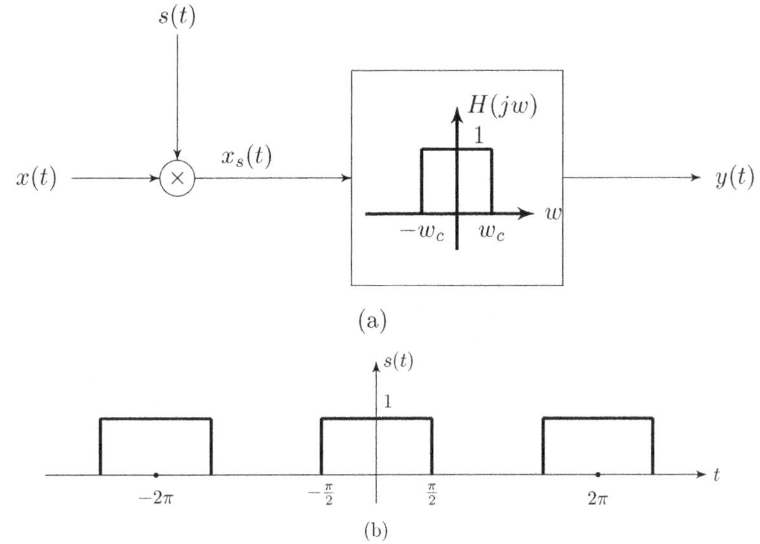

Figure 8.10: Problem 153

Solution.
From Fig. 8.10, we have

$$x_s(t) = x(t) \cdot s(t).$$

Applying the multiplication property of the continues-time Fourier transform, we know that

$$X_s(jw) = \frac{1}{2\pi}[X(jw) * S(jw)].$$

Since $s(t)$ is periodic with period 2π, then its Fourier transform can be directly obtained from its Fourier coefficients, that is,

$$\begin{aligned} S(jw) &= 2\pi \sum_{k=-\infty}^{+\infty} a_k \delta(w - k\frac{2\pi}{T_s}) \\ &= 2\pi \sum_{k=-\infty}^{+\infty} a_k \delta(w - k). \end{aligned} \tag{8.12}$$

We know from Chapter 4 that the Fourier series coefficients of the pulse train is given as

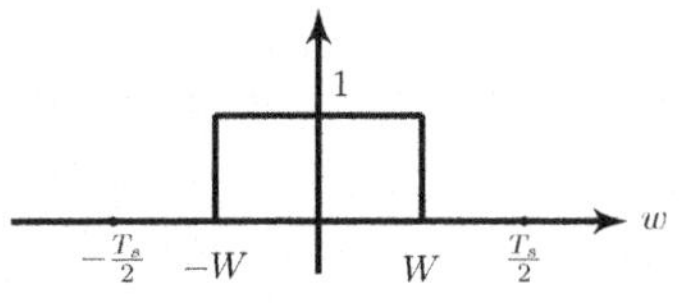

$$\xrightarrow{\mathcal{F.S.}} \quad a_k = \frac{\sin(k\pi 2W/T_s)}{k\pi} = \frac{\sin(kw_s W)}{k\pi}$$

Substituting $T_s = 2\pi$, $W = \pi/2$ and $w_s = 1$, the Fourier series coefficients of $s(t)$ are obtained as

$$a_k = \frac{\sin(k\pi/2)}{k\pi}. \tag{8.13}$$

Substituting (8.13) into (8.12), we get

$$S(jw) = 2\pi \sum_{k=-\infty}^{+\infty} \frac{\sin(k\pi/2)}{k\pi}\delta(w-k). \tag{8.14}$$

That is, $X_s(jw)$ is a periodic function of w consisting of a superposition of shifted replicas of $X(jw)$, scaled by $S(jw)$, as illustrated in Fig. 8.11. Based on Fig. 8.11(b), to avoid aliasing, we must have $w_m < (w_s - w_m)$, or equivalently, $w_s/2 > w_m$. Since

$$w_s = \frac{2\pi}{T} = 1,$$

we have

$$w_m < \frac{1}{2}.$$

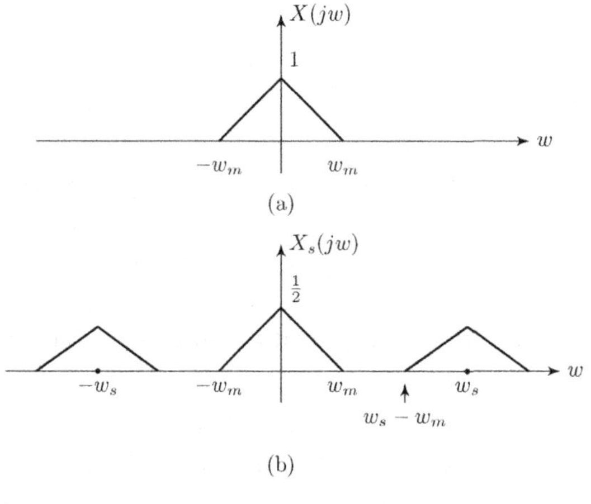

Figure 8.11: (a) representative spectrum for $x(t)$; (b) corresponding spectrum for $x_s(t)$.

Problem 154. In the system shown in Fig. 8.12, the signal $x[n]$ is the input to the system S_1. The system S_1 is defined by

$$S_1 : w[n] = x[2n].$$

The output $w[n]$ of the system S_1 is sampled by a periodic impulse train. The signal $x[n]$ is band limited to $w = \frac{\pi}{4}$; that is,

$$X(e^{jw}) = 0, \ |w| \geq \frac{\pi}{4}.$$

Determine the maximum value of N such that $w[n]$ is recoverable from $w_p[n]$ through the use of an ideal low-pass filter.

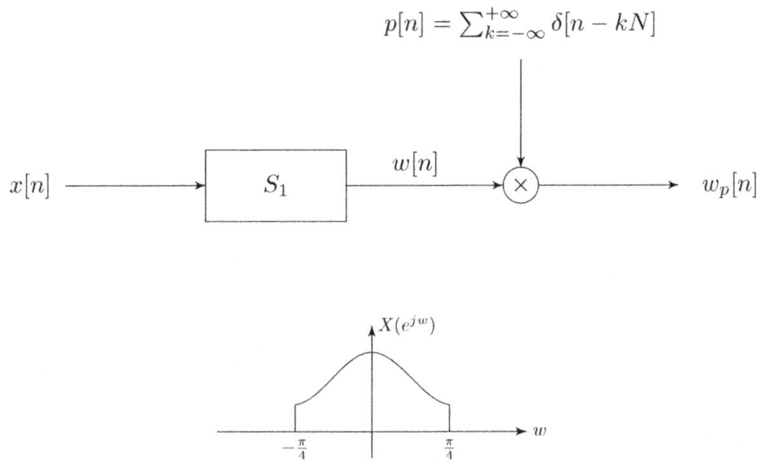

Figure 8.12: Problem 154

Solution.

From Chapter 7 and using the time compression property, the Fourier transform of $x[2n]$ is given by

$$x[mn] \xrightarrow{\mathcal{F.T.}} \frac{1}{m}\sum_{k=0}^{m-1} X(e^{j(\frac{w}{m}-\frac{2k\pi}{m})})$$

$$W(e^{jw}) = \frac{1}{2}X(e^{j\frac{w}{2}}) + \frac{1}{2}X(e^{j(\frac{w}{2}-\pi)}). \tag{8.15}$$

where $X(e^{jw})$ is the Fourier transform of $x[n]$.

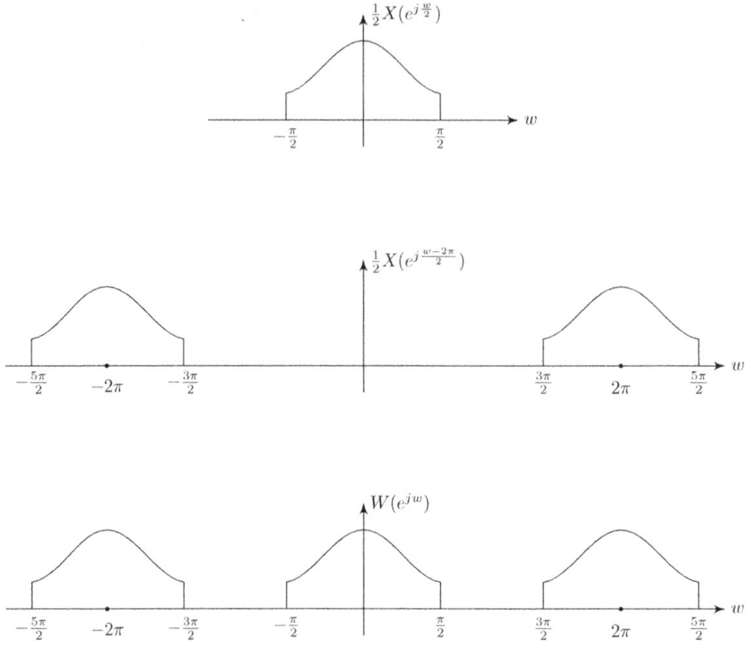

Figure 8.13: Obtaining The Fourier transform $W(e^{jw})$ from $X(e^{jw})$, Problem 154.

The aperiodic Fourier transform $W(e^{jw})$ is shown in Fig. 8.14.

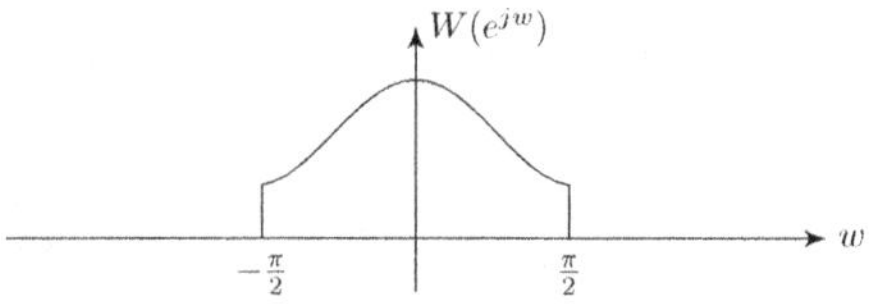

Figure 8.14: The aperiodic Fourier transform $W(e^{jw})$, Problem 154.

From Chapter 7, the Fourier transform of the impulse train with period N is

$$P(e^{jw}) = \frac{2\pi}{N} \sum_{k=-\infty}^{+\infty} \delta[w - k\frac{2\pi}{N}]. \tag{8.16}$$

From Fig. 8.12, we have

$$w_p[n] = w[n] \cdot p[n].$$

Thus, using the multiplication property and from (8.15) and (8.16), the Fourier transform of $w_p[n]$ is given by

$$W_p(e^{jw}) = \frac{1}{2\pi}\left[W(e^{jw}) * P(e^{jw})\right],$$

which is depicted in Fig. 8.15 for $k = -N,\ 0,\ N$.

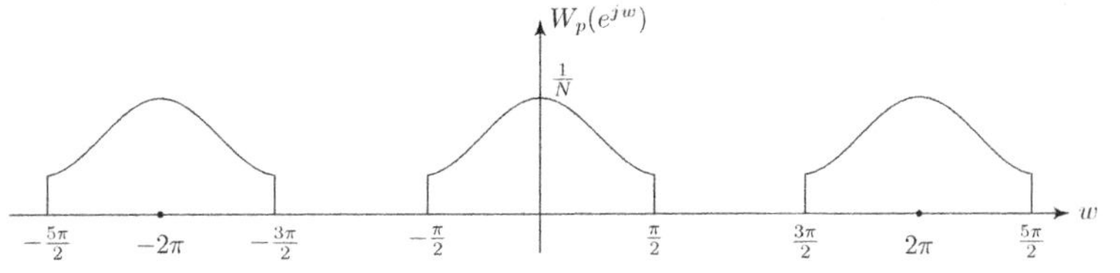

Figure 8.15: The Fourier transform $W_p(e^{jw})$, Problem 154.

From Fig. 8.15, the highest frequency present in signal $w_p[n]$ is $\frac{\pi}{2}$. Thus we must have

$$w_m = \frac{\pi}{2}, \quad w_s \geq 2w_m,$$

so

$$\frac{2\pi}{N} \geq \frac{2\pi}{2} \quad \Rightarrow \quad N \leq 2 \quad \Rightarrow \quad N_{max} = 2.$$

Problem 155. Consider the system shown in Fig. 8.16. The system consisting of a casual continuous-time LTI system followed by a sampler, Continuous-to-Discrete (C/D) Converter, and an LTI discrete-time system. The relationship between $x(t)$ and $y(t)$ is given by:

$$\frac{dy(t)}{dt} + 2y(t) = x(t). \tag{8.17}$$

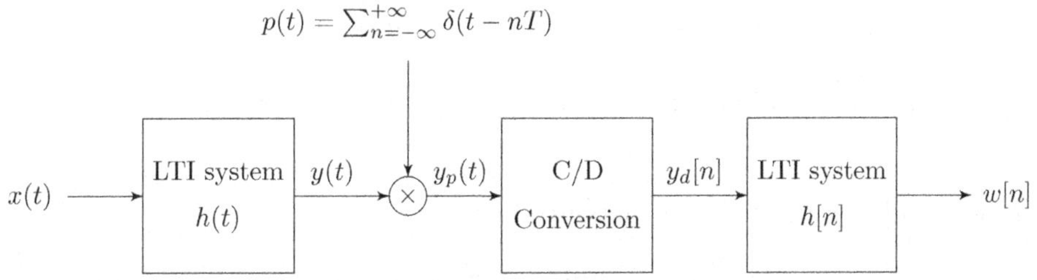

Figure 8.16: Problem 155

Determine $h[n]$ such that if the input of the system is $x(t) = \delta(t)$, then the output is equal to $w[n] = \delta[n]$.

Solution. Taking the Fourier transform of both sides of eq.(8.17), we have

$$(jw)Y(jw) + 2Y(jw) = X(jw),$$

then

$$H(jw) = \frac{Y(jw)}{X(jw)} = \frac{1}{jw + 2}. \tag{8.18}$$

If $x(t) = \delta(t)$, thus $y(t)$ is given by

$$y(t) = e^{-2t}u(t).$$

From Fig. 8.16, we can see that

$$y_p(t) = y(t) \cdot p(t),$$

where

$$\begin{aligned} y_p(t) &= e^{-2t}u(t) \sum_{n=-\infty}^{+\infty} \delta(t - nT) \\ &= \sum_{n=0}^{+\infty} e^{-2nT}\delta(t - nT). \end{aligned} \tag{8.19}$$

$y_p(t)$ is fed into the Analog to Digital converter to get $y_d[n]$:

$$\begin{aligned} y_d[n] = y_p(nT) &= \sum_{k=0}^{+\infty} e^{-2kT}\delta[nT - kT] \\ &= \sum_{k=0}^{+\infty} e^{-2kT}\delta[n - k] = e^{-2nT}u[n]. \end{aligned} \tag{8.20}$$

Thus the input to the discrete-time filter $h[n]$ is $y_d[n]$ in (8.20). If $w[n] = \delta[n]$, then we must have:

$$y_d[n] * h[n] = \delta[n].$$

Taking the Fourier transform of both sides of above equation, we have

$$Y_d(e^{jw}).H(e^{jw}) = 1 \;\Rightarrow\; H(e^{jw}) = \frac{1}{Y_d(e^{jw})} = 1 - e^{-2T}e^{-jw}.$$

Then the impulse response $h[n]$ is obtained as

$$h[n] = \delta[n] - e^{-2T}\delta[n-1].$$

Problem 156. Consider the continuous-time filter with the frequency response as follows:

$$H(jw) = \frac{1}{jw+1}.$$

We want to implement the discrete-time version of this filter by setting $h[n] = h(nT)$. Determine the value of T such that the half power bandwidth (3dB bandwidth) is equal to $w = \frac{\pi}{2}$.

Solution. We know that the peak value for this filter happens at $w = 0$, then we have

$$10\log\left(\frac{|H(e^{j\pi/2})|}{|H(e^{j0})|}\right)^2 = -3 \text{ dB}.$$

The impulse response of the continuous-time system is

$$h(t) = e^{-t}u(t).$$

So

$$h[n] = h(nT) = e^{-nT}u[n],$$

and the frequency response of the discrete-time system is given by

$$H(e^{jw}) = \sum_{n=0}^{+\infty} e^{-nT}e^{-jnw} = \frac{1}{1 - e^{-T}e^{-jw}}. \tag{8.21}$$

Substituting $w = \frac{\pi}{2}$ and $w = 0$ into (8.21), we obtain

$$|H(e^{j\pi/2})|^2 = \frac{1}{1 + e^{-2T}}, \quad |H(e^{j0})|^2 = \frac{1}{(1 - e^{-T})^2}.$$

Then we have

$$20\log(\frac{|H(e^{j\pi/2})|}{|H(e^{j0})}|) = 10\log(\frac{|H(e^{j\pi/2})|^2}{|H(e^{j0})|^2}) = -3$$

$$\Rightarrow \frac{(1 - e^{-T})^2}{1 + e^{-2T}} = 10^{-0.3} \simeq 0.5,$$

and

$$0.5e^{-2T} - 2e^{-T} + 0.5 = 0.$$

By solving the above quadratic equation, we have 2 solutions:

$$e^{-T} = 3.7, \Rightarrow T = -1.3$$

$$e^{-T} = 0.3, \Rightarrow T = 1.2$$

The only acceptable solution is $T = 1.2$.

Problem 157. Let $x(t)$ be a continuous-time signal whose Fourier transform has the property that $X(jw) = 0$ for $|w| > w_0$. Signal $x(t)$ is sampled at sampling frequency $w_s > 2w_0$, to obtain signal $x_d[n]$.

Determine the relationship between the energy E_d of $x_d[n]$ and the energy E of the original signal, if the sampling period is T. Note that the energy in a continuous-time function $x(t)$ is defined as

$$E = \int_{-\infty}^{+\infty} |x(t)|^2 dt,$$

and the energy of a sequence $x[n]$ is defined as

$$E_d = \sum_{n=-\infty}^{+\infty} |x[n]|^2.$$

Solution. Using Parseval's identity, we have

$$\begin{aligned} E &= \int_{-\infty}^{+\infty} |x(t)|^2 dt \\ &= \frac{1}{2\pi}\int_{-\infty}^{+\infty} |X(jw)|^2 dw \\ &= \frac{1}{2\pi}\int_{-w_0}^{+w_0} |X(jw)|^2 dw. \end{aligned} \tag{8.22}$$

The Fourier transform of $x_d[n]$ is given by

$$X_d(e^{jw}) = \begin{cases} \frac{1}{T_s}X(\frac{jw}{T_s}) & |w| \leq w_0 T_s \\ 0 & w_0 T_s < |w| < \pi \end{cases}.$$

Again based on Parseval's identity, the energy of $x_d[n]$ is given by

$$\begin{aligned} E_d &= \sum_{n=-\infty}^{+\infty} |x[n]|^2 \\ &= \frac{1}{2\pi}\int_{-\pi}^{+\pi} |X(e^{jw})|^2 dw \\ &= \frac{1}{2\pi}\int_{-w_0T_s}^{+w_0T_s} \frac{1}{T_s^2}|X(\frac{jw}{T_s})|^2 dw, \end{aligned}$$

where we used the above relation between $X_d(e^{jw})$ and $X(jw)$ to obtain the last equality. Now letting

$$\frac{w}{T_s} = u \quad \text{and} \quad \frac{dw}{T_s} = du,$$

we obtain

$$E_d = \frac{1}{T_s 2\pi}\int_{-w_0}^{+w_0} |X(ju)|^2 du. \tag{8.23}$$

Comparing (8.22) with (8.23), we conclude that

$$E_d = \frac{1}{T_s}E \quad \Rightarrow \quad E = T_s E_d.$$

Problem 158. Shown in Fig. 8.17(a), is a system that processes continuous-time signals using a discrete-time filter $h[n]$. The Fourier transform of the input signal is as indicated in Fig. 8.17 (b).

$$h[n] = \delta[n],$$

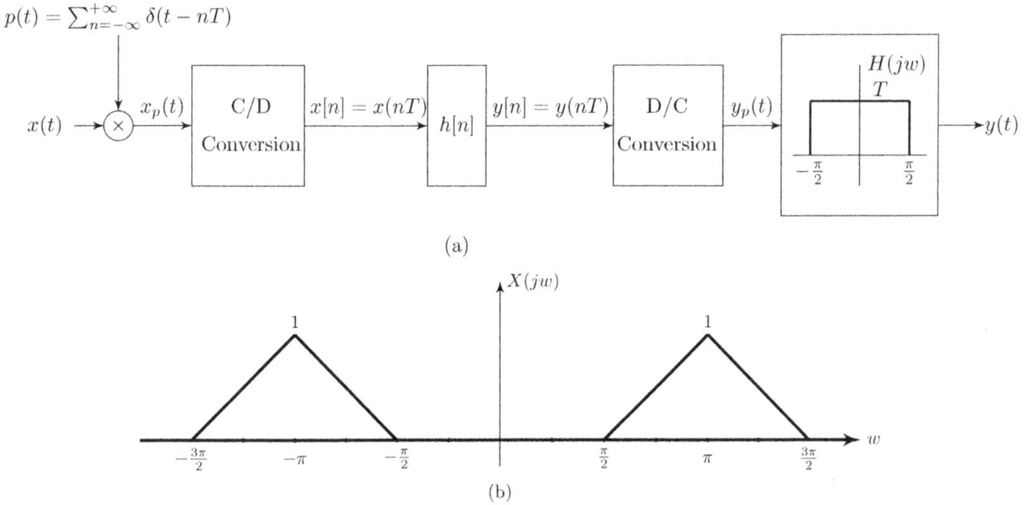

Figure 8.17: Problem 158

(a) Determine the maximum sampling interval T such that $x(t)$ is recoverable from $x[n]$ through the use of a low-pass filter.

(b) If $T = 2$, sketch $X_p(jw)$.

(c) If $T = 2$, sketch $Y(e^{jw})$ and $Y(jw)$.

Solution.

(a) To determine the maximum sampling interval T, we must calculate the highest frequency present in the signal. From Fig. 8.17(b), we see that the highest frequency present in signal $x(t)$ is

$$w_{max} = \frac{3\pi}{2} \quad \Rightarrow \quad w_s > 2w_{max} = 3\pi.$$

Then, the maximum sampling interval is given by

$$T_s = \frac{2\pi}{w_s} \quad \Rightarrow \quad T < \frac{2}{3}.$$

(b) The spectrum $X_p(jw)$ is a periodic function of w consisting of superposition of shifted replicas of $X(jw)$ as illustrated in Fig. 8.18.

$$x_p(t) = x(t) \times \sum_{n=-\infty}^{+\infty} \delta(t - nT)$$

Substituting $T = 2$, we write

$$X_p(jw) = \frac{1}{2\pi} X(jw) * \frac{2\pi}{T} \sum_{n=-\infty}^{+\infty} \delta(w - \frac{2\pi k}{T})$$

$$= \frac{1}{2} X(jw) * \sum_{n=-\infty}^{+\infty} \delta(w - k\pi)$$

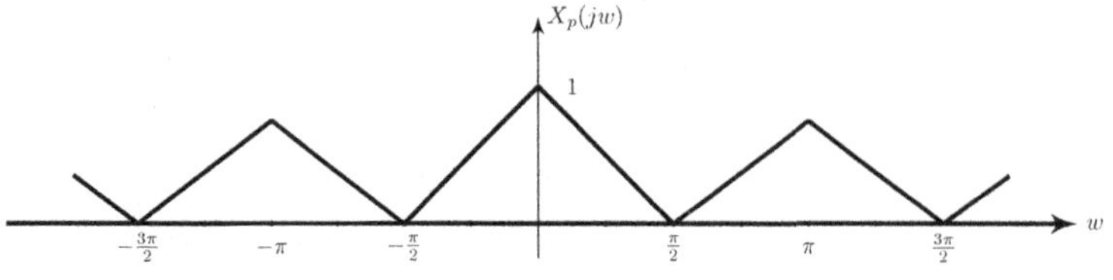

Figure 8.18: Spectrum of $x_p(t)$ in Problem 158

(c) First note that

$$x[n] = x_p(nT) \quad \Rightarrow \quad X(e^{jw}) = X_p(\frac{jw}{T}).$$

Now we see that the discrete-time filter $h[n]$ is in fact zero for all values of n except $n = 0$. So we have

$$y[n] = x[n],$$

and so

$$Y(e^{jw}) = X_p(\frac{jw}{T}),$$

where $Y(e^{jw})$ is depicted in Fig. 8.19.

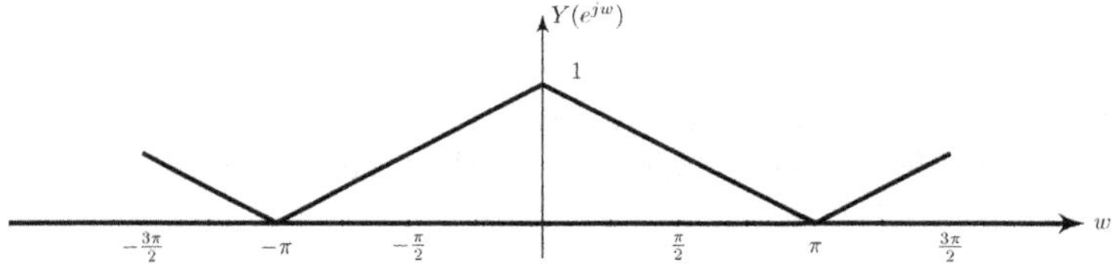

Figure 8.19: Fourier transform of $y[n]$, Problem 158

After passing through D/C conversion, we obtain $Y_p(jw)$ as the figure below.

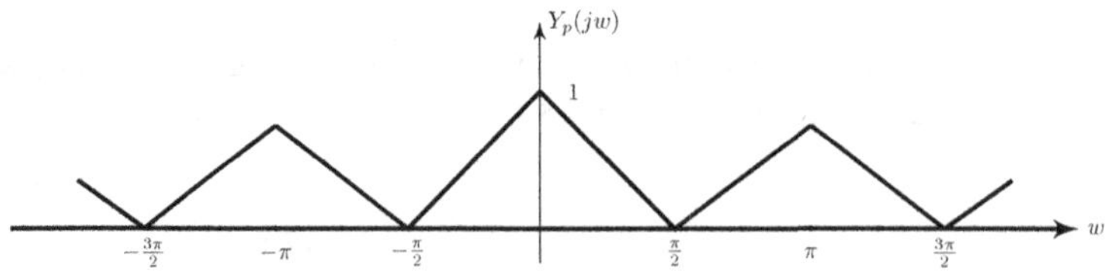

Figure 8.20: Spectrum of $y_p(t)$ in Problem 158.

From Fig. 8.17(a), we see that $H(jw)$ is an ideal low-pass filter with gain T and cutoff frequency $w_c = \frac{\pi}{2}$. Then, the Fourier transform of $y(t)$ is obtained as depicted in Fig. 8.21.

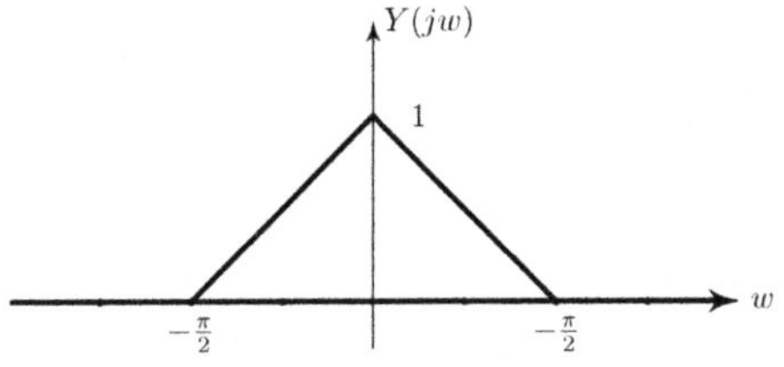

Figure 8.21: Spectrum of $y(t)$, Problem 158.

Chapter 9

Laplace Transform

9.1 Introduction

In this chapter, we will introduce the Laplace transform and its applications in representing the continuous-time signals and systems.

9.2 The Laplace Transform

Consider a continuous-time LTI system with impulse response $h(t)$. It is a well-known fact that the output of the system to an input in the form of e^{st} is

$$y(t) = H(s)e^{st}, \tag{9.1}$$

where s in general is a complex number

$$s = \sigma + j\omega,$$

and

$$H(s) = \int_{-\infty}^{+\infty} h(t)e^{-st}dt. \tag{9.2}$$

In (9.2), $H(s)$ is defined as the Laplace transform of the impulse response $h(t)$. In general, the Laplace transform of a continuous-time signal $x(t)$ is defined as

$$X(s) = \int_{-\infty}^{+\infty} x(t)e^{-st}dt. \tag{9.3}$$

where

$$s = \sigma + j\omega.$$

The relationship between the Laplace transform and Fourier transform is then given by

$$X(s) = \int_{-\infty}^{+\infty} x(t)e^{-(\sigma+j\omega)t}dt$$

$$= \int_{-\infty}^{+\infty} x(t)e^{-\sigma t}e^{-j\omega t}dt$$
$$= \mathcal{F}.\left\{x(t)e^{-\sigma t}\right\}. \tag{9.4}$$

From (9.4), we see that the Laplace transform of $x(t)$ is the Fourier transform of $x(t)e^{-\sigma t}$. According to the convergence conditions for the continuous-time Fourier transform, $x(t)$ has a Laplace transform, if $x(t)e^{-\sigma t}$ is absolutely integrable; that is,

$$\int_{-\infty}^{+\infty} x(t)e^{-\sigma t}dt < \infty.$$

9.3 The Region of Convergence for Laplace Transform

The range of values of the complex variables s for which the Laplace transform is defined for a continuous-time signal, is called the *region of convergence* (ROC). This is equivalent to the range of values of σ for which the Fourier transform of $|x(t)e^{-\sigma t}|$ converges.

9.3.1 Definition of Poles and Zeros of $X(s)$

Usually, the Laplace transform of a signal, $X(s)$, is in the form of a rational function of s as

$$X(s) = \frac{N(s)}{D(s)} = \frac{a_0 s^m + a_1 s^{m-1} + \cdots + a_m}{b_0 s^n + b_1 s^{n-1} + \cdots + b_n} = \frac{a_0}{b_0}\frac{(s-z_1)\cdots(s-z_m)}{(s-p_1)\cdots(s-p_n)}, \tag{9.5}$$

where $N(s)$ and $D(s)$ are called the numerator polynomial and denominator polynomial, respectively. The roots of $N(s)$ are called the *zeros* of $X(s)$ and for these values $X(s) = 0$. Similarly, the roots of $D(s)$ are called the *poles* of $X(s)$ and for these values $X(s) = \infty$. Since $X(s)$ does not converge at the poles, then the ROC of $X(s)$ does not include these values. The location of each pole is shown by "×" and the location of each zero is illustrated by "o" on the s-plane:

$$X(s) = 0 \Rightarrow N(s) = 0 \Rightarrow s = z_1,\ z_2, \cdots, z_m.$$
$$X(s) = \infty \Rightarrow D(s) = 0 \Rightarrow s = p_1,\ p_2, \cdots, p_n.$$

9.3.2 Properties of the ROC:

To determine a complete specification of a Laplace transform, we need to know both the algebraic expression for X(s) and the associated region of convergence. In this subsection, we introduce some properties of the ROC that are helpful to determine ROC of a given $X(s)$.

Property 1: The ROC does not contain any poles.

Property 2: If $x(t)$ is a *finite-duration* signal and absolutely integrable, then the ROC is the entire s-plane. Note that a finite-duration signal is a signal for which $x(t) = 0$ except in a finite interval $t_1 \leq t \leq t_2$.

Property 3: If $x(t)$ is a *right-sided* signal, i.e., $x(t) = 0$ for $t < t_0$, for some $t_0 \in$ R, and $\Re\{s\} = \sigma_0$ is in the ROC, then all values of s for which $\Re\{s\} > \sigma_0$ will also be in the ROC. In other words, ROC contains $\sigma = +\infty$.

Property 4: If $x(t)$ is a *left-sided* signal, $x(t) = 0$ for $t > t_0$, for some $t_0 \in \mathrm{R}$, and $\Re\{s\} = \sigma_0$ is in the ROC, then all values of s for which $\Re\{s\} < \sigma_0$ will also be in the ROC. In other words, ROC contains $\sigma = -\infty$.

Property 5: If $x(t)$ is *two-sided*, and $\Re\{s\} = \sigma_0$ is in the ROC, then the ROC will include a strip in the s-plane that contains the line $\Re\{s\} = \sigma_0$.

9.4 Properties of the Laplace Transform

(a) Linearity

If

$$x_1(t) \overset{\mathcal{L}}{\longleftrightarrow} X_1(s), \quad \text{with ROC} = R_1$$

$$x_2(t) \overset{\mathcal{L}}{\longleftrightarrow} X_2(s), \quad \text{with ROC} = R_2$$

then we have

$$y(t) = ax_1(t) + bx_2(t) \overset{\mathcal{L}}{\longleftrightarrow} Y(s) = aX_1(s) + bX_2(s), \quad \text{with ROC of at least } R_1 \cap R_2. \tag{9.6}$$

(b) Time Shifting

If

$$x(t) \overset{\mathcal{L}}{\longleftrightarrow} X(s), \quad \text{with ROC} = R,$$

then

$$x(t - t_0) \overset{\mathcal{L}}{\longleftrightarrow} e^{-st_0}X(s), \quad \text{with ROC} = R. \tag{9.7}$$

From (9.7), it is clear that the ROC does not change after the time-shift operation.

(c) Shifting in the s-Domain

If

$$x(t) \overset{\mathcal{L}}{\longleftrightarrow} X(s), \quad \text{with ROC} = R,$$

then

$$e^{s_0 t}x(t) \overset{\mathcal{L}}{\longleftrightarrow} X(s - s_0), \quad \text{with ROC} = R \text{ horizontally shifted by } \Re\{s_0\}. \tag{9.8}$$

(d) Conjugation

If

$$x(t) \overset{\mathcal{L}}{\longleftrightarrow} X(s), \quad \text{with ROC} = R,$$

then

$$x^*(t) \overset{\mathcal{L}}{\longleftrightarrow} X^*(s^*), \quad \text{with ROC} = R. \tag{9.9}$$

Therefore,

$$X(s) = X^*(s^*), \quad \text{when } x(t) \text{ is real.} \tag{9.10}$$

(e) Time Scaling

If

$$x(t) \overset{\mathcal{L}}{\longleftrightarrow} X(s), \quad \text{with ROC} = R,$$

then for $a \in R^+$

$$x(at) \overset{\mathcal{L}}{\longleftrightarrow} \frac{1}{|a|}X(\frac{s}{a}), \quad \text{with ROC} = R \text{ scaled horizontally by factor } a. \tag{9.11}$$

From (9.11), it is clear that when $a > 1$, the time scaling causes ROC to be expanded, and for $0 < a < 1$, the ROC is compressed.

(f) Convolution Property

If

$$x_1(t) \overset{\mathcal{L}}{\longleftrightarrow} X_1(s), \quad \text{with ROC} = R_1$$
$$x_2(t) \overset{\mathcal{L}}{\longleftrightarrow} X_2(s), \quad \text{with ROC} = R_2$$

Then we have

$$x_1(t) * x_2(t) \overset{\mathcal{L}}{\longleftrightarrow} X_1(s)X_2(s), \quad \text{with ROC of at least } R_1 \cap R_2. \tag{9.12}$$

(g) Laplace Transform of Periodic Signals

Consider a periodic signal $y(t)$ with Period T, where one period of this signal is denoted by $f(t), 0 < t < T$. We already know the Fourier transform of a periodic signal is in the form of Impulse functions. A periodic signal is in general a 2-sided signal defined over $(-\infty\ +\infty)$. If we consider only one side of the signal, say, for $t > 0$, and with some negligence, we still call it a periodic signal named $x(t)$, i.e.,

$$x(t) = \begin{cases} y(t) & t > 0 \\ 0 & \text{else,} \end{cases}$$

then, the Laplace transform of $x(t)$ can be obtained based on the Laplace transform of $f(t)$ as follows:

$$x(t) \overset{\mathcal{L}}{\longleftrightarrow} \frac{F(s)}{1 - e^{-Ts}}, \quad \text{with ROC} = Re\{s\} > 0. \tag{9.13}$$

9.5 Table of Laplace Properties and of Common Laplace Transform Pairs

Some of the important properties of the Laplace transform are given in Table 9.1 In Table 9.2, we have summarized some common Laplace transform pairs.

Table 9.1: Properties of the Laplace Transform

Property	**signal**	**Laplace transform**	ROC
	$x(t)$ $x_1(t)$ $x_2(t)$	$X(s)$ $X_1(s)$ $X_2(s)$	R R_1 R_2
Linearity	$ax_1(t) + bx_2(t)$	$aX_1(s) + bX_2(s)$	At least $R_1 \cap R_2$
Time Shifting	$x(t - t_0)$	$e^{-st_0}X(s)$	R
Shifting in the s-Domain	$e^{s_0 t}x(t)$	$X(s - s_0)$	Shifted version of R
Conjugation	$x^*(t)$	$X^*(s^*)$	R
Time Reversal	$x(-t)$	$X(-s)$	Flipped R
Time scaling	$x(at)$	$\frac{1}{\lvert a\rvert}X\left(\frac{s}{a}\right)$	Scaled version of R
Convolution	$x_1(t) * x_2(t)$	$X_1(s)X_2(s)$	At least $R_1 \cap R_2$
Differentiation in Time	$\frac{d}{dt}x(t)$	$sX(s)$	At least R
Integration in Time	$\int_{-\infty}^{t} x(\tau)d\tau$	$\frac{1}{s}X(s)$	At lst $R \cap \{\Re\{s\} > 0\}$
Differentiation in the s-Domain	$-tx(t)$	$\frac{d}{ds}X(s)$	R
Integration in the s-Domain	$\frac{1}{t}x(t)$	$\int_{s}^{+\infty} X(\tau)d\tau$ $\lim_{t\to 0} \frac{x(t)}{t} \neq \infty$	R

Table 9.2: Common Laplace Transforms Pairs

Signal	**Transform**	ROC
$\delta(t)$	1	the whole s-plane
$\delta(t-t_0)$	e^{-st_0}	the whole s-plane
$u(t)$	$\frac{1}{s}$	$\Re\{s\} > 0$
$-u(-t)$	$\frac{1}{s}$	$\Re\{s\} < 0$
$\frac{t^{n-1}}{(n-1)!}u(t)$	$\frac{1}{s^n}$	$\Re\{s\} > 0$
$-\frac{t^{n-1}}{(n-1)!}u(-t)$	$\frac{1}{s^n}$	$\Re\{s\} < 0$
$e^{-at}u(t)$	$\frac{1}{s+a}$	$\Re\{s\} > -a$
$-e^{-at}u(-t)$	$\frac{1}{s+a}$	$\Re\{s\} < -a$
$\frac{t^{n-1}}{(n-1)!}e^{-at}u(t)$	$\frac{1}{(s+a)^n}$	$\Re\{s\} > -a$
$-\frac{t^{n-1}}{(n-1)!}e^{-at}u(-t)$	$\frac{1}{(s+a)^n}$	$\Re\{s\} < -a$
$\cos(w_0t)u(t)$	$\frac{s}{s^2+w_0^2}$	$\Re\{s\} > 0$
$\sin(w_0t)u(t)$	$\frac{w_0}{s^2+w_0^2}$	$\Re\{s\} > 0$
$e^{-at}\cos(w_0t)u(t)$	$\frac{s+a}{(s+a)^2+w_0^2}$	$\Re\{s\} > -a$
$e^{-at}\sin(w_0t)u(t)$	$\frac{w_0}{(s+a)^2+w_0^2}$	$\Re\{s\} > -a$
$u_n(t) = \frac{d^n\delta(t)}{dt^n}$	s^n	All s

9.6 The Initial and Final Value Theorems

9.6.1 The Initial-Value Theorem

If $x(t) = 0$ for $t < 0$, and $x(t)$ contains no impulses or higher order singularities at $t = 0$, then

$$x(0^+) = \lim_{s\to\infty} sX(s). \tag{9.14}$$

9.6.2 The Final-Value Theorem

If $x(t) = 0$ for $t < 0$, and all poles of $sX(s)$ are located in the left-side of the jw-axis (i.e, the ROC contains the jw-axis), then the final-value theorem states that

$$\lim_{t \to \infty} x(t) = \lim_{s \to 0} sX(s). \tag{9.15}$$

Extended Final-Value Theorem:
Let σ be a real positive number. The following points are important when using this theorem:
1) If ROC contains the jw-axis, we necessarily must have $x(\infty) = x(-\infty) = 0$
2) If ROC does not contain the jw-axis and lies in its right side, i.e., $0 < Re\{s\} < \sigma$, then $x(\infty) = \lim_{s \to 0} sX(s)$ and $x(-\infty) = 0$.
3) If ROC does not contain the jw-axis and lies in its left side, i.e., $-\sigma < Re\{s\} < 0$, then $x(-\infty) = \lim_{s \to 0} - sX(s)$ and $x(\infty) = 0$.

9.7 The Inverse Laplace Transform

If $X(S)$ is the Laplace transform of $x(t)$, $x(t)$ can be obtained based on $X(s)$ as follows:

$$x(t) = \int_{-\infty}^{+\infty} X(S)e^{st}ds. \tag{9.16}$$

Instead of direct integration, there are some simpler methods to obtain a signal $x(t)$ from its Laplace transform $X(s)$. In this section, we introduce two methods for obtaining the inverse Laplace transform.

9.7.1 Use of Tables of Laplace Transform Pairs

In this method we try to express $X(s)$ as a sum of some functions whose transform we already know:

$$X(s) = X_1(s) + X_2(s) + \cdots + X_n(s). \tag{9.17}$$

Knowing that signals $x_1(t),\ x_2(t), \cdots,\ x_n(t)$ are the inverse transforms of

$$X_1(s),\ X_2(s), \cdots,\ X_n(s)$$

respectively, from the linearity property, we have

$$x(t) = x_1(t) + x_2(t) + \cdots + x_n(t). \tag{9.18}$$

9.7.2 Partial-Fraction Expansion

Let $X(s)$ be a rational function as

$$X(s) = \frac{N(s)}{D(s)} = k\frac{(s - z_1)\cdots(s - z_m)}{(s - p_1)\cdots(s - p_n)}; \tag{9.19}$$

(a) If all poles of $X(s)$ are simple and $m < n$, then we can write

$$X(s) = \frac{c_1}{s - p_1} + \cdots + \frac{c_n}{s - p_n}, \tag{9.20}$$

where coefficients c_k are given by

$$c_k = (s - p_k)X(s)\Big|_{s=p_k}. \tag{9.21}$$

(b) If the poles of $X(s)$ are multiple and $m < n$, then for each multiple pole of $X(s)$, the expansion of $X(s)$ will consist of terms of the form

$$\frac{\lambda_1}{s - p_i} + \frac{\lambda_2}{(s - p_i)^2} \cdots + \frac{\lambda_r}{(s - p_i)^r} \tag{9.22}$$

where

$$\lambda_{r-k} = \frac{1}{k!}\frac{d^k}{ds^k}\left[(s - p_i)^r X(s)\right]\Big|_{s=p_i}. \tag{9.23}$$

(c) If $m > n$, we first divide the numerator by the denominator to comp up with a polynomial plus a rational function for which the polynomial degree of the numerator is less than the denominator. Now for the rational part, we end up with one of the above cases and for the polynomial part, we use the last row of Table 9.2 to obtain the inverse.

For all the above cases, we can now use the method of Section 9.7.1 to obtain the inverse.

9.8 Analysis of LTI Systems Using the Laplace Transform

Consider an LTI system with the impulse response $h(t)$. Let the input be a continuous-time signal $x(t)$. The Laplace transform of the output signal $y(t)$, and eventually $y(t)$ itself, can be obtained based on the Laplace transform of $x(t)$ and the Laplace transform of $h(t)$. That is, if

$$y(t) = h(t) * x(t) \quad \overset{\mathcal{L}}{\longleftrightarrow} \quad Y(s) = H(s)X(s) \tag{9.24}$$

then

$$H(s) = \frac{Y(s)}{X(s)}. \tag{9.25}$$

where $H(s)$ is referred to as the system function.

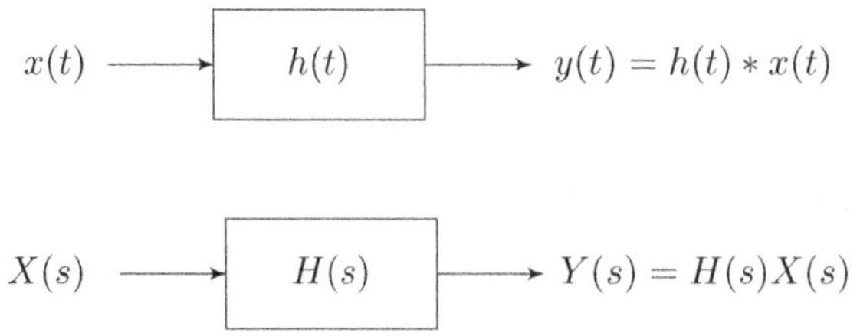

Hint 1: Based on the convolution property stated in Table 9.1, when 2 time-domain signals are convolved, the ROC of the Laplace transform of the resulting signals is the intersection of the ROC's of 2 convolved signals Laplace transforms. As such, for the above formulation

to hold, the ROC of the input function Laplace transform, $X(s)$, the ROC of output function Laplace transform, $Y(s)$, and the ROC of the system transfer function $H(s)$, should be have non-empty intersection:

$$\text{LTI system} \overset{Y(s)=X(s)H(s)}{\Longleftrightarrow} [R_X \cap R_Y \cap R_H] \neq \emptyset. \tag{9.26}$$

Hint 2:

We already know that the output $y(t)$ of an LTI system to the input e^{jw_0t} is equal to $H(w_0)e^{jw_0t}$. In the Laplace domain, the same relationship exists for the input e^{s_0t}. However, s_0 must be in the ROC of $H(S)$, otherwise,the output is unbounded, i.e.,

$$y(t) = \begin{cases} H(s_0)e^{s_0t}, & \forall t \ (s_0 \in \text{ROC}) \\ \\ \infty, & \forall t \ (s_0 \notin \text{ROC}) \end{cases}. \tag{9.27}$$

9.9 Characterization of LTI Systems Using the Laplace Transform

9.9.1 Causality

For a causal LTI system, we know that

$$h(t) = 0, \quad t < 0.$$

It is clear that $h(t)$ is a right-sided signal. Consequently, the ROC of the system transfer function for a causal LTI system spans a region such that $\sigma_0 < \Re\{s\} < +\infty$ for some real σ_0.

9.9.2 Stability

From Chapter 3 we know that a continuous-time LTI system is stable if the impulse response is absolutely integrable, i.e., if

$$\int_{-\infty}^{+\infty} |h(t)|dt < \infty. \tag{9.28}$$

Therefore, if an LTI system is stable, the ROC of its system transfer function necessarily contains the jw-axis.

9.9.3 Causal and Stable Systems

If an LTI system is both causal and stable, the ROC must contain both the jw-axis and $s = +\infty$. This means that all poles of $H(s)$ must be located in the left half of the s-plane, i.e., the real part of all poles is negative.

9.10 LTI Systems Characterized by Constant-Coefficient Difference Equations

Consider a continuous-time LTI system with the input $x(t)$ and output $y(t)$ which satisfy the relationship of the form

$$\sum_{k=0}^{N} a_k \frac{d^k y(t)}{dt^k} = \sum_{k=0}^{M} b_k \frac{d^k x(t)}{dt^k}. \tag{9.29}$$

Applying the Laplace transform and using the differentiation property of the Laplace transform we obtain

$$\sum_{k=0}^{N} a_k s^k Y(s) = \sum_{k=0}^{M} b_k s^k X(s), \tag{9.30}$$

or

$$Y(s) \sum_{k=0}^{N} a_k s^k = X(s) \sum_{k=0}^{M} b_k s^k. \tag{9.31}$$

Thus we have

$$H(s) = \frac{Y(s)}{X(s)} = \frac{\sum_{k=0}^{M} b_k s^k}{\sum_{k=0}^{N} a_k s^k}. \tag{9.32}$$

9.11 Systems Interconnection

For Laplace transforms associated to different system interconnections, we can consider the following cases:

(a) The impulse response of a parallel combination of two LTI systems:

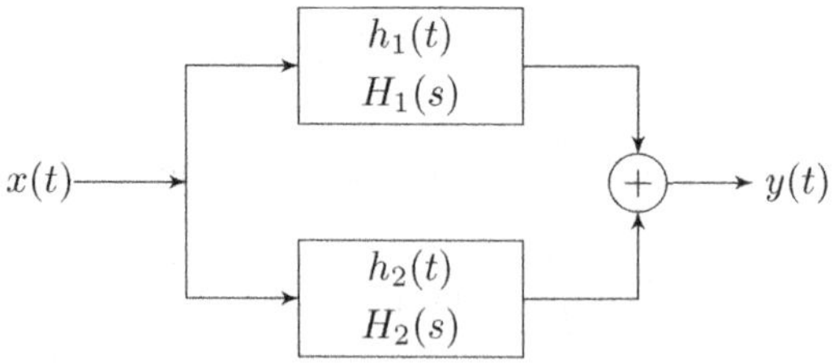

Figure 9.1: A parallel interconnection of two systems.

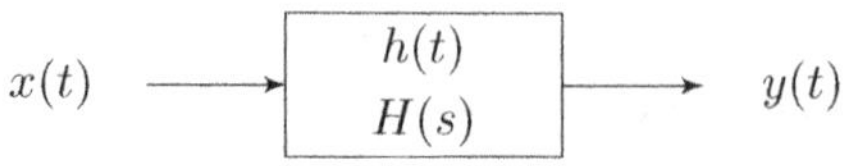

$$h(t) = h_1(t) + h_2(t), \tag{9.33}$$

$$H(s) = H_1(s) + H_2(s). \tag{9.34}$$

(b) The impulse response of a cascade combination of two LTI systems:

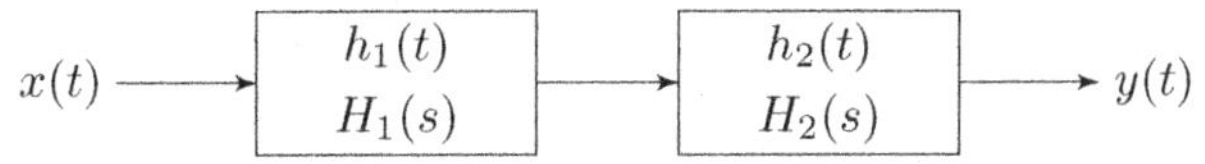

Figure 9.2: A series or cascade interconnection of two systems.

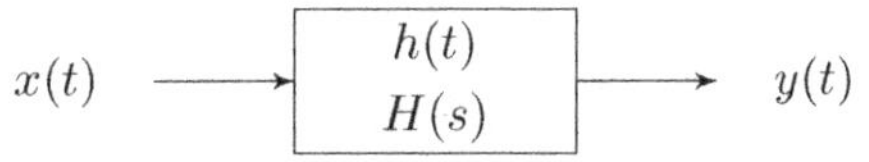

$$h(t) = h_1(t) * h_2(t), \tag{9.35}$$

$$H(s) = H_1(s)H_2(s). \tag{9.36}$$

(c) The impulse response of a feedback interconnection:

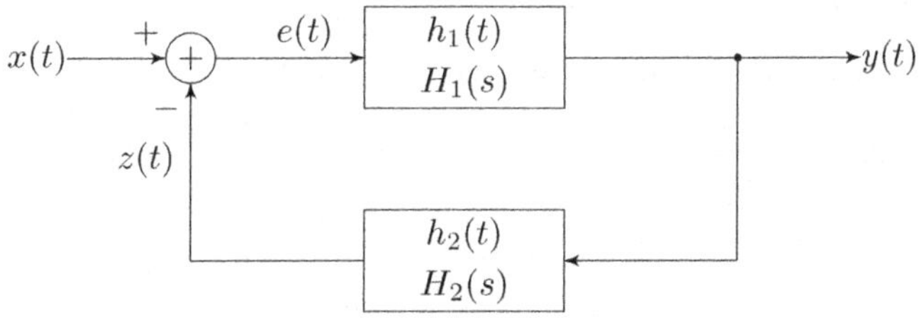

Figure 9.3: A feedback interconnection.

$$\begin{aligned} Y(s) = H_1(s)E(s), \quad E(s) &= X(s) - Y(s)H_2(s), \\ \Rightarrow Y(s) &= H_1(s)\left[X(s) - Y(s)H_2(s)\right], \\ H(s) = \frac{Y(s)}{X(s)} &= \frac{H_1(s)}{1 + H_1(s)H_2(s)}. \end{aligned} \tag{9.37}$$

9.12 Problems

Problem 159. Use the definition to obtain the Laplace transform of the following signals:

(a) $x(t) = 3e^{2t}u(t)$

(b) $x(t) = 3e^{-2t}u(t) + te^{-t}u(t)$

(c) $x(t) = e^{-2|t|}$

(d) $x(t) = u(t+2)$

Solution.

(a) $x(t) = 3e^{2t}u(t)$

Using the definition of the Laplace transform we write

$$\begin{aligned} X(s) &= \int_{-\infty}^{\infty} 3e^{2t}e^{-st}u(t)dt \\ &= \int_{0}^{\infty} 3e^{2t}e^{-st}dt \\ &= \frac{3}{2-s}e^{(2-s)t}|_0^{\infty}. \end{aligned}$$

The integral is finite only for $Re\{s\} > 2$ as for these values of s, $e^{(2-s)\infty} = 0$. So the Laplace transform and the ROC are obtained as:

$$X(s) = \frac{3}{s-2}, \qquad ROC: Re\{s\} > 2$$

We use similar argument to find the ROC for other parts of this question.

(b) $x(t) = 3e^{-2t}u(t) + te^{-t}u(t)$

Using the definition of the Laplace transform we write

$$\begin{aligned} X(s) &= \int_{-\infty}^{\infty} 3e^{-2t}e^{-st}u(t)dt + \int_{-\infty}^{\infty} te^{-t}e^{-st}u(t)dt \\ &= \int_{0}^{\infty} 3e^{-2t}e^{-st}dt + \underbrace{\int_{0}^{\infty} te^{-t}e^{-st}dt}_{\text{evaluated using integral by part}} \\ &= \frac{3}{s+2} + \frac{e^{-(1+s)t}}{(1+s)^2}(-ts-t-1)|_0^{\infty} = \frac{3}{s+2} + \frac{1}{(1+s)^2} \\ &= \frac{3s^2+7s+5}{(s+1)^2(s+2)}, \qquad ROC: Re\{s\} > -1. \end{aligned}$$

(c) $x(t) = e^{-2|t|}$

Using the definition of the Laplace transform we write

$$\begin{aligned} X(s) &= \int_{-\infty}^{0} e^{2t}e^{-st}dt + \int_{0}^{\infty} e^{-2t}e^{-st}dt \\ &= \frac{1}{2-s}e^{(2-s)t}|_{-\infty}^{0} - \frac{1}{2+s}e^{-(2+s)t}|_0^{\infty} \\ &= \frac{1}{2-s} + \frac{1}{2+s} = \frac{4}{2-s^2}, \qquad ROC: -2 < Re\{s\} < 2. \end{aligned}$$

(d) $x(t) = u(t+2)$

By definition of the Laplace transform, we can write

$$X(s)=\int_{-\infty}^{\infty}u(t+2)e^{-st}dt=\int_{-2}^{\infty}e^{-st}dt=-\frac{1}{s}e^{-ts}|_{-2}^{\infty}=\frac{1}{s}e^{2s}, \qquad ROC: Re\{s\}>0.$$

Problem 160. For each of the following signals, use Laplace transform properties to determine Laplace transforms.

(a) $x(t)=te^{-t}u(t)$

(b) $x(t)=te^{-t}\cosh(t)u(t)$

(c) $x(t)=\int_0^t e^{-2\tau}\cos(\tau)d\tau$

(d) $x(t)=\int_0^t (t-\lambda)^4\sin(\lambda)d\lambda$

(e) $x(t)=\sin^2(wt)u(t)$

(f) $x(t)=te^{-2|t|}$

(g) $x(t)=\cos(t)u(t)+e^{-t}u(-t)$

Solution.

(a) $x(t)=te^{-t}u(t)$

We know that $\mathcal{L}\{u(t)\}=\frac{1}{s}$, and using the Laplace-domain derivation property, we can write

$$u(t)\ \overset{\mathcal{L}}{\to}\ \frac{1}{s},\quad tu(t)\ \overset{\mathcal{L}}{\to}\ \frac{-d}{ds}(\frac{1}{s})=\frac{1}{s^2}.$$

Now using the frequency shifting property we have

$$\mathcal{L}\{e^{-t}tu(t)\}=\frac{1}{(s+1)^2}.$$

(b) $x(t)=te^{-t}\cosh(t)u(t)$

First let's obtain the Laplace transform of $\cosh(t)u(t)=(\frac{e^t}{2}+\frac{e^{-t}}{2})u(t)$:

$$\mathcal{L}\{\cosh(t)u(t)\}=\frac{1}{2}(\frac{1}{s-1}+\frac{1}{s+1})=\frac{s}{s^2-1}.$$

Now we use the Laplace-domain derivation property to get

$$\mathcal{L}\{t\cosh(t)u(t)\}=\frac{-d}{ds}(\frac{s}{s^2-1})=\frac{s^2+1}{(s^2-1)^2},$$

and using frequency shifting property we have

$$\mathcal{L}\{te^{-t}\cosh(t)u(t)\}=\frac{(s+1)^2+1}{((s+1)^2-1)^2}=\frac{s^2+2s+2}{s^2(s+2)^2}.$$

(c) $x(t) = \int_0^t e^{-2\tau}\cos(\tau)d\tau$

Let $a(t) = e^{-2t}\cos(t)u(t)$. We have

$$\mathcal{L}\{\cos(t)\} = \frac{s}{s^2+1},$$

and using frequency shifting property we have

$$\mathcal{L}\{e^{-2t}\cos(t)u(t)\} = \frac{s+2}{(s+2)^2+1}.$$

Thus, the Laplace transform of $a(t)$ is obtained as below

$$\mathcal{L}\{a(t)\} = \frac{s+2}{(s+2)^2+1} = A(s).$$

Now based on the time-domain integration property we have

$$\mathcal{L}\{x(t)\} = \frac{1}{s}A(s) + \underbrace{\frac{1}{s}\int_{-\infty}^{0^-} a(\tau)d\tau}_{integral\ is\ equal\ zero} = \frac{s+2}{s((s+2)^2+1)}.$$

(d) $x(t) = \int_0^t (t-\lambda)^4 \sin(\lambda)d\lambda$

$x(t)$ can be seen as the Convolution of $t^4u(t)$ and $\sin(t)u(t)$, so we have

$$\mathcal{L}\{t^4u(t)\} = \frac{4!}{s^5} \quad \text{and} \quad \mathcal{L}\{\sin(t)u(t)\} = \frac{1}{s^2+1},$$

and using the Convolution property, we simply multiply the 2 above transforms to obtain the corresponding Laplace Transform:

$$\mathcal{L}\{x(t)\} = \frac{4!}{s^5}\cdot\frac{1}{s^2+1} = \frac{24}{s^7+s^5}.$$

(e) $x(t) = \sin^2(wt)u(t)$

Let's rewrite $x(t)$ as

$$x(t) = \sin^2(wt)u(t) = \sin(wt)\sin(wt)u(t).$$

We also know that

$$\sin(wt) = \frac{1}{2j}\{e^{jwt} - e^{-jwt}\} = \frac{j}{2}\{e^{-jwt} - e^{jwt}\}.$$

Now we write

$$\mathcal{L}\{\sin^2(wt)u(t)\} = \mathcal{L}\{\sin(wt)u(t)\frac{1}{2j}\{e^{jwt} - e^{-jwt}\}\}.$$

The Laplace transform can be obtained by applying the frequency shifting property to the sin function, once to the left and once to the right, and adding the result together:

$$\mathcal{L}\{\sin(wt)u(t)\frac{1}{2j}\{e^{jwt} - e^{-jwt}\}\} = \frac{j}{2}[\frac{w}{w^2+(s+jw)^2} - \frac{w}{w^2+(s-jw)^2}]$$
$$= \frac{j}{2}[\frac{w}{s^2+2sjw} - \frac{w}{s^2-2sjw}] = \frac{jw}{2}\frac{-4jsw}{(s^2+2sjw)(s^2-2sjw)}$$
$$= \frac{2sw^2}{s^2(s+2jw)(s-2jw)} = \frac{2w^2}{s(s^2+4w^2)}.$$

(f) $x(t) = te^{-2|t|}$

We rewrite $x(t)$ as:

$$x(t) = te^{-2t}u(t) + te^{2t}u(-t) = te^{-2t}u(t) - [-te^{-(-2t)}u(-(-t))].$$

Now if we let $f(t) = te^{-2t}u(t)$ then $x(t) = f(t) - f(-t)$. Concentrating on $f(t)$ we obtain

$$\mathcal{L}\{e^{-2t}u(t)\} = \frac{1}{s+2}$$
$$\mathcal{L}\{f(t) = te^{-2t}u(t)\} = \frac{-d}{ds}\frac{1}{s+2} = \frac{1}{(s+2)^2}.$$

We know that:

$$\text{if} \quad \mathcal{L}\{x(t)\} = X(s) \quad ROC = R \qquad \text{then} \qquad \mathcal{L}\{x(-t)\} = X(-s) \quad ROC = -R.$$

As such

$$\mathcal{L}\{f(t)\} = \frac{1}{(s+2)^2} \quad ROC : Re\{s\} > -2, \quad \mathcal{L}\{f(-t)\} = \frac{1}{(-s+2)^2} \quad ROC : Re\{s\} < 2.$$

So we can obtain the Laplace transform of $x(t)$ as follows:

$$\mathcal{L}\{x(t)\} = \mathcal{L}\{f(t) - f(-t)\} = \frac{1}{(s+2)^2} - \frac{1}{(-s+2)^2} \qquad ROC : -2 < Re\{s\} < 2.$$

(g) $x(t) = \cos(t)u(t) + e^{-t}u(-t)$

Based on Table 9.1,

$$\underbrace{\mathcal{L}\{\cos(t)u(t)\}}_{part\ 1} = \frac{s}{(s^2+1)}, \qquad ROC : Re\{s\} > 0$$
$$\underbrace{\mathcal{L}\{e^{-t}u(-t)\}}_{part\ 2} = \frac{-1}{(s+1)}, \qquad ROC : Re\{s\} < -1.$$

Since the intersection of ROC's of Part 1 and Part 2 is empty, $x(t)$ does not have a Laplace transform.

Problem 161. For each of the following signals find the Laplace transform.

(a) $x(t)$ as in the figure.

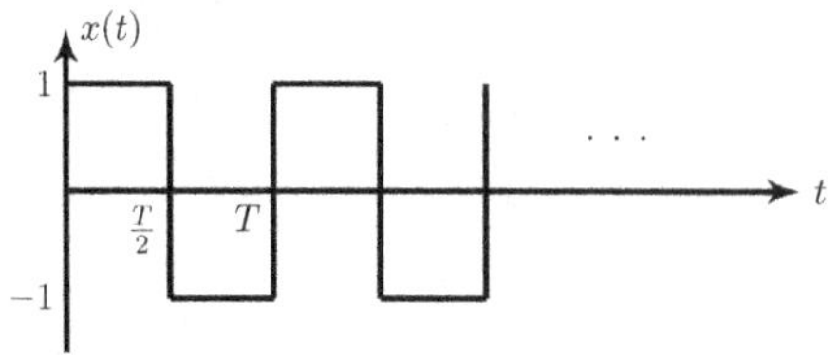

Figure 9.4: Problem 161 - Part (a)

(b) $x(t)$ as in the figure.

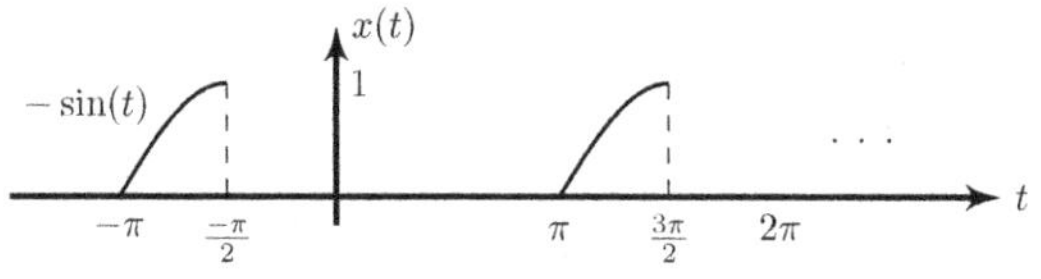

Figure 9.5: Problem 161 - Part (b)

(c) $x(t)$ as in the figure.

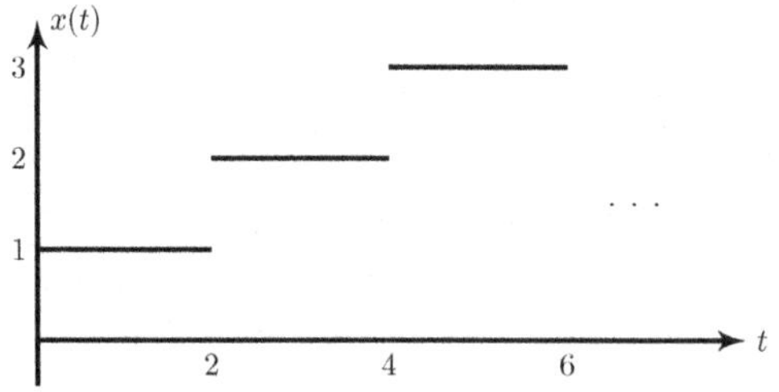

Figure 9.6: Problem 161 - Part (c)

Solution.

(a) As can be seen, the signal is periodic with period T for $t > 0$. As such, we can use (9.13) where $f(t)$ is given in the figure below:

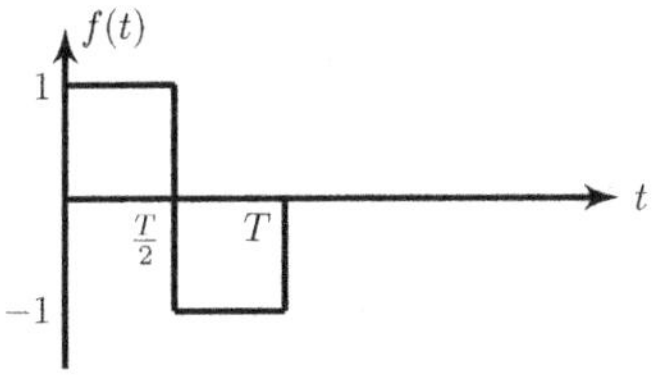

Figure 9.7: Problem 161 - Part (a), function $f(t)$

According to the above figure, we can write the $f(t)$ as:

$$f(t) = u(t) - 2u(t - \frac{T}{2}) + u(t - T).$$

Now we have to obtain $F(S)$. Using the Laplace transform time shifting property, $F(s)$ can be written as:

$$F(s) = \frac{1}{s} - \frac{2}{s}e^{\frac{-T}{2}s} + \frac{1}{s}e^{-Ts} = \frac{1}{s}(1 - 2e^{\frac{-T}{2}s} + e^{-Ts}).$$

Thus using (9.13) we have

$$\begin{aligned}\mathcal{L}\{x(t)\} &= \frac{F(s)}{1 - e^{-Ts}} = \frac{\frac{1}{s}(1 - 2e^{\frac{-T}{2}s} + e^{-Ts})}{1 - e^{-Ts}} \\ &= \frac{\frac{1}{s}(1 - e^{\frac{-T}{2}s})^2}{(1 + e^{\frac{-T}{2}s})(1 - e^{\frac{-T}{2}s})} = \frac{\frac{1}{s}(1 - e^{\frac{-T}{2}s})}{(1 + e^{\frac{-T}{2}s})}.\end{aligned}$$

(b) The signal is periodic with period 2π. We let $f(t)$ be one period of this signal from 0 to 2π and obtain $F(s)$ as follows:

$$\begin{aligned}F(s) &= \int_{\pi}^{\frac{3\pi}{2}} -\sin(t)e^{-st}dt \\ &= -\frac{e^{-st}}{s^2+1}(s\sin(t) - \cos(t))|_{\pi}^{\frac{3\pi}{2}} \\ &= \frac{e^{-\pi s} - se^{-\frac{3\pi}{2}s}}{s^2+1}.\end{aligned}$$

As the signal does not start from $t = 0$, we can obtain the Laplace transform of the shifted version of this signal, i.e., $x(t + 2\pi)$ using (9.13):

$$\mathcal{L}\{x(t + 2\pi)\} = \frac{F(s)}{1 - e^{-Ts}} = \frac{\frac{e^{-\pi s} - se^{-\frac{3\pi}{2}s}}{s^2+1}}{1 - e^{-2\pi s}}.$$

But

$$\mathcal{L}\{x(t+2\pi)\} = X(s)e^{2\pi s},$$

and so $X(s)$ is obtained as

$$X(s) = \frac{e^{-\pi s} - se^{-\frac{3\pi}{2}s}}{(1-e^{-2\pi s})(s^2+1)e^{2\pi s}} = \frac{e^{-\pi s} - se^{-\frac{3\pi}{2}s}}{(e^{2\pi s}-1)(s^2+1)}.$$

(c) At the first glance, the signal does not seem to be periodic. However, consider the following signal:

$$y(t) = \sum_{k=-\infty}^{+\infty} u(t-2k).$$

In this case, we can see that:

$$y(t+2) = \sum_{k=-\infty}^{+\infty} u(t+2-2k) = \sum_{k=-\infty}^{+\infty} u(t+2(1-k)) = \sum_{k'=-\infty}^{+\infty} u(t-2k') = y(t)$$

Thus, $y(t)$ is periodic with period $T = 2$. Now we observe that $x(t)$ is in fact the same as $y(t)$ for $t > 0$ and as such, we can use (9.13) where we set $f(t) = u(t)$ and $\mathcal{L}\{f(t)\} = \frac{1}{s}$. So the Laplace transform of $x(t)$ is given as

$$\mathcal{L}\{x(t)\} = \frac{\frac{1}{s}}{1-e^{-2s}} = \frac{1}{s(1-e^{-2s})}.$$

Problem 162. Let $\mathcal{L}\{x(t)\} = X(s)$. For each of the following signals find the Laplace transforms based on $X(s)$.

(a) $y(t) = x(3t)$

(b) $y(t) = x(t-2)$

(c) $y(t) = x(t) * \frac{d}{dt}x(t)$

(d) $y(t) = e^{-t}x(t)$

(e) $y(t) = 2tx(t)$

(f) $y(t) = \int_{-\infty}^{t} x(3\tau)d\tau$

(g) $y(t) = x^*(-2t+1)$

(h) $y(t) = x(t-3) + x^*(-t+2)$

(i) $y(t) = x(-3t)e^{(1+3j)t}$, $-1 < Re(s) < 2$

Solution.

(a) $y(t) = x(3t)$
By the time scaling property, the Laplace transform is obtained as below

$$\mathcal{L}\{y(t)\} = \frac{1}{3}X(\frac{s}{3}).$$

(b) $y(t) = x(t-2)$
By the time shifting property we have

$$\mathcal{L}\{y(t)\} = e^{-2s}X(s).$$

(c) $y(t) = x(t) * \frac{d}{dt}x(t)$
As we know $\mathcal{L}\{x'(t)\} = sX(s)$, and convolution in time domain is equivalent to multiplication in Laplace domain, so we have

$$\mathcal{L}\{y(t)\} = X(s) \times sX(s) = sX^2(s).$$

(d) $y(t) = e^{-t}x(t)$
By the frequency shifting property, $\mathcal{L}\{e^{s_0 t}x(t)\} = X(s - s_0)$, where s_0 is -1 here. So we have

$$\mathcal{L}\{y(t)\} = X(s-(-1)) = X(s+1).$$

(e) $y(t) = 2tx(t)$
Using the Laplace-domain differentiation property we have

$$\mathcal{L}\{y(t)\} = 2(-X'(s)) = -2\frac{dX(s)}{ds}.$$

(f) $y(t) = \int_{-\infty}^{t} x(3\tau)d\tau$
Using the time scaling property, the Laplace transform of $x(3\tau)$ is $\frac{X(\frac{s}{3})}{3}$ and by the time-domain integration property for a given signal $y(t)$, $\mathcal{L}\{\int_{-\infty}^{t} y(\tau)d\tau\} = \frac{Y(s)}{s}$. As such we have

$$\mathcal{L}\{y(t)\} = \frac{X(\frac{s}{3})}{3s}.$$

(g) $y(t) = x^*(-2t+1)$
For the systems below, if the input is $z(t)$, we have indicated the output and based on that, we can track the changes of $x(t)$:

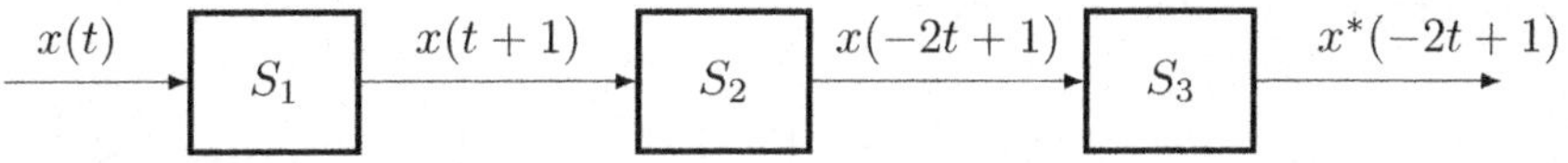

$$\begin{cases} S_1: & z(t) \to z(t+1) \\ S_2: & z(t) \to z(-2t) \\ S_3: & z(t) \to z^*(t) \end{cases}$$

In the Laplace domain, the systems are shown in the figure below:

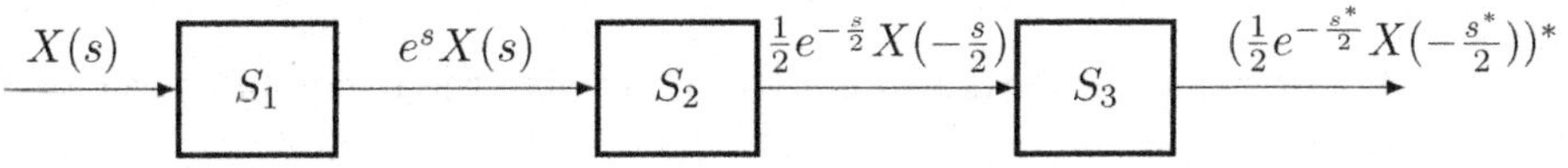

Note that in the last step, to find the Laplace transform of the conjugate of a signal, at first s is changed to s^* and then the whole transform is also conjugated as stated in the fifth row of Table 9.1. Finally, $Y(s)$ is obtained as

$$\mathcal{L}\{y(t)\} = \mathcal{L}\{x^*(-2t+1)\} = \frac{1}{2}e^{-\frac{s}{2}}X^*(-\frac{s^*}{2})$$

(h) $y(t) = x(t-3) + x^*(-t+2)$

$$\mathcal{L}\{x(t-3)\} = e^{-3s}X(s) \qquad \text{(by the time shifting property)}$$

and

$$\mathcal{L}\{x(-t+2)\} = e^{-2s}X(-s) \qquad \text{(by the time shifting and time reversal property)}$$

and using the hint given in part **g**, we have

$$\mathcal{L}\{x^*(-t+2)\} = (e^{-2s^*}X(-s^*))^* = e^{-2s}X^*(-s^*).$$

The overall transform is

$$\mathcal{L}\{y(t)\} = e^{-3s}X(s) + e^{-2s}X^*(-s^*).$$

(i) $y(t) = x(-3t)e^{(1+3j)t}, \qquad ROC: -1 < Re(s) < 2$
By the time scaling property, for $x(-3t)$ we have

$$\mathcal{L}\{x(-3t)\} = \frac{1}{3}X(\frac{-s}{3}),$$

and using frequency shifting property,

$$\mathcal{L}\{x(-3t)e^{(1+3j)t}\} = \frac{1}{3}X(\frac{s-(1+3j)}{-3}).$$

Now we obtain the ROC for $y(t)$:

$$-1 < Re(s) < 2 \xrightarrow{s\to\frac{-s}{3}} -1 < Re(\frac{-s}{3}) < 2 \Rightarrow -6 < Re(s) < 3,$$

$$-6 < Re(s) < 3 \xrightarrow{s\to s-(1+3j)} -6 < Re(s-(1+3j)) < 3 \Rightarrow -5 < Re(s) < 4.$$

So we have

$$Y(s) = \frac{1}{3}X(\frac{s-(1+3j)}{-3}), \qquad -5 < Re(s) < 4.$$

Problem 163. For signals in Parts (a) and (b), determine the region of convergence for each of the following cases:

(1) If $x(t)$ is a right-sided signal.

(2) If $x(t)$ is a left-sided signal.

(3) If $x(t)$ is a two-sided signal and $\int_{-\infty}^{\infty} x(t)e^{-1.5t}dt = \infty$.

(a) $x(t)$ is a signal with Laplace transform $X(s) = \frac{s^2-4}{(s+1)(s^2-9)}$.

(b) $x(t)$ is a signal with Laplace transform $X(s) = \frac{1-e^s}{(s+2)(s+1)}$.

Solution.

(a) Poles of $X(s)$ are located at $s = -3, s = 3$ and $s = -1$, so with the properties of regions of convergence, we can consider 4 regions of convergence for $X(s)$: $Re(s) < -3$, $-3 < Re(s) < -1$, $-1 < Re(s) < 3$ and $Re(s) > 3$. Now by considering the condition of Parts **1**, **2** and **3** for $x(t)$, the regions of convergence can be found as below:

(1) In this case since $x(t)$ is right-sided, using Property 3 of Section 9.3, ROC has to cover the region which lies in the right side of the rightmost pole as ROC does not contain any poles. So the ROC is $Re(s) > 3$.

(2) In this case since $x(t)$ is left-sided, using Property 4 of Section 9.3, ROC has to cover the region which lies in the left side of the leftmost pole as ROC does not contain any poles. So the ROC is $Re(s) < -3$.

(3) In this case since $x(t)$ is 2-sided, using Property 5 of Section 9.3, ROC is in the form of a vertical strip that does not contain any poles. So the ROC can either be $-3 < Re(s) < -1$ or $-1 < Re(s) < 3$. Meanwhile based on the question assumption, $\int_{-\infty}^{\infty} x(t)e^{-1.5t}dt$ which is equivalent to $X(1.5)$ is unbounded. This means that $s = 1.5$ can not be in the region of convergence. So the region of convergence is the one that does not contain 1.5, i.e., $-3 < Re(s) < -1$.

(b) Poles of $X(s)$ are located at $s = -1$ and $s = -2$, so with the properties of region of convergence we can consider 3 regions: $Re(s) < -2$, $-2 < Re(s) < -1$ and $Re(s) > -1$, now by considering the condition of Parts **1**, **2** and **3** for $x(t)$, the regions of convergence can be found as below:

(1) In this case since $x(t)$ is right-sided, using Property 3 of Section 9.3, ROC has to cover the region which lies in the right side of the rightmost pole as ROC does not contain any poles. So the ROC is $Re(s) > -1$.

(2) In this case since $x(t)$ is left-sided, using Property 4 of Section 9.3, ROC has to cover the region which lies in the left side of the leftmost pole as ROC does not contain any poles. So the ROC is $Re(s) < -2$.

(3) As the $x(t)$ is a two-sided signal, the ROC is a vertical strip between two poles, i.e., $-2 < Re(s) < -1$.

Problem 164. Determine the inverse Laplace transform for each of the following cases.

(a) $F(s) = \frac{2(s+6)}{(s+1)(s^2+5s+6)}$

(b) $F(s) = \frac{100(s+4)}{s^2(s^2+2s+10)}$

(c) $F(s) = \frac{4s^2}{(s^2+2s+3)(s^2+1)}$

(d) $F(s) = \frac{5}{(s^2+6s+10)(s+1)}$

(e) $F(s) = \frac{2s^3+8s^2+11s+3}{(s+2)(s+1)^3}$

Solution.

(a) $F(s) = \frac{2(s+6)}{(s+1)(s^2+5s+6)}$

We expand the $F(s)$ to partial fractions:

$$F(s) = \frac{2(s+6)}{(s+1)(s^2+5s+6)} = \frac{k_1}{(s+1)} + \frac{k_2}{(s+2)} + \frac{k_3}{(s+3)},$$

where k_1, k_2 and k_3 are computed as follows:

$$k_1 = [(s+1)F(s)]_{s=-1} = \frac{2\times 5}{1\times 2} = 5$$
$$k_2 = [(s+2)F(s)]_{s=-2} = \frac{2\times 4}{-1\times 1} = -8$$
$$k_3 = [(s+3)F(s)]_{s=-3} = \frac{2\times 3}{-2\times -1} = 3$$

So the inverse Laplace transform is obtained as

$$f(t) = [5e^{-t} - 8e^{-2t} + 3e^{-3t}]u(t).$$

(b) $F(s) = \frac{100(s+4)}{s^2(s^2+2s+10)}$

We expand $F(s)$ into partial fractions as

$$F(s) = \frac{100(s+4)}{s^2(s^2+2s+10)} = \frac{k_0}{s^2} + \frac{k_1}{s} + \frac{As+B}{s^2+2s+10}.$$

Now the values of k_0 and k_1 are obtained as

$$k_0 = [s^2(F(s))]_{s=0} = 40,$$
$$k_1 = \frac{d}{ds}[s^2(F(s))]_{s=0} = \frac{50(10-8)}{100} = 2.$$

To obtain A and B, we tend s to 0 as follows:

$$\lim_{s\to 0} sF(s) = 0 \;\Rightarrow\; k_1 + A = 0 \Rightarrow A = -2.$$

For $s = 1$ we have

$$\frac{100\times 5}{13} = 40 + 2 + \frac{B-2}{13} \Rightarrow B = -44.$$

Finally, $F(s)$ is obtained as

$$F(s) = \frac{40}{s^2} + \frac{2}{s} - \frac{2s+44}{s^2+2s+10} = \frac{40}{s^2} + \frac{2}{s} - \frac{2(s+1)+42}{(s+1)^2+9}$$

$$= \underset{\substack{\Downarrow \\ 40t}}{\frac{40}{s^2}} + \underset{\substack{\Downarrow \\ 2}}{\frac{2}{s}} - \underset{\substack{\Downarrow \\ 2e^{-t}\cos(3t)}}{\frac{2(s+1)}{(s+1)^2+9}} - 14\cdot \underset{\substack{\Downarrow \\ 14e^{-t}\sin(3t)}}{\frac{3}{(s+1)^2+9}}.$$

As such, the inverse Laplace transform is given as

$$f(t) = [40t + 2 - 2e^{-t}\cos(3t) - 14e^{-t}\sin(3t)]u(t).$$

(c) $F(s) = \frac{4s^2}{(s^2+2s+3)(s^2+1)}$

We have two quadratic (second order) polynomials in the denominator of $F(s)$, so the expansion is as follows:

$$F(s) = \frac{4s^2}{(s^2+2s+3)(s^2+1)} = \frac{As+B}{s^2+2s+3} + \frac{Cs+D}{s^2+1}.$$

To obtain the coefficients, we act as follows; we first find $F(0)$:

$$F(0) = 0 = \frac{B}{3} + D. \qquad (1)$$

Now we obtain the limit of $sF(s)$ when s goes to ∞:

$$\lim_{s\to\infty} sF(s) = 0 = A + C. \qquad (2)$$

By obtaining the values of $F(1)$ and $F(-1)$ (any other finite value could have been used), we come up with the following equation,

$$\begin{aligned} F(s) &= \frac{3-s}{s^2+2s+3} + \frac{s-1}{s^2+1} \\ &= \frac{4}{s^2+2s+3} - \frac{s+1}{s^2+2s+3} + \frac{s}{s^2+1} - \frac{1}{s^2+1} \\ &= \frac{4}{(s+1)^2+2} - \frac{s+1}{(s+1)^2+2} + \frac{s}{s^2+1} - \frac{1}{s^2+1} \end{aligned}$$

So the inverse transform is obtained as

$$f(t) = [2\sqrt{2}e^{-t}\sin(\sqrt{2}t) - e^{-t}\cos(\sqrt{2}t) + \cos(t) - \sin(t)]u(t).$$

(d) $F(s) = \frac{5}{(s^2+6s+10)(s+1)}$

We first expand $F(s)$ as

$$F(s) = \frac{5}{(s^2+6s+10)(s+1)} = \frac{As+B}{s^2+6s+10} + \frac{A_1}{s+1}.$$

To get the coefficients, we have

$$A_1 = [(s+1)F(s)]_{s=-1} = \frac{5}{1-6+10} = 1$$

$$F(0) = \frac{5}{10} = \frac{B}{10} + 1 \quad \Rightarrow \quad B = -5$$

$$\lim_{s\to\infty} sF(s) = 0 = A + A_1 \quad \Rightarrow \quad A = -1$$

so we can write the $F(s)$ as

$$F(s) = \underbrace{\frac{-s-5}{s^2+6s+10}}_{Part1} + \underbrace{\frac{1}{s+1}}_{Part2}.$$

Part 2 is already in Table 9.2. For Part 1, we have to rewrite it in a from that exists in that Table. To do so we have

$$\begin{aligned}\frac{-s-5}{s^2+6s+10} &= \frac{-s-5}{(s+3)^2+1} \\ &= \frac{-s-3-2}{(s+3)^2+1} \\ &= -\frac{s+3}{(s+3)^2+1} - \frac{2}{(s+3)^2+1}.\end{aligned}$$

So we write

$$F(s) = -\underset{\substack{\Downarrow \\ -e^{-3t}\cos(t)}}{\frac{s+3}{(s+3)^2+1}} - \underset{\substack{\Downarrow \\ -2e^{-3t}\sin(t)}}{\frac{2}{(s+3)^2+1}} + \underset{\substack{\Downarrow \\ e^{-t}}}{\frac{1}{s+1}},$$

and the inverse Laplace transform is obtained as

$$f(t) = [-e^{-3t}\cos(t) - 2e^{-3t}\sin(t) + e^{-t}]u(t).$$

(e) $F(s) = \frac{2s^3+8s^2+11s+3}{(s+2)(s+1)^3}$

As we have multiple roots at $s = -1$, the expansion into partial fractions for $F(s)$ is as follows:

$$F(s) = \frac{A}{s+2} + \frac{B_1}{s+1} + \frac{B_2}{(s+1)^2} + \frac{B_3}{(s+1)^3}.$$

For the simple root $s = -2$, we readily find B as

$$A = [(s+2)F(s)]_{s=-2} = 3.$$

For B_1, B_2 and B_3 we act as follows:

$$B_3 = [(s+1)^3 F(s)]_{s=-1} = -2$$

$$B_2 = [\frac{d}{ds}((s+1)^3 F(s))]_{s=-1}$$
$$= [\frac{(6s^2+16s+11)(s+2) - (2s^3+8s^2+11s+3)}{(s+2)^2}]_{s=-1} = 3$$

$$B_1 = \frac{1}{2}[\frac{d^2}{ds^2}((s+1)^3 F(s))]_{s=-1}$$
$$= [\frac{1}{2}\frac{4s^3+24s^2+48s+26}{(s+2)^3}]_{s=-1} = -1.$$

We thus have

$$f(t) = [3e^{-2t} - e^{-t} + 3te^{-t} - t^2e^{-t}]u(t).$$

As an alternative way, one can easily come up with the values of A and B_3 as above but for B_1 and B_2, we can simply tend s to 0 and obtain $1.5 = 1.5 + B_1 + B_2 - 2$. We also compute the value of $sF(s)$ as s tends to infinity and obtain $B_1 + 3 = 2$. Solving the system of equations, we get

$$B_1 = -1 \text{ and } B_2 = 3$$

as expected.

Problem 165. Find the values of $f(0)$ and $f(\infty)$ for the following Laplace transforms using the initial and final value theorems.

(a) $F(s) = \frac{10s-2}{s^2+6s+10}$

(b) $F(s) = \frac{2s^3-s^2-3s-10}{s^3+3s^2+6s}$

(c) $F(s) = \frac{10s-2}{s^2-6s+10}$

(d) $F(s) = \frac{10s^2-2}{(s+2)^2(s+1)(s^2+5s+10)}$

Solution.

(a) $f(0^+) = \lim_{s\to\infty} sF(s) = \lim_{s\to\infty} \frac{s(10s-2)}{s^2+6s+10} = 10 \quad \Leftarrow \quad$ initial value theorem

$f(\infty) = \lim_{s\to 0} sF(s) = \lim_{s\to 0} \frac{s(10s-2)}{s^2+6s+10} = 0 \quad \Leftarrow \quad$ final value theorem

(b) $f(0^+) = \lim_{s\to\infty} sF(s) = \lim_{s\to\infty} \frac{s(2s^3-s^2-3s-10)}{s^3+3s^2+6s} = \infty \quad \Leftarrow \quad$ initial value theorem

$f(\infty) = \lim_{s\to 0} sF(s) = \lim_{s\to 0} \frac{s(2s^3-s^2-3s-10)}{s^3+3s^2+6s} = -\frac{10}{6} = -\frac{5}{3} \quad \Leftarrow \quad$ final value theorem

(c) $f(0^+) = \lim_{s\to\infty} sF(s) = \lim_{s\to\infty} \frac{s(10s-2)}{s^2-6s+10} = 10 \quad \Leftarrow \quad$ initial value theorem

Since the poles of $F(s)$ ($s = 3 \pm j$) are in the right hand side of the jw-axis, the final value theorem can not be applied.

(d) $f(0^+) = \lim_{s\to\infty} sF(s) = \lim_{s\to\infty} \frac{s(10s^2-2)}{(s+2)^2(s+1)(s^2+5s+10)} = 0 \quad \Leftarrow \quad$ initial value theorem

$f(\infty) = \lim_{s\to 0} sF(s) = \lim_{s\to 0} \frac{s(10s^2-2)}{(s+2)^2(s+1)(s^2+5s+10)} = 0 \quad \Leftarrow \quad$ final value theorem

Problem 166. **(a)** Let $y(t)$ be the output of a causal LTI system with transfer function $H(s) = \frac{3}{s-3}$ when the input is $x(t) = u(t)$. Find the value $y(t)$ when t tends to infinity.

(b) Solve the problem when the system is stable instead of being causal.

Solution.

Since $H(s)$ has a pole in $s = 3$, it can have two possible convergence regions, $Re\{s\} < 3$ and $Re\{s\} > 3$.

(a) As it is assumed that the system is causal, the ROC has to contain $Re\{s\} = \infty$, so the region of convergence of $H(s)$ is $Re\{s\} > 3$. Since the ROC of $X(s)$ is $Re\{s\} > 0$, and the ROC of $Y(s) = X(s)H(s)$ is the intersection of the ROC's of $X(s)$ and $H(s)$, the ROC of $Y(s)$ is $Re\{s\} > 3$.

Since the ROC does not contain the jw-axis, as stated in Section 9.6.2, we cannot use the final value theorem. So we have to first obtain $y(t)$:

$$X(s) = \frac{1}{s}, \quad ROC: Re\{s\} > 0, \qquad H(s) = \frac{3}{s-3}, \quad ROC: Re\{s\} > 3,$$

$$Y(s) = X(s)H(s) = \frac{3}{s(s-3)} \quad ROC: Re\{s\} > 3,$$

$$Y(s) = \frac{1}{s-3} - \frac{1}{s} \quad ROC: Re\{s\} > 3.$$

So we have

$$y(t) = e^{3t}u(t) - u(t) \Rightarrow y(\infty) = \infty.$$

(b) If the system is stable, the ROC contains the jw-axis and so the ROC of $H(s)$ is $Re\{s\} < 3$. This means that the ROC of $Y(s)$ is $0 < Re\{s\} < 3$ and from Section (9.6.2), Extended Final-Value Theorem, we obtain

$$y(\infty) = \lim_{s \to 0} sY(s) = \lim_{s \to 0} \frac{3}{s-3} = -1.$$

To check the validity of the answer, let's obtain it directly in the time domain:

$$Y(s) = X(s)H(s) = \frac{3}{s(s-3)} \qquad ROC: 0 < Re\{s\} < 3$$

$$Y(s) = \frac{1}{s-3} - \frac{1}{s} \qquad ROC: 0 < Re\{s\} < 3.$$

So $y(t)$ is computed as

$$y(t) = -e^{3t}u(-t) - u(t) \Rightarrow y(\infty) = -1.$$

Problem 167. Using Laplace transform, determine which of the following systems are LTI.

(a) $x(t) = e^{t}u(t), \qquad y(t) = -e^{2t}u(-t)$

(b) $x(t) = e^{3t}u(t), \qquad y(t) = -e^{t}u(-t)$

Solution.

(a) First we obtain the Laplace transforms of $x(t)$ and $y(t)$:

$$x(t) = e^{t}u(t) \quad \overset{\mathcal{L}}{\rightarrow} \quad X(s) = \frac{1}{s-1}, \qquad ROC_X: Re\{s\} > 1,$$

$$y(t) = -e^{2t}u(-t) \quad \overset{\mathcal{L}}{\rightarrow} \quad Y(s) = \frac{1}{s-2}, \qquad ROC_Y: Re\{s\} < 2.$$

So we obtain $H(s)$ as

$$H(s) = \frac{Y(s)}{X(s)} = \frac{\frac{1}{s-2}}{\frac{1}{s-1}} = \frac{s-1}{s-2}. \tag{9.38}$$

$H(s)$ has two possible ROC's; $Re\{s\} < 2$ and $Re\{s\} > 2$. $H(s)$ with the ROC_H: $Re\{s\} < 2$ satisfies the condition $[ROC_X \cap ROC_Y \cap ROC_H] \neq \emptyset$ (see Section 9.8). So this system can be an LTI system if we assume $H(s) = \frac{s-1}{s-2}$, $ROC_H: Re\{s\} < 2$.

(b) First we obtain the Laplace transforms of $x(t)$ and $y(t)$:

$$x(t) = e^{3t}u(t) \quad \overset{\mathcal{L}}{\rightarrow} \quad X(s) = \frac{1}{s-3}, \qquad ROC_X: Re\{s\} > 3,$$

$$y(t) = -e^{t}u(-t) \quad \overset{\mathcal{L}}{\rightarrow} \quad Y(s) = \frac{1}{s-1}, \qquad ROC_Y: Re\{s\} < 1.$$

In this case, since ROC_X and ROC_Y do not intersect, it is obvious that that $[ROC_X \cap ROC_Y \cap ROC_H] = \emptyset$. As the relationship $Y(s) = H(s)X(s)$ can not be established in this case (see Section 9.8), there is no LTI system that can relate $x(t)$ and $y(t)$.

Problem 168. Find the inverse Laplace transform of following functions by considering their region of convergence.

(a) $X(s) = \dfrac{s+1}{(s+1)^2+9} - \dfrac{5}{s+2} - \dfrac{1}{s-2}, \qquad -1 < Re\{s\} < 2$

(b) $X(s) = \dfrac{s+2}{s^2+7s+12}, \qquad -4 < Re\{s\} < -3$

(c) $X(s) = \dfrac{1-e^{-s}}{s(1-e^{-3s})}, \qquad Re\{s\} > 0$

(d) $X(s) = \dfrac{1}{s^2(s+1)^2}, \qquad -1 < Re\{s\} < 0$

(e) $X(s) = \dfrac{s-5}{s^2-6s+10}, \qquad Re\{s\} > 3$

Solution.

(a) In this part $X(s)$ is formed by the summation of three polynomial fractions. To find the inverse Laplace transform, the region of convergence for all parts should be first determined. The first fraction, $\frac{s+1}{(s+1)^2+9}$, has two poles in $s = -1 \pm 3j$ and possible ROCs are $Re\{s\} > -1$ and $Re\{s\} < -1$. We pick up the one that has non-empty intersection with the assumed ROC ($-1 < Re\{s\} < 2$). So the region of convergence for this transform is $Re\{s\} > -1$. The second fraction, $\frac{5}{s+2}$, has a pole in $s = -2$ and possible ROCs are $Re\{s\} > -2$ and $Re\{s\} < -2$. We pick up the one that has non-empty intersection with the assumed ROC. So the ROC for this transform is $Re\{s\} > -2$. Finally, the third fraction, $\frac{-1}{s-2}$, has a pole in $s = 2$ and possible ROCs are $Re\{s\} > 2$ and $Re\{s\} < 2$. We pick up the one that has non-empty intersection with the assumed ROC. So the ROC for this transform is $Re\{s\} < 2$.

Now, the inverse Laplace transform of the first is obtained using Table 9.2:

$$\frac{s^2}{s^2+9}, \qquad Re\{s\} > 0 \quad \xrightarrow{\mathcal{L}^{-1}} \quad \cos(3t)u(t),$$

and using the frequency shifting property we get

$$\frac{(s+1)^2}{(s+1)^2+9}, \qquad Re\{s+1\} > 0 \triangleq Re\{s\} > -1 \quad \xrightarrow{\mathcal{L}^{-1}} \quad e^{-t}\cos(3t)u(t).$$

For the second part we have

$$\frac{-5}{s+2}, \qquad Re\{s\} > -2 \quad \xrightarrow{\mathcal{L}^{-1}} \quad -5e^{-2t}u(t),$$

and for the third part we get

$$-\frac{1}{s-2}, \qquad Re\{s\} < 2 \quad \xrightarrow{\mathcal{L}^{-1}} \quad e^{2t}u(-t).$$

So we can write

$$x(t) = e^{-t}\cos(3t)u(t) - 5e^{-2t}u(t) + e^{2t}u(-t).$$

(b) $X(s) = \dfrac{s+2}{s^2+7s+12}, \qquad -4 < Re\{s\} < -3$

Let's first decompose $X(s)$ into partial fractions:

$$X(s) = \frac{s+2}{s^2+7s+12} = \frac{A}{s+3} + \frac{B}{s+4}.$$

$$A = (s+3)X(s)|_{s=-3} = -1$$

$$B = (s+4)X(s)|_{s=-4} = 2$$

Now we have to deal with $\frac{-1}{s+3}$ and $\frac{2}{s+4}$. The first fraction has two possible regions of convergence, $Re\{s\} > -3$ and $Re\{s\} < -3$. Looking at the assumed ROC, we have to pick $Re\{s\} < -3$. For the second fraction, again there are two possible regions of convergence, $Re\{s\} > -4$ and $Re\{s\} < -4$. Looking at the assumed ROC, we have to pick $Re\{s\} > -4$. So we have

$$\frac{-1}{s+3}, \qquad Re\{s\} < -3 \quad \xrightarrow{\mathcal{L}^{-1}} \quad e^{-3t}u(-t),$$

$$\frac{2}{s+4}, \qquad Re\{s\} > -4 \quad \xrightarrow{\mathcal{L}^{-1}} \quad 2e^{-4t}u(t),$$

and we get

$$x(t) = e^{-3t}u(-t) + 2e^{-4t}u(t).$$

(c) $X(s) = \dfrac{1-e^{-s}}{s(1-e^{-3s})}, \qquad Re\{s\} > 0$

We rewrite $X(s)$ as

$$X(s) = \frac{\frac{1}{s}(1-e^{-s})}{(1-e^{-3s})} = \frac{F(s)}{(1-e^{-3s})}, \qquad Re\{s\} > 0,$$

where $F(s)$ and $f(t)$ are

$$F(s) = \frac{1}{s}(1-e^{-s}) = \frac{1}{s} - \frac{1}{s}e^{-s} \quad \xrightarrow{\mathcal{L}^{-1}} \quad f(t) = u(t) - u(t-1),$$

and $f(t)$ is plotted below.

Figure 9.8: Problem 168 - Part (c) - Signal $f(t)$

As we have $X(s) = \frac{F(s)}{(1-e^{-3s})}$ and comparing with (9.13), the time-domain signal $x(t)$ is the periodic version of $f(t)$ with period $T = 3$ for $t > 0$. We have depicted $x(t)$ in tg following figure.

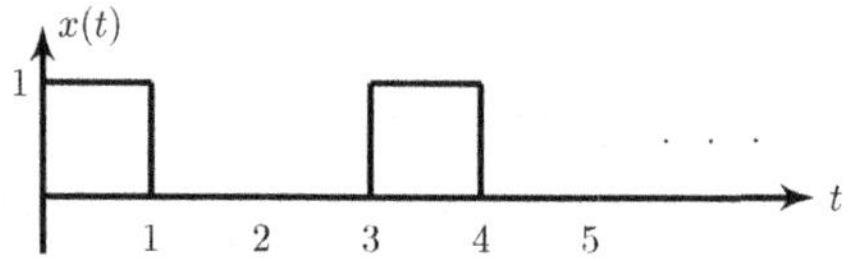

Figure 9.9: Problem 168 - Part (c), Signal $x(t)$

(d) $X(s) = \dfrac{1}{s^2(s+1)^2}, \qquad -1 < Re\{s\} < 0$

We first decompose $X(s)$ into partial fractions, bearing in mind that both poles are multiple:

$$X(s) = \frac{1}{s^2(s+1)^2} = \frac{A}{s^2} + \frac{B}{s} + \frac{C}{(s+1)^2} + \frac{D}{(s+1)}$$

where

$$A = s^2 X(s)|_{s=0} = 1,$$
$$C = (s+1)^2 X(s)|_{s=-1} = 1.$$

By setting $s = 1$ and $s = 2$ we have

$$X(1) = \frac{1}{1^2(1+1)^2} = 1 + B + \frac{1}{2^2} + \frac{D}{2} \quad \Rightarrow \quad B + \frac{D}{2} = -1 \quad \rightarrow ①$$

$$X(2) = \frac{1}{2^2(2+1)^2} = \frac{1}{2^2} + \frac{B}{2} + \frac{1}{3^2} + \frac{D}{2+1} \quad \Rightarrow \quad \frac{B}{2} + \frac{D}{3} = \frac{-1}{3} \quad \rightarrow ②$$

Based on ① and ② we obtain $B = -2$ and $D = 2$. Now we obtain the inverse Laplace transform of each part using the determined ROCs:

$$\frac{1}{s^2} \overset{s=0}{\Longrightarrow} \Big| \begin{smallmatrix} Re\{s\}>0 \\ Re\{s\}<0\surd \end{smallmatrix} \quad \overset{\mathcal{L}^{-1}}{\longrightarrow} -tu(-t),$$

$$\frac{-2}{s} \overset{s=0}{\Longrightarrow} \Big| \begin{smallmatrix} Re\{s\}>0 \\ Re\{s\}<0\surd \end{smallmatrix} \quad \overset{\mathcal{L}^{-1}}{\longrightarrow} 2u(-t),$$

$$\frac{1}{(s+1)^2} \overset{s=-1}{\Longrightarrow} \Big| \begin{smallmatrix} Re\{s\}>-1\surd \\ Re\{s\}<-1 \end{smallmatrix} \quad \overset{\mathcal{L}^{-1}}{\longrightarrow} te^{-t}u(t),$$

$$\frac{2}{(s+1)} \overset{s=-1}{\Longrightarrow} \Big|^{Re\{s\}>-1 \checkmark}_{Re\{s\}<-1} \quad \xrightarrow{\mathcal{L}^{-1}} 2e^{-t}u(t).$$

Finally we obtain $x(t)$ as

$$x(t) = -tu(-t) + 2u(-t) + te^{-t}u(t) + 2e^{-t}u(t).$$

(e) $X(s) = \dfrac{s-5}{s^2-6s+10}, \qquad Re\{s\} > 3$

Since the poles are complex, the time domain signal contains sin or cos functions. So we write

$$s^2 - 6s + 10 = s^2 - 6s + 9 + 1 = (s-3)^2 + 1,$$

$$X(s) = \frac{s-5}{s^2-6s+10} = \frac{s-3-2}{(s-3)^2+1} = \frac{s-3}{(s-3)^2+1} - \frac{2}{(s-3)^2+1}.$$

We know that

$$\cos(at)u(t) \xrightarrow{\mathcal{L}} \frac{s}{s^2+a^2}, \qquad Re\{s\} > 0,$$

$$\sin(at)u(t) \xrightarrow{\mathcal{L}} \frac{a}{s^2+a^2}, \qquad Re\{s\} > 0.$$

Thus by frequency shifting property we can obtain the inverse Laplace transform as below

$$\frac{s-3}{(s-3)^2+1}, \quad Re\{s\} > 3 \quad \xrightarrow{\mathcal{L}^{-1}} e^{3t}\cos(t)u(t),$$

$$\frac{-2}{(s-3)^2+1}, \quad Re\{s\} > 3 \quad \xrightarrow{\mathcal{L}^{-1}} -e^{3t}\sin(t)u(t).$$

Finally, $x(t)$ is obtained as

$$x(t) = [e^{3t}\cos(t) - e^{3t}\sin(t)]u(t).$$

Problem 169. Consider a stable and LTI system which it is described with the following differential equation. Which of the following inputs: $x_1(t) = e^{-t}, x_2(t) = e^{-2t}, x_3(t) = e^{8t}$, could result in a bounded output?

$$9y(t) - \frac{d^2y(t)}{dt^2} = x(t) \tag{9.39}$$

Solution.

$$9y(t) - \frac{d^2y(t)}{dt^2} = x(t) \quad \xrightarrow{\mathcal{L}} \quad 9Y(s) - s^2Y(s) = X(s)$$

$$9Y(s) - s^2Y(s) = X(s) \quad \Rightarrow \quad H(s) = \frac{Y(s)}{X(s)} = \frac{1}{9-s^2}.$$

$H(s)$ has two poles in $s = \pm 3$, so we can consider 3 regions of convergence, but the system is stable so the ROC should contain the jw-axis. Consequently, the ROC is obtained as $-3 < Re\{s\} < 3$. The input signals are a bunch of exponentials and we can use (9.27) to see if the output is bounded or not:

$$x_1(t) = e^{-t} \quad \xrightarrow{-1\in ROC} \quad y_1(t) = H(-1)e^{-t} = \frac{1}{8}e^{-t}$$

$$x_2(t) = e^{-2t} \quad \xrightarrow{-2\in RoC} \quad y_2(t) = H(-2)e^{-2t} = \frac{1}{5}e^{-2t}$$

$$x_3(t) = e^{8t} \quad \xrightarrow{8\notin RoC} \quad y_3(t) = \infty, \quad \forall t$$

Therefor, the $y_3(t)$ is unbounded and so do $y_1(t)$ and $y_2(t)$ when $t \to -\infty$.

Problem 170. Transfer function of an LTI system is $H(s) = \frac{4s}{s^2+s-6}$, $-3 < Re\{s\} < 2$. Determine the correctness of the following propositions:

(a) System is stable.

(b) System is non-causal.

(c) If $h(t)$ is the impulse response of the system, then $\int_{-\infty}^{\infty} h(t)dt = 0$.

Solution.

(a) The jw-axis (s = 0) is in the ROC of $H(s)$, and $h(t)$ does not contain derivations of the impulse function because the order of the numerator is lower than the order of the denominator. Thus the system is stable.

(b) ROC does not contain $s = \infty$ and thus the system is not causal.

(c) By the definition of the Laplace transform and setting $s = 0$ we get

$$H(s) = \int_{-\infty}^{\infty} h(t)e^{-st}dt \quad \xrightarrow{s=0} \quad H(0) = \int_{-\infty}^{\infty} h(t)dt.$$

As $s = 0$ is in the region of convergence, we simply evaluate $H(0)$ which is equal to 0. Thus the proposition is correct.

Problem 171. Let $X(s)$ be the Laplace transform of $x(t)$. The following information is available on $X(s)$.

(1) $X(0) = 4$

(2) $X(s)$ has a pole in $s = -1 + j$

(3) $s = 3$ is not in the ROC.

What can be said about the ROC of $X(s)$?

Solution.

Since $X(s)$ is defined at $s = 0$, it implies that the ROC contains the jw-axis. The fact that $s = -1 + j$ is a pole, states that ROC is either located on $s > -1$ or $s < -1$. Since ROC should span a continuous region and contains $s = 0$, it will be a subset of a strip between -1 and 3 as shown in the following figure.

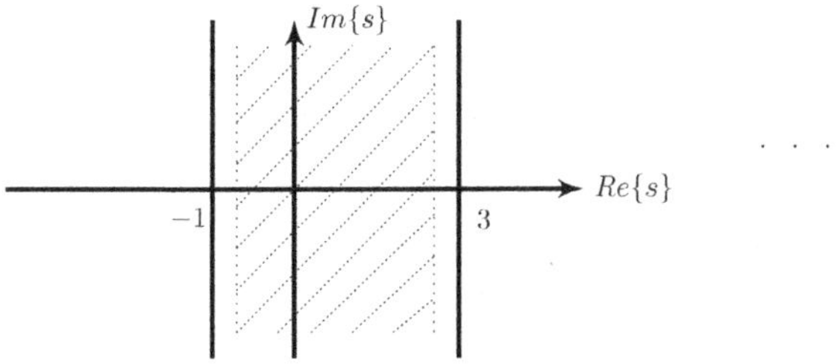

Figure 9.10: Problem 171, a typical sketch of the ROC of $X(s)$

Problem 172. Transfer function of an LTI system is $H(s) = \frac{s+3}{s^2-4s+4-\alpha^2}$. For all values of α, investigate the causality and stability of this system.

Solution.

To investigate the stability and causality, we first have to locate poles of the system by finding the roots of the denominator of $H(s)$:

$$s^2 - 4s + 4 - \alpha^2 = 0 \quad \Rightarrow \quad (s-2)^2 - \alpha^2 = 0 \quad \Rightarrow \quad s = 2 \pm \alpha.$$

Now, we should determine the position of poles $s = 2 + \alpha$ and $s = 2 - \alpha$ with respect to the jw-axis for different values of α. If $|\alpha| < 2$, then both poles are placed in the right side of the jw-axis, and if $|\alpha| > 2$ then one pole is placed in the right side of the jw-axis an the other pole in the left side of the jw-axis, as shown in the figures below.

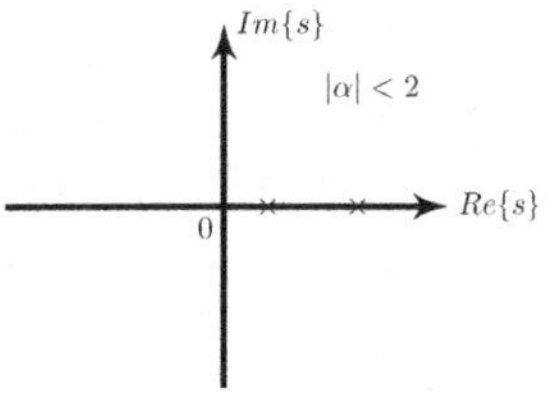

Figure 9.11: Problem 172 - Pole locations for $|\alpha| < 2$

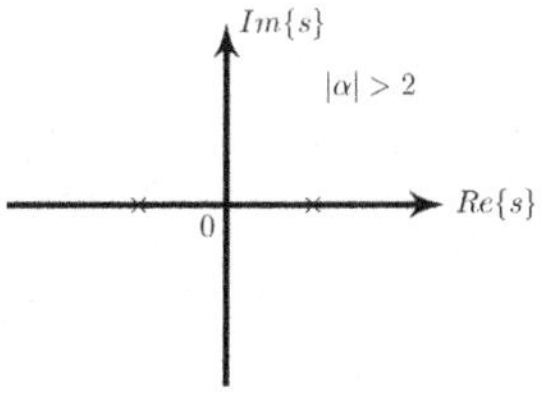

Figure 9.12: Problem 172 - Pole locations for $|\alpha| > 2$

As in both cases, at least one pole is in the right side of the jw-axis, this system cannot be causal and stable simultaneously for any value of α.

Problem 173. The Laplace transform of $x(t)$ has the following pole-zero diagram. If the region of convergence is $-1 < Re\{s\} < 1$, find $x(t)$.

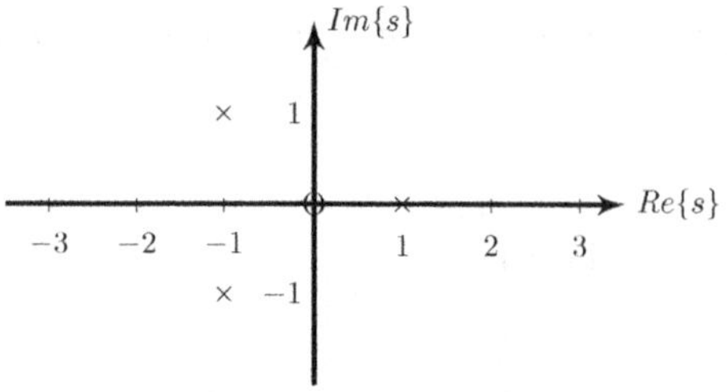

Figure 9.13: Problem 173 - The location of poles and zeros

Solution. With the given poles and zeros, we write the $X(s)$ as follows where we set $k = 1$ for simplicity:

$$X(s) = \frac{ks}{(s-1)(s+1+j)(s+1-j)} = \frac{s}{(s-1)[(s+1)^2+1]}.$$

By the partial fraction decomposition we have

$$X(s) = \frac{s}{(s-1)[(s+1)^2+1]} = \frac{A}{(s-1)} + \frac{Bs+C}{(s+1)^2+1},$$

where

$$A = (s-1)X(s)|_{s=1} = \frac{1}{5}.$$

So we can write

$$X(s) = \frac{\frac{1}{5}}{(s-1)} + \frac{Bs+C}{(s+1)^2+1}.$$

To find B and C, we set $s = 0$ and $s = -1$ in both sides of the above equality and obtain $B = \frac{-1}{5}$ and $C = \frac{2}{5}$. Thus we have

$$\begin{aligned} X(s) &= \frac{\frac{1}{5}}{(s-1)} + \frac{\frac{1}{5}(2-s)}{(s+1)^2+1} \\ &= \frac{\frac{1}{5}}{(s-1)} + \frac{\frac{3}{5}}{(s+1)^2+1} - \frac{1}{5}\cdot\frac{s+1}{(s+1)^2+1} \end{aligned}$$

All the 3 terms are well-known Laplace transforms and all we have to do to obtain their inverse, is to decide the ROC with respect ot the original ROC $(-1 < Re\{s\} < 1)$:

$$\frac{\frac{1}{5}}{(s-1)} \overset{s=1}{\Longrightarrow} \Big| {}^{Re\{s\}>1}_{Re\{s\}<1\surd} \quad \overset{\mathcal{L}^{-1}}{\longrightarrow} -\frac{1}{5}e^{t}u(-t),$$

$$\frac{\frac{3}{5}}{(s+1)^2+1} \overset{s=-1}{\Longrightarrow} \Big| {}^{Re\{s\}>-1\surd}_{Re\{s\}<-1} \quad \overset{\mathcal{L}^{-1}}{\longrightarrow} \frac{3}{5}e^{-t}\sin(t)u(t),$$

$$-\frac{1}{5}\frac{s+1}{(s+1)^2+1} \overset{s=-1}{\Longrightarrow} \Big| {}^{Re\{s\}>-1\surd}_{Re\{s\}<-1} \quad \overset{\mathcal{L}^{-1}}{\longrightarrow} \frac{-1}{5}e^{-t}\cos(t)u(t).$$

Finally, $x(t)$ is obtained as

$$x(t) = -\frac{1}{5}e^{t}u(-t) + \frac{3}{5}e^{-t}\sin(t)u(t) - \frac{1}{5}e^{-t}\cos(t)u(t).$$

Problem 174. Consider an LTI system whose response to the input

$$x(t) = te^{-3t}u(t)$$

is the output below:

$$y(t) = e^{-2t}[\cos(t) + \sin(t)]u(t) - e^{-3t}u(t).$$

(a) Find the system transfer function and its region of convergence. Is the system causal? Is the system stable?

(b) Find the differential equation corresponding to this system.

Solution.

(a) At first the Laplace transform of the input and output is obtained:

$$X(s) = \frac{1}{(s+3)^2}, \quad ROC : Re\{s\} > -3,$$

$$Y(s) = \frac{s+2}{(s+2)^2+1} + \frac{1}{(s+2)^2+1} - \frac{1}{s+3}, \quad ROC = [Re\{s\} > -3] \cap [Re\{s\} > -2]$$

$$\Longrightarrow Y(s) = \frac{2(s+2)}{(s^2+4s+5)(s+3)}, \quad ROC : Re\{s\} > -2.$$

So the transfer function is obtained as

$$H(s) = \frac{Y(s)}{X(s)} = \frac{2(s+2)(s+3)}{s^2+4s+5}.$$

Looking at the poles of $H(s)$, there are two possible ROCs for $H(s)$: $Re\{s\} < -2$ and $Re\{s\} > -2$. We have to pick the one for which the intersection between the 3 available ROCs is not empty. As such, we have to pick $Re\{s\} > -2$.

From the obtained ROC we conclude that the system is causal as the region of convergence contains $Re\{s\} = \infty$. Moreover, the system is stable as the ROC contains the jw-axis.

(b) Using the system transfer function, the differential equation is obtained as below

$$\frac{d^2y(t)}{dt^2} + 4\frac{dy(t)}{dt} + 5y(t) = 2\frac{d^2x(t)}{dt^2} + 10\frac{dx(t)}{dt} + 12x(t).$$

Problem 175. Evaluate the following integrals using the Laplace transform properties and pairs in Tables 9.1 and 9.2 and in particular the Laplace-domain integration property.

(a) $\int_0^\infty \frac{\sin(t)}{t}dt$

(b) $\int_0^\infty te^{-3t}\cos(t)dt$

(c) $\int_0^\infty \frac{\cos(2t)-\cos(3t)}{t}dt$

(d) $x(t) = e^{-2|t|} \Rightarrow \begin{cases} A = \int_{-\infty}^{\infty} x(t)e^{t}dt \\ B = \int_{-\infty}^{\infty} x(t)e^{-5t}dt \end{cases}$

Solution.

(a) We know that the integral can be evaluated if we let $s = 0$ in the Laplace transform of $\frac{sin(t)}{t}$:

$$\mathcal{L}\{\frac{sin(t)}{t}\} = \int_0^\infty e^{-st}\frac{sin(t)}{t}dt.$$

To obtain the Laplace transform, we know that $\mathcal{L}\{\sin(t)\} = \frac{1}{1+s^2}$ and $\lim_{t\to 0}\frac{sin(t)}{t} = 1$ so using the integration property in the Laplace domain, we have

$$\mathcal{L}\{\frac{sin(t)}{t}\} = \int_s^\infty \frac{du}{u^2+1} = \tan^{-1}(u)|_s^\infty = \frac{\pi}{2} - \tan^{-1}(s).$$

By setting $s = 0$ in the above equation, we get:

$$\int_0^\infty \frac{sin(t)}{t}dt = \frac{\pi}{2} - \tan^{-1}(s)|_{s=0} = \frac{\pi}{2}.$$

(b) The Laplace transform of $t\cos(t)$ is

$$\mathcal{L}\{t\cos(t)\} = -\frac{d}{ds}\frac{s}{s^2+1} = \frac{s^2-1}{(s^2+1)^2}, \tag{9.40}$$

At the same time,

$$\mathcal{L}\{t\cos(t)\} = \int_0^\infty te^{-st}\cos(t)dt.$$

Comparing between the above Laplace transform and the integral in question, we can see that they are the same if we let $s = 3$. As such we have

$$\int_0^\infty te^{-3t}\cos(t)dt = \mathcal{L}\{t\cos(t)\}|_{s=3} = \frac{s^2-1}{(s^2+1)^2}|_{s=3} = \frac{2}{25}.$$

(c) Since

$$\lim_{t\to 0}\frac{\cos(2t)-\cos(3t)}{t} = 0,$$

we can use the integration in Laplace-domain property in the row of Table 9.1 and write

$$\mathcal{L}\{\frac{\cos(2t)-\cos(3t)}{t}\} = \mathcal{L}\{\frac{\cos(2t)}{t} - \frac{\cos(3t)}{t}\} = \int_s^\infty [\frac{udu}{u^2+2^2} - \frac{udu}{u^2+3^2}]$$
$$= \left[\frac{1}{2}\ln(u^2+2^2) - \frac{1}{2}\ln(u^2+3^2)\right]\Big|_s^\infty = \frac{1}{2}\ln(\frac{u^2+2^2}{u^2+3^2})\Big|_s^\infty$$

$$= \frac{1}{2}\ln(1) + \frac{1}{2}\ln(\frac{s^2+4}{s^2+9}) = \frac{1}{2}\ln(\frac{s^2+4}{s^2+9}).$$

The integral in question is the above Laplace transform with $s = 0$. Thus,

$$\int_0^{\infty} \frac{\cos(2t) - \cos(3t)}{t} dt = \frac{1}{2}\ln(\frac{s^2+4}{s^2+9})\Big|_{s=0} = \frac{1}{2}\ln(\frac{4}{9}) = \ln(\frac{2}{3}).$$

(d) Based on the Laplace transform definition, the integrals A and B are equal to $X(s)$ at $s = -1$ and $s = 5$, respectively. Thus we just need to obtain the Laplace transform of $x(t) = e^{-2|t|}$.

$$x(t) = e^{-2t}u(t) + e^{2t}u(-t)$$

Using Table 9.2 we get

$$X(s) = \frac{1}{s+2} - \frac{1}{s-2} = \frac{-4}{s^2-4}, \quad ROC: -2 < Re\{s\} < 2.$$

In other words,

$$X(s) = \begin{cases} \frac{-4}{s^2-4} & -2 < Re\{s\} < 2 \\ \infty & O.W. \end{cases}$$

As $s = -1$ is in the ROC ($-2 < Re\{s\} < 2$), we have

$$A = X(-1) = \frac{-4}{(-1)^2 - 4} = \frac{4}{3},$$

and as the $s = 5$ is not in the ROC, we have

$$B = X(5) = \infty.$$

Problem 176. Consider the following causal system:

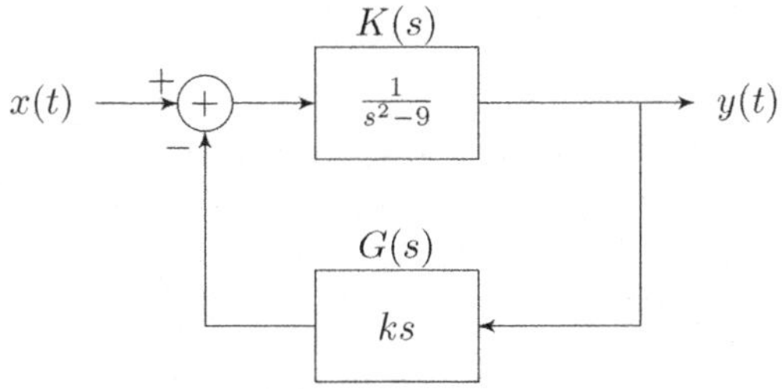

(a) Find the transfer function of the system.

(b) For $k = 0$, find the impulse response of the system, and determine if the system is stable or not.

(c) Find the value of k for which the system has two poles at $s = 1$ and $s = -9$.

Solution.

(a) As we know the relation between the input and output in a system with feedback is as below

$$H(s) = \frac{Y(s)}{X(s)} = \frac{K(s)}{1 + G(s)K(s)},$$

where $K(s) = \frac{1}{s^2-9}$ and $G(s) = ks$. Therefore

$$H(s) = \frac{1}{s^2 - 9 + ks}.$$

(b) In this case we have $H(s) = K(s)$ and by decomposition into partial fractions, we can obtain $h(t)$:

$$\begin{aligned} H(s) &= \frac{1}{s^2 - 9} \\ &= \frac{A}{s-3} + \frac{B}{s+3} \\ &= \frac{1}{6} \times \frac{1}{s-3} - \frac{1}{6} \times \frac{1}{s+3} \end{aligned}$$

$$\xRightarrow{\mathcal{L}^{-1}} h(t) = \frac{1}{6}(e^{3t} - e^{-3t})u(t).$$

As $\lim_{t\to\infty} h(t) \to \infty$, this system is not stable.

(c) To have two poles in $s = 1$ and $s = -9$, we should have

$$s^2 - 9 + ks = (s-1)(s+9) = s^2 + 8s - 9,$$

so we must set $k = 8$.

Problem 177. The Laplace transform of the impulse response for a system is given as follows:

$$H(s) = \frac{1}{(s+2)(s-3)}.$$

(a) Find the region of convergence for the transfer function for which the system is stable and causal.

(b) Find the impulse response assuming the system is stable.

(c) Find the impulse response when the system is neither causal nor stable.

Solution.

(a) This system has two poles in $s = -2$ and $s = 3$ and so there are 3 possible ROCs:

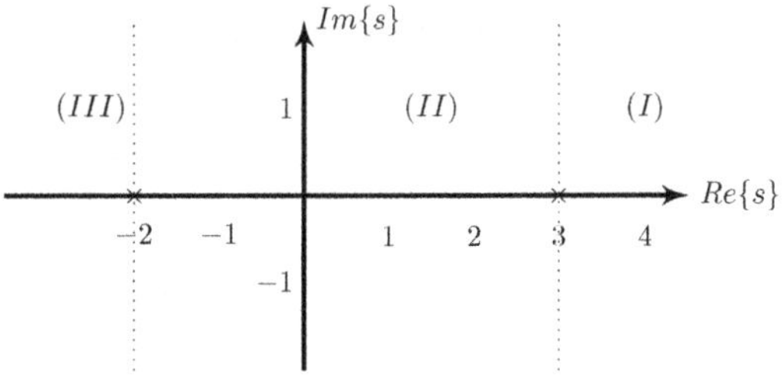

Figure 9.14: Problem 177 - Possible ROCs

Part (I): The system is unstable and causal (as ROC does not contain the jw-axis but contains $s = \infty$).

Part (II): The system is stable and non-causal (as ROC contains the jw-axis but does not contain $s = \infty$).

Part (III): The system is unstable and non-causal (as ROC neither contains the jw-axis, nor $s = \infty$).

Therefore, the system cannot be stable and causal at the same time.

(b) We write

$$H(s) = \frac{1}{(s+2)(s-3)} = \frac{\frac{1}{5}}{s-3} - \frac{\frac{1}{5}}{s+2}.$$

Based on the analysis in Part (a), the only ROC for which the system is stable is $-2 < Re\{s\} < 3$. Thus,

$$\frac{\frac{1}{5}}{(s-3)} \overset{s=3}{\Longrightarrow} \Big| \begin{matrix} Re\{s\}>3 \\ Re\{s\}<3\surd \end{matrix} \quad \overset{\mathcal{L}^{-1}}{\longrightarrow} -\frac{1}{5}e^{3t}u(-t),$$

$$\frac{-\frac{1}{5}}{(s+2)} \overset{s=-2}{\Longrightarrow} \Big| \begin{matrix} Re\{s\}>-2\surd \\ Re\{s\}<-2 \end{matrix} \quad \overset{\mathcal{L}^{-1}}{\longrightarrow} -\frac{1}{5}e^{-2t}u(t),$$

Thus we have

$$h(t) = -\frac{1}{5}(e^{3t}u(-t) + e^{-2t}u(t)).$$

(c) Based on the analysis in Part (a), the only ROC for which the system is neither stable nor causal is for $Re\{s\} < -2$. Thus,

$$\frac{\frac{1}{5}}{(s-3)} \overset{s=3}{\Longrightarrow} \Big| \begin{matrix} Re\{s\}>3 \\ Re\{s\}<3\surd \end{matrix} \quad \overset{\mathcal{L}^{-1}}{\longrightarrow} -\frac{1}{5}e^{3t}u(-t),$$

$$\frac{-\frac{1}{5}}{(s+2)} \overset{s=-2}{\Longrightarrow} \Big| \begin{matrix} Re\{s\}>-2 \\ Re\{s\}<-2\surd \end{matrix} \quad \overset{\mathcal{L}^{-1}}{\longrightarrow} \frac{1}{5}e^{-2t}u(-t),$$

Thus we have

$$h(t) = \frac{1}{5}(e^{-2t} - e^{3t})u(-t).$$

Problem 178. The following information about an LTI system with input $x(t)$ and output $y(t)$ is at hand:

$$X(s) = \frac{s+2}{s+3} \quad \text{and} \quad x(t) = 0 \text{ for } \quad t < 0$$

and

$$y(t) = \frac{1}{5}e^{-3t}u(t) - \frac{4}{5}e^{2t}u(-t).$$

Find the response of the system to the input $x_1(t) = e^{-2t}$ and $x_2(t) = e^{3t}$.

Solution. From (9.27), we know that the response to input $x(t) = e^{-2t}$ is equal to $y(t) = H(-2)e^{-2t}$, if $s = -2$ is in the ROC of $H(s)$. So we have to determine $H(s)$. Since $x(t)$ is zero for $t < 0$, it is a right-sided signal and consequently its ROC contains $+\infty$:

$$X(s) = \frac{s+2}{s+3}, \quad ROC : Re\{s\} > -3.$$

Also, the Laplace transform of $y(t)$ is obtained as below:

$$Y(s) = \frac{\frac{1}{5}}{s+3} + \frac{\frac{4}{5}}{s-2}, \quad ROC = \{Re\{s\} > -3\} \cap \{Re\{s\} < 2\}$$
$$\Rightarrow Y(s) = \frac{s+2}{(s+3)(s-2)}, \quad -3 < Re\{s\} < 2.$$

Now with $X(s)$ and $Y(s)$ at hand, $H(s)$ is computed as

$$H(s) = \frac{Y(s)}{X(s)} = \frac{\frac{s+2}{(s+3)(s-2)}}{\frac{s+2}{s+3}} = \frac{1}{s-2}.$$

As $H(s)$ has one pole in $s = 2$, it has two possible ROCs: $Re\{s\} < 2$ and $Re\{s\} > 2$. However, only the ROC which has non-empty intersection with ROC of $X(s)$ and $Y(s)$ is acceptable. Thus, the ROC is $Re\{s\} < 2$. As $s = -2$ is in the ROC, the response of system to the input $x_1(t) = e^{-2t}$ is computed as

$$y_1(t) = H(-2)e^{-2t} = -\frac{1}{4}e^{-2t}.$$

For $x_2(t) = e^{3t}$, since $s = 3$ is not in the ROC, the response of system is $y_2(t) = \infty$.

Problem 179. Step response of an LTI system is $s(t) = (1 - e^{-2t} - 2te^{-2t})u(t)$. If the system response to the input $x(t)$ is

$$y(t) = (1 + e^{-4t} - 2e^{-2t})u(t),$$

find $x(t)$.

Solution.

We know that

$$h(t) = \frac{ds(t)}{dt} \quad \Rightarrow \quad H(s) = sS(s).$$

Thus

$$S(s) = \frac{1}{s} - \frac{1}{s+2} - \frac{2}{(s+2)^2} = \frac{4}{s(s+2)^2} \text{ and}$$
$$H(s) = sS(s) = \frac{4}{(s+2)^2}, \quad ROC : Re\{s\} > -2.$$

Now we obtain $Y(s)$ as

$$Y(s) = \frac{1}{s} + \frac{1}{s+4} - \frac{2}{s+2} = \frac{8}{s(s+4)(s+2)}, \quad ROC : Re\{s\} > 0.$$

Therefore,

$$X(s) = \frac{Y(s)}{H(s)} = \frac{\frac{8}{s(s+4)(s+2)}}{\frac{4}{(s+2)^2}}$$
$$= \frac{2(s+2)}{s(s+4)} = \frac{1}{s} + \frac{1}{s+4}.$$

To determine $x(t)$, its corresponding ROC has to be found. $X(s)$ has two poles in $s = 0$ and $s = -4$, which constitute 3 possible ROCs: $Re\{s\} < -4$, $-4 < Re\{s\} < 0$, and $Re\{s\} > 0$. We have to choose the one with non-empty intersection with ROC of $Y(s)$ and $H(s)$ which is indeed $Re\{s\} > 0$. Thus we write

$$X(s) = \frac{1}{s} + \frac{1}{s+4}, \quad ROC : Re\{s\} > 0$$
$$\Rightarrow x(t) = (1 + e^{-4t})u(t).$$

Problem 180. If an LTI system where transfer function $H(s)$ is invertible, the transfer function of the inverse system, $H_i(s)$, is given by $H_i(s) = \frac{1}{H(s)}$. Using this fact, work out the following questions.

(a) The impulse response for systems S_1 is as below. Investigate reversibility of this system.

$$h_1(t) = e^{-t}u(t).$$

(b) The pole-zero diagram of an LTI system, and its ROC is indicated in the following figure. Find the region of convergence of the inverse system $H_i(s) = \frac{1}{H(s)}$.

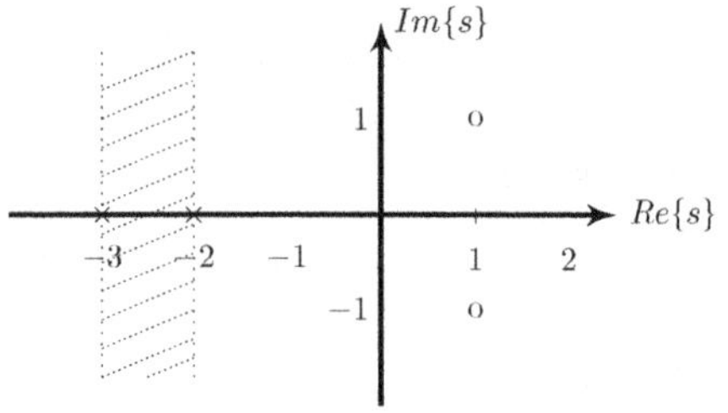

Figure 9.15: Problem 180 - Part (b), The pole-zero diagram of $H(s)$

Solution.

(a) Note that if the transfer function contains zeros in its ROC, say at $S = z_0$, the system is not invertible. The reason is that the response of the system to both inputs $x_1(t) = 0$ and $x_2(t) = e^{z_0 t}$ is $y(t) = 0$, and this does not coincide with the definition of invertibility. Otherwise, the system can be invertible. So let's obtain $H_1(s)$:

$$h_1(t) = e^{-t}u(t) \quad \xrightarrow{\mathcal{L}} \quad H_1(s) = \frac{1}{s+1}, \quad Re\{s\} > -1.$$

As can be seen, $H_1(s)$ does not have any zeros in its ROC and we obtain $H_i(s) = s + 1$.

(b) Based on the poles and zeros in the figure we obtain

$$H(s) = \frac{(s-1+j)(s-1-j)}{(s+2)(s+3)} = \frac{(s-1)^2+1}{(s+2)(s+3)}, \quad ROC: -3 < Re\{s\} < -2.$$

The zeros of $H(s)$ are located at $s = 1 \pm j$ which are outside of the ROC. So the system can be invertible and we obtain:

$$H_i(s) = \frac{1}{H(s)} = \frac{(s+2)(s+3)}{(s-1)^2+1}.$$

Two different ROCs can be imagined for $H_i(s)$: $Re\{s\} < 1$, $Re\{s\} > 1$. Among these, we have to choose the one that has non-empty intersection with the ROC of $H(s)$, i.e., $Re\{s\} < 1$.

Problem 181. There is an LTI system with the input and output as below. Find the impulse response of this system.

$$x(t) = 2e^{-4t}u(t) + 2e^{-2t}u(-t),$$
$$y(t) = e^{-4t}u(t) - e^{-3t}u(t).$$

Solution.
First, $X(s)$ and $Y(s)$ are obtained as follows:

$$x(t) = 2e^{-4t}u(t) + 2e^{-2t}u(-t)$$
$$\xRightarrow{\mathcal{L}} X(s) = \frac{2}{s+4} - \frac{2}{s+2} = \frac{-4}{(s+4)(s+2)}, \quad ROC: -4 < Re\{s\} < -2,$$

$$y(t) = e^{-4t}u(t) - e^{-3t}u(t)$$

$$\stackrel{\mathcal{L}}{\Longrightarrow} Y(s) = \frac{1}{s+4} - \frac{1}{s+3} = \frac{-1}{(s+4)(s+3)}, \quad ROC: Re\{s\} > -3.$$

Now, $H(s)$ is computed as

$$H(s) = \frac{Y(s)}{X(s)} = \frac{\frac{-1}{(s+4)(s+3)}}{\frac{-4}{(s+4)(s+2)}} = \frac{s+2}{4(s+3)}.$$

$H(s)$ has one pole in $s = -3$, resulting in two possible ROCs, $Re\{s\} < -3$ and $Re\{s\} > -3$, but the ROC should be such that its intersection with ROCs of $X(s)$ and $Y(s)$ is not empty. Thus, $Re\{s\} > -3$ is the proper ROC. So we have

$$H(s) = \frac{s+2}{4(s+3)} = \frac{1}{4}(1 - \frac{1}{s+3}), \quad ROC: Re\{s\} > -3$$

$$\stackrel{\mathcal{L}^{-1}}{\Longrightarrow} h(t) = \frac{1}{4}\delta(t) - \frac{1}{4}e^{-3t}u(t).$$

Chapter 10

Z-Transform

10.1 Introduction

In this chapter, we will introduce the z-transform and its basic concepts to represent discrete-time signals, and also describe the applications of the z-transform in analyzing discrete-time LTI systems.

10.2 The Z-Transform

Consider a discrete-time LTI system with impulse response $h[n]$. From Chapter 3, we know that the output of the system to a complex exponential input of the form z^n is

$$y[n] = H(z)z^n, \tag{10.1}$$

where

$$H(z) = \sum_{n=-\infty}^{+\infty} h[n]z^{-n}. \tag{10.2}$$

For an arbitrary complex variable z in (10.2), $H(z)$ is defined as the z-transform of the impulse response $h[n]$. The variable z is expressed in polar form as

$$z = re^{jw}, \tag{10.3}$$

where r is the magnitude of z and w is the angle of z.

Similarly, the z-transform of a discrete-time signal $x[n]$ is defined as

$$X(z) = \sum_{n=-\infty}^{+\infty} x[n]z^{-n}, \tag{10.4}$$

and shown as

$$x[n] \overset{\mathcal{Z}}{\longleftrightarrow} X(z).$$

The z-transform and the discrete-time Fourier transform are closely related. For the z-transform we write:

$$X(z) = X(re^{jw}) = \sum_{n=-\infty}^{+\infty} x[n]z^{-n}$$
$$= \sum_{n=-\infty}^{+\infty} \left[x[n]r^{-n}\right] e^{-jwn}. \tag{10.5}$$

On the other hand,

$$x[n] \overset{\mathcal{F.T.}}{\longleftrightarrow} \sum_{n=-\infty}^{+\infty} x[n]e^{-jwn}.$$

Thus

$$X(z) = \mathcal{F}\{x[n]r^{-n}\}. \tag{10.6}$$

From (10.6), we see that the z-transform of $x[n]$ is the Fourier transform of the signal $x[n]r^{-n}$. Alternatively, if $|z| = 1$, from (10.6) we have

$$X(z)\Big|_{z=e^{jw}} = X(e^{jw}) = \mathcal{F}\{x[n]\}. \tag{10.7}$$

10.3 Convergence of the Z-Transform

From Chapter 7, we know that the sufficient condition for the convergence of discrete-time Fourier transform is that $x[n]$ must be absolutely summable. Therefore, for convergence of the z-transform, it requires that the Fourier transform of $x[n]r^{-n}$ converges. As a result, $x[n]$ has z-transform, if $x[n]r^{-n}$ is absolutely summable:

$$\sum_{n=-\infty}^{+\infty} |x[n]|r^{-n} < \infty. \tag{10.8}$$

10.4 The Region of Convergence for the Z-Transform

The range of values of the complex variables z for which the z-transform is defined for a discrete-time signal is called the *region of convergence* (ROC). This is equivalent to the range of values of r for which $|x[n]|r^{-n}$ absolutely summable.

10.4.1 Definition of Poles and Zeros of $X(z)$:

As in the case of the Laplace transform in many cases, the z-transform of a signal $X(z)$ is a rational function in z,

$$X(z) = \frac{N(z)}{D(z)}, \tag{10.9}$$

where $N(z)$ and $D(z)$ are the numerator polynomial and denominator polynomial, respectively. By definition, the roots of $N(z)$ are called the *zeros* of $X(z)$ and for these values $X(z) = 0$. Similarly, the roots of $D(z)$ are called the *poles* of $X(z)$ and for these values $X(z) = \infty$:

$$X(z) = 0 \Rightarrow N(z) = 0 \Rightarrow z = z_1,\ z_2, \cdots, z_m.$$

$$X(z) = \infty \Rightarrow D(z) = 0 \Rightarrow z = p_1,\ p_2, \cdots, p_n.$$

In general, the poles and zeros of $X(z)$ are complex numbers. Since $X(z)$ does not converge at the poles, the ROC of $X(z)$ can not include any poles.

10.4.2 Properties of the ROC

To have a complete identification of z-transform of a given signal, we need to know both of the algebraic expression for $X(z)$ and the associated region of convergence. In this subsection, we introduce some properties of the ROC that are helpful to determine ROC of $X(Z)$.

Property 1: The ROC of $X(z)$ consists of a ring in the z-plane centered around the origin.

Property 2: The ROC does not contain any poles.

Property 3: If $x[n]$ is a *finite-duration* signal and absolutely summable, then the ROC is the entire z-plane, except possibly $z = 0$ and/or $z = \infty$. Note that a finite-duration signal is a signal for which $x[n] = 0$ except in a finite interval $N_1 \leq n \leq N_2$.

Property 4: If $x[n]$ is a *right-sided* signal, and if the circle $|z| = r_0$ is in the ROC, then all values of z for which $|z| > r_0$ will also be in the ROC. Note that a right-sided signal is a signal for which $x[n] = 0$ for $n < N_1 < \infty$ for some N_1.

Property 5: If $x[n]$ is a *left-sided* signal, and if the circle $|z| = r_0$ is in the ROC, then all values of z for which $0 < |z| < r_0$ will also be in the ROC. Note that a left-sided signal is a signal for which $x[n] = 0$ for $-\infty < N_2 < n$ for some N_2.

Property 6: If $x[n]$ is *two-sided*, and if the circle $|z| = r_0$ is in the ROC, then the ROC will include a ring in the z-plane that contains the circle $|z| = r_0$.

Property 7: If the z-transform $X(z)$ of $x[n]$ is rational, then its ROC is bounded by circles passing through poles or extends to infinity.

Property 8: If the z-transform $X(z)$ of $x[n]$ is rational, and if $x[n]$ is right-sided, then the ROC is the region in the z-plane outside the outermost pole, i.e., outside the circle with radius equal to the largest magnitude of the poles of $X(z)$. This implies that the ROC includes $z = \infty$.

Property 9: If the z-transform $X(z)$ of $x[n]$ is rational, and if $x[n]$ is left-sided, then the ROC is the region in the z-plane inside the innermost nonzero pole, i.e., inside the circle of radius equal to the smallest magnitude of the poles of $X(z)$ other than any at $z = 0$ and extending inward to and possibly including $z = 0$.

10.5 Properties of the Z-Transform

(a) Linearity

If

$$x_1[n] \overset{\mathcal{Z}}{\longleftrightarrow} X_1(z), \quad \text{with ROC} = R_1,$$
$$x_2[n] \overset{\mathcal{Z}}{\longleftrightarrow} X_2(z), \quad \text{with ROC} = R_2,$$

then we have

$$y[n] = ax_1[n] + bx_2[n] \overset{\mathcal{Z}}{\longleftrightarrow} Y(z) = aX_1(z) + bX_2(z), \quad \text{with ROC containing } R_1 \cap R_2. \tag{10.10}$$

(b) Time Shifting

If

$$x[n] \overset{\mathcal{Z}}{\longleftrightarrow} X(z), \quad \text{with ROC} = R,$$

then

$$x[n-n_0] \overset{\mathcal{Z}}{\longleftrightarrow} z^{-n_0}X(z), \quad \text{with ROC} = R, \text{ except for the possible addition or exclusion of the origin or infinity.} \tag{10.11}$$

(c) Scaling in the z-Domain

If

$$x[n] \overset{\mathcal{Z}}{\longleftrightarrow} X(z), \quad \text{with ROC} = R,$$

then

$$z_0^n x[n] \overset{\mathcal{Z}}{\longleftrightarrow} X(\frac{z}{z_0}), \quad \text{with ROC scaled by } |z_0|. \tag{10.12}$$

If $X(z)$ has a pole (or zero) at $z = z_k$, then $X(z/z_0)$ has a pole (or zero) at $z = z_0 z_k$ and the ROC expands or shrinks by factor $|z_0|$.

(d) Conjugation

If

$$x[n] \overset{\mathcal{Z}}{\longleftrightarrow} X(z), \quad \text{with ROC} = R,$$

then

$$x^*[n] \overset{\mathcal{Z}}{\longleftrightarrow} X^*(z^*), \quad \text{with ROC} = R. \tag{10.13}$$

Therefore,

$$X(z) = X^*(z^*), \quad \text{when } x[n] \text{ is real.} \tag{10.14}$$

(e) Time Expansion

If

$$x[n] \overset{\mathcal{Z}}{\longleftrightarrow} X(z), \quad \text{with ROC} = R,$$

then

$$x_k[n] = \begin{cases} x[n/k] & \text{if } n = mk \\ 0 & \text{if } n \neq mk \end{cases} \overset{\mathcal{Z}}{\longleftrightarrow} X(z^k), \text{ with ROC scaled by the power of } 1/k. \tag{10.15}$$

(f) Convolution Property

If

$$x_1[n] \overset{\mathcal{Z}}{\longleftrightarrow} X_1(z), \quad \text{with ROC} = R_1,$$
$$x_2[n] \overset{\mathcal{Z}}{\longleftrightarrow} X_2(z), \quad \text{with ROC} = R_2,$$

then we have

$$x_1[n] * x_2[n] \overset{\mathcal{Z}}{\longleftrightarrow} X_1(z)X_2(z), \text{ with ROC containing } R_1 \cap R_2. \tag{10.16}$$

10.6 Table of Z-Transform Properties and Common Z-Transform Pairs

Some of the important properties of the z-transform are given in Table 10.1, some of which have been discussed above. In Table 10.2, we have summarized some common z-transform pairs.

Table 10.1: Properties of the z-Transform

Property	**signal**	**Laplace transform**	ROC
	$x[n]$ $x_1[n]$ $x_2[n]$	$X(z)$ $X_1(z)$ $X_2(z)$	R R_1 R_2
Linearity	$ax_1[n] + bx_2[n]$	$aX_1(z) + bX_2(z)$	At least $R_1 \cap R_2$
Time Shifting	$x[n - n_0]$	$z^{-n_0}X(z)$	$R \cap \{0 < \lvert z\rvert < \infty\}$
Scaling in the z-domain	$z_0^n x[n]$	$X(\frac{z}{z_0})$	Scaled R
Conjugation	$x^*[n]$	$X^*(z^*)$	R
Time Reversal	$x[-n]$	$X(z^{-1})$	Flipped R
Convolution	$x_1[n] * x_2[n]$	$X_1(z)X_2(z)$	At least $R_1 \cap R_2$
First difference	$x[n] - x[n-1]$	$(1 - z^{-1})X(z)$	At least $R \cap \{\lvert z\rvert > 0\}$
Accumulation	$\sum_{k=-\infty}^{n} x[k]$	$\frac{1}{1-z^{-1}}X(z)$	At least $R \cap \{\lvert z\rvert > 1\}$
Differentiation in the z-Domain	$nx[n]$	$-z\frac{d}{dz}X(z)$	R

Table 10.2: Common z-Transforms Pairs

Signal	**Transform**	ROC
$\delta[n]$	1	All z
$\delta[n-m]$	z^{-m}	All z except 0 if $(m>0)$ or ∞ if $(m<0)$
$u[n]$	$\frac{1}{1-z^{-1}}$	$\|z\|>1$
$-u[-n-1]$	$\frac{1}{1-z^{-1}}$	$\|z\|<1$
$a^n u[n]$	$\frac{1}{1-az^{-1}}$	$\|z\|>\|a\|$
$-a^n u[-n-1]$	$\frac{1}{1-az^{-1}}$	$\|z\|<\|a\|$
$na^n u[n]$	$\frac{az^{-1}}{(1-az^{-1})^2}$	$\|z\|>\|a\|$
$-na^n u[-n-1]$	$\frac{az^{-1}}{(1-az^{-1})^2}$	$\|z\|<\|a\|$
$[\cos w_0 n]u[n]$	$\frac{1-[\cos w_0]z^{-1}}{1-[2\cos w_0]z^{-1}+z^{-2}}$	$\|z\|>1$
$[\sin w_0 n]u[n]$	$\frac{1-[\sin w_0]z^{-1}}{1-[2\cos w_0]z^{-1}+z^{-2}}$	$\|z\|>1$
$[r^n \cos w_0 n]u[n]$	$\frac{1-[r\cos w_0]z^{-1}}{1-[2r\cos w_0]z^{-1}+r^2z^{-2}}$	$\|z\|>r$
$[r^n \sin w_0 n]u[n]$	$\frac{1-[r\sin w_0]z^{-1}}{1-[2r\cos w_0]z^{-1}+r^2z^{-2}}$	$\|z\|>r$

10.7 The Initial- and Final-Value Theorems

10.7.1 The Initial-Value Theorem

From (10.4), $X(z)$ can be rewritten as

$$X(z) = \cdots + x[-2]z^2 + x[-1]z^1 + x[0] + x[1]z^{-1} + x[2]z^{-2} + \cdots . \tag{10.17}$$

If $x[n] = 0$ for $n < 0$, then we have

$$X(z) = x[0] + x[1]z^{-1} + x[2]z^{-2} + \cdots . \tag{10.18}$$

The initial-value theorem states that

$$x[0] = \lim_{z\to\infty} X(z). \tag{10.19}$$

At the same time, we can write

$$x[1] = \lim_{z \to \infty} z\left(X(z) - x[0]\right). \tag{10.20}$$

We can continue similarly for other values of $x[n]$.

10.7.2 The Final-Value Theorem

According to the ROC of $X(z)$, 3 cases of the Final-Value Theorem can be stated:

1. ROC $= 1 < |z| < r_2$:

$$x[+\infty] = \lim_{z \to 1}(1 - z^{-1})X(z), \quad x[-\infty] = 0. \tag{10.21}$$

2. ROC $= r_1 < |z| < 1$:

$$x[-\infty] = \lim_{z \to 1} -(1 - z^{-1})X(z), \quad x[+\infty] = 0. \tag{10.22}$$

3. If the ROC does not contain the unit circle :

$$x[+\infty] = x[-\infty] = 0. \tag{10.23}$$

10.8 The Inverse Z-Transform

Signal $x[n]$ can be obtained from $X(z)$ through taking the inverse Z-Transform as follows:

$$x[n] = \mathcal{Z}^{-1}\{X(z)\} = \frac{1}{2\pi j}\oint_C X(z)z^{n-1}dz,$$

where C is a counterclockwise closed path encircling the origin and entirely in the ROC.

Instead of direct integration, there are some simpler methods to obtain a signal $x[n]$ from its z-transform $X(z)$. In this section, we introduce two methods for obtaining the inverse z-transform.

10.8.1 Use of Tables of Z-Transform Pairs:

In this method we try to express $X(z)$ as a sum

$$X(z) = X_1(z) + X_2(z) + \cdots + X_n(z), \tag{10.24}$$

and assuming we know the signals $x_1[n]$, $x_2[n], \cdots,$ $x_n[n]$ are the inverse transforms of

$$X_1(z),\ X_2(z), \cdots,\ X_n(z)$$

respectively, from the linearity property, we have

$$x[n] = x_1[n] + x_2[n] + \cdots + x_n[n]. \tag{10.25}$$

10.8.2 Partial Fraction Expansion:

If $X(z)$ is a rational function

$$X(z) = \frac{N(z)}{D(z)} = k\frac{(z-z_1)\cdots(z-z_m)}{(z-p_1)\cdots(z-p_n)}, \tag{10.26}$$

then:

(a) If all poles of $X(z)$ are simple and $m < n$, then we can write

$$\frac{X(z)}{z} = \frac{c_0}{z} + \frac{c_1}{z-p_1} + \cdots + \frac{c_n}{z-p_n}, \tag{10.27}$$

where coefficients c_k are given by

$$c_0 = X(z)\Big|_{z=0}, \qquad c_k = (z-p_k)X(z)\Big|_{z=p_k}. \tag{10.28}$$

(b) If the poles of $X(z)$ are multiple and $m < n$, then for each multiple pole of $X(z)$ the expansion of $X(z)/z$ will consist of terms of the form

$$\frac{\lambda_1}{z-p_i} + \frac{\lambda_2}{(z-p_i)^2} \cdots + \frac{\lambda_r}{(z-p_i)^r}, \tag{10.29}$$

where

$$\lambda_{r-k} = \frac{1}{k!}\frac{d^k}{dz^k}\left[(z-p_i)^r \frac{X(z)}{z}\right]\Bigg|_{z=p_i}. \tag{10.30}$$

(c) If $m > n$, we first divide the numerator by the denominator to comp up with a polynomial plus a rational function for which the polynomial degree of the numerator is less than the denominator. Now for the rational part, we end up with one of the above cases and for the polynomial part, we use the second row of Table 10.2 to obtain the inverse.

10.9 Analysis of Discrete-Time LTI Systems Using the Z-Transform

Consider an LTI system with the impulse response $h[n]$. Let the input be a discrete-time signal $x[n]$. The z-transform of the output signal $y[n]$, and eventually $y[n]$ itself, can be obtained based on the z-transforms of $x[n]$ and $h[n]$. That is, if

$$y[n] = h[n] * x[n] \quad \overset{\mathcal{Z}}{\longleftrightarrow} \quad Y(z) = H(z)X(z), \tag{10.31}$$

then

$$H(z) = \frac{Y(z)}{X(z)}. \tag{10.32}$$

where $H(z)$ is referred to as the system transfer function.

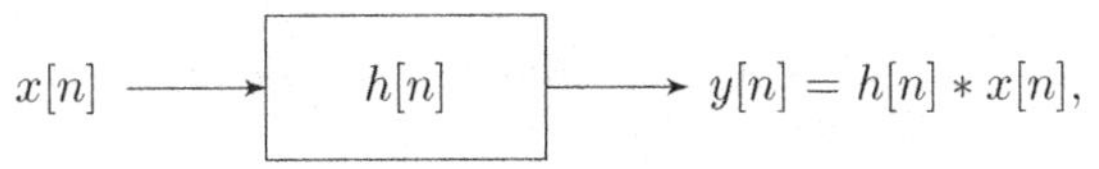

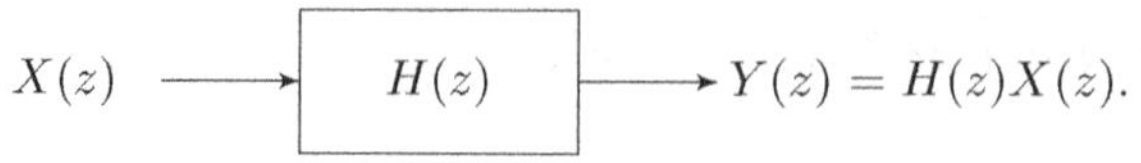

The output of an LTI system with input z^n is given by

$$y[n] = H(z)z^n. \tag{10.33}$$

10.10 Characterization of LTI Systems Using the Z-Transform

10.10.1 Causality

From Chapter 3, we know that a causal LTI system satisfies following condition:

$$h[n] = 0, \quad n < 0.$$

It is clear that $h[n]$ is a right-sided signal. Consequently, the ROC of the system function for a causal LTI system includes $z = +\infty$.

10.10.2 Stability

From Chapter 3, we know that a discrete-time LTI system is stable if the impulse response is absolutely summable, i.e., if

$$\sum_{n=-\infty}^{+\infty} |h[n]| < \infty. \tag{10.34}$$

Therefore, if an LTI system is stable, the ROC of its transfer function contains the unit circle:

$$|z| = 1 \in \text{ROC of } H(z).$$

10.10.3 Causal and Stable Systems

If an LTI system is both causal and stable, then the ROC must contain both the unit circle $|z| = 1$ and $|z| = +\infty$. It means that all poles of $H(z)$ must be inside of the unit circle.

10.11 LTI Systems Characterized by Constant-Coefficient Difference Equations

Consider a discrete-time LTI system with the input $x[n]$ and output $y[n]$ which satisfy the linear relationship of the form

$$\sum_{k=0}^{N} a_k y[n-k] = \sum_{k=0}^{M} b_k x[n-k]. \tag{10.35}$$

Applying the z-transform and using the differentiation property of the z-transform, we obtain

$$\sum_{k=0}^{N} a_k z^{-k} Y(z) = \sum_{k=0}^{M} b_k z^{-k} X(z), \tag{10.36}$$

or

$$Y(z) \sum_{k=0}^{N} a_k z^{-k} = X(s) \sum_{k=0}^{M} b_k z^{-k}. \tag{10.37}$$

Then,

$$H(z) = \frac{Y(z)}{X(z)} = \frac{\sum_{k=0}^{M} b_k z^{-k}}{\sum_{k=0}^{N} a_k z^{-k}}. \tag{10.38}$$

10.12 Systems Interconnection

(a) The impulse response of a parallel combination of two LTI systems:

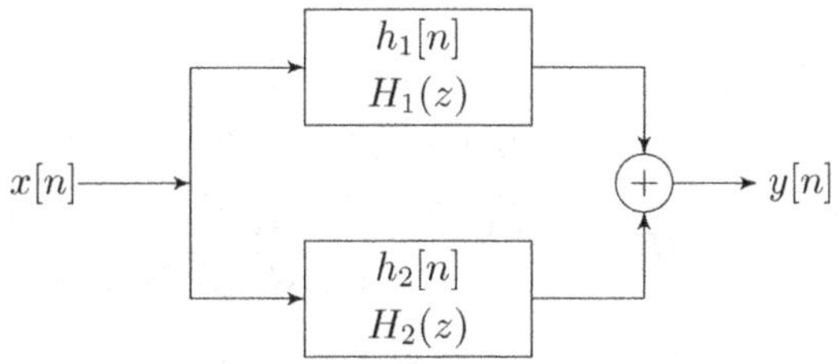

Figure 10.1: A parallel interconnection of two systems.

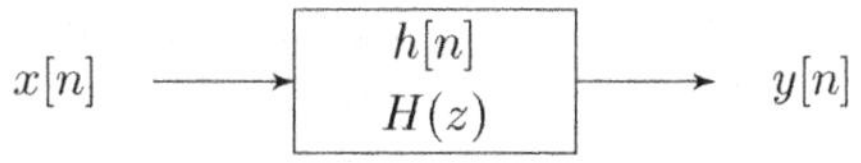

$$h[n] = h_1[n] + h_2[n], \tag{10.39}$$

$$H(z) = H_1(z) + H_2(z). \tag{10.40}$$

(b) The impulse response of a cascade combination of two LTI systems:

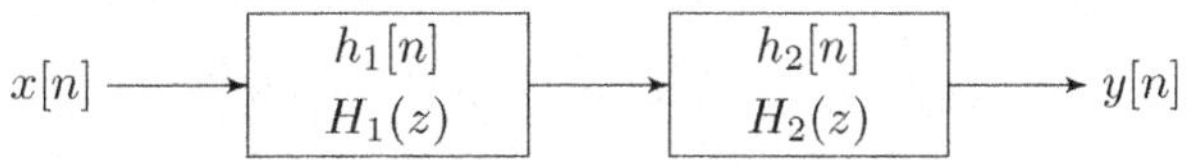

Figure 10.2: A series or cascade interconnection of two systems.

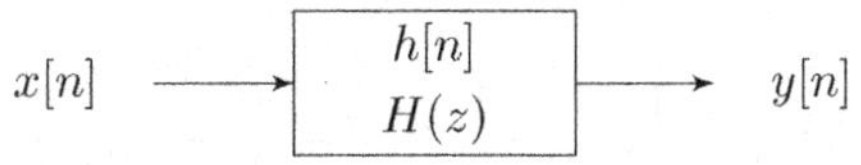

$$h[n] = h_1[n] * h_2[n], \tag{10.41}$$
$$H(z) = H_1(z)H_2(z). \tag{10.42}$$

(c) The impulse response of a feedback interconnection:

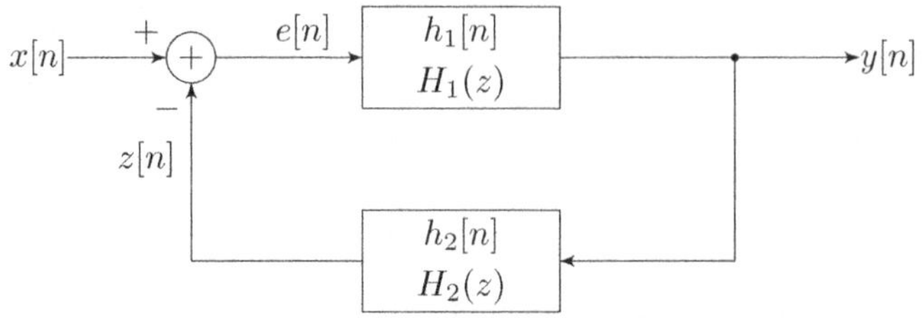

Figure 10.3: A feedback interconnection.

$$Y(z) = H_1(z)E(z), \quad E(z) = X(z) - Y(z)H_2(z),$$
$$\Rightarrow Y(z) = H_1(z)\left[X(z) - Y(z)H_2(z)\right],$$
$$H(z) = \frac{Y(z)}{X(z)} = \frac{H_1(z)}{1 + H_1(z)H_2(z)}. \tag{10.43}$$

10.13 Problems

Problem 182. For each of the following signals find the z-transform.

(a) $x[n] = \cos(nw_0)u[n]$

(b) $x[n] = -a^n u[-n-1]$

(c) $x[n] = (\frac{1}{2})^n u[3-n]$

(d) $x[n] = a^{|n|}, \quad |a| < 1$

(e) $x[n] = (\frac{1}{4})^{n-1}u[n-1] - 3(\frac{1}{2})^n u[-n-1]$

Solution.

(a) $x[n] = \cos(nw_0)u[n]$

$$x[n] = \frac{1}{2}(e^{jw_0 n} + e^{-jw_0 n})u[n].$$

As we know

$$u[n] \xrightarrow{Z} \frac{1}{1 - z^{-1}}, \quad ROC : |z| > 1.$$

Thus we have

$$X(z) = \frac{1}{2}\left(\frac{1}{1 - e^{jw_0}z^{-1}} + \frac{1}{1 - e^{-jw_0}z^{-1}}\right)$$

$$= \frac{1}{2}\left(\frac{2 - z^{-1}(e^{jw_0} + e^{-jw_0})}{(1 - e^{jw_0}z^{-1})(1 - e^{-jw_0}z^{-1})}\right)$$

Finally,

$$X(z) = \frac{1 - (\cos w_0)z^{-1}}{1 - 2(\cos w_0)z^{-1} + z^{-2}}, \quad ROC : |z| > 1.$$

(b) $x[n] = -a^n u[-n-1]$

This is already in Table 10.2, but we want to derive it using the definition:

$$X(z) = \sum_{n=-\infty}^{\infty} x[n]z^{-n} = \sum_{n=-\infty}^{\infty} -a^n u[-n-1]z^{-n} = -\sum_{n=-\infty}^{-1} (az^{-1})^n.$$

Substituting $m = -n$ into the last summation, we obtain

$$X(z) = -\sum_{m=1}^{\infty} (a^{-1}z)^m = -(a^{-1}z + a^{-2}z^2 + a^{-3}z^3 ...)$$

$$= -\frac{a^{-1}z}{1 - a^{-1}z}, \quad |a^{-1}z| < 1.$$

By multiplying the denominator and numerator by az^{-1}, we obtain

$$X(z) = -\frac{1}{az^{-1} - 1} = \frac{1}{1 - az^{-1}}, \quad ROC : |z| < a$$

(c) $x[n] = (\frac{1}{2})^n u[3-n]$
As we know

$$a^n u[n] \xrightarrow{Z} \frac{1}{1 - az^{-1}}, \quad ROC : |z| > a,$$

$$-(a)^n u[-n-1] \xrightarrow{Z} \frac{1}{1 - az^{-1}}, \quad ROC : |z| < a$$

Thus,

$$-(\frac{1}{2})^n u[-n-1] \xrightarrow{Z} \frac{1}{1 - \frac{1}{2}z^{-1}}, \quad ROC : |z| < \frac{1}{2}.$$

Substituting n with $n-4$, we have

$$-(\frac{1}{2})^{n-4} u[-n+3] \xrightarrow{Z} \frac{z^{-4}}{1 - \frac{1}{2}z^{-1}}, \quad |z| < \frac{1}{2}$$

$$\Rightarrow -(\frac{1}{2})^{-4}(\frac{1}{2})^n u[-n+3] \xrightarrow{Z} \frac{z^{-4}}{1 - \frac{1}{2}z^{-1}}, \quad |z| < \frac{1}{2}$$

$$\Rightarrow (\frac{1}{2})^n u[-n+3] \xrightarrow{Z} \frac{\frac{-z^{-4}}{16}}{1-\frac{1}{2}z^{-1}} = \frac{-z^{-4}}{8(2-z^{-1})}, \quad ROC: |z| < \frac{1}{2}.$$

(d) $x[n] = a^{|n|}, \quad |a| < 1$

At first $x[n]$ is rewritten as

$$x[n] = a^{|n|} = a^n u[n] + a^{-n} u[-n] - \delta[n].$$

We know that

$$a^n u[n] \xrightarrow{Z} \frac{1}{1-az^{-1}}, \quad ROC: |z| > a$$

and

$$x[-n] \xrightarrow{Z} X(z^{-1}).$$

Therefore

$$a^{-n} u[-n] \xrightarrow{Z} \frac{1}{1-az}, \quad ROC: |z| < \frac{1}{a}$$

Thus we write

$$X(z) = \frac{1}{1-az^{-1}} + \frac{1}{1-az} - 1, \quad ROC: a < |z| < \frac{1}{a}$$

$$\Rightarrow X(z) = \frac{1-a^2}{(1-az^{-1})(1-az)}, ROC: \quad a < |z| < \frac{1}{a}.$$

(e) $x[n] = (\frac{1}{4})^{n-1} u[n-1] - 3(\frac{1}{2})^n u[-n-1]$

The signal can be seen as the summation of two signals. To deal with the first part we have:

$$(\frac{1}{4})^n u[n] \xrightarrow{Z} \frac{1}{1-\frac{1}{4}z^{-1}}, \quad ROC: |z| > \frac{1}{4},$$

substituting n with $n-1$, we get:

$$(\frac{1}{4})^{n-1} u[n-1] \xrightarrow{Z} \frac{z^{-1}}{1-\frac{1}{4}z^{-1}}, \quad ROC: |z| > \frac{1}{4}.$$

For the 2nd signal we readily have:

$$-3(\frac{1}{2})^n u[-n-1] \xrightarrow{Z} \frac{3}{1-\frac{1}{2}z^{-1}}, \quad ROC: |z| < \frac{1}{2}.$$

Finally, the answer is obtained as below where the ROC is the intersection of the above two ROCs:

$$X(z) = \frac{z^{-1}}{1-\frac{1}{4}z^{-1}} + \frac{3}{1-\frac{1}{2}z^{-1}}, \quad ROC: \frac{1}{4} < |z| < \frac{1}{2}.$$

Problem 183. Determine the z-transform for each of following signals using z-transform properties.

(a) $x[n] = (2)^n \cos(\frac{n\pi}{3})u[-n-1]$

(b) $x[n] = n(\frac{1}{2})^n u[n-2]$

(c) $x[n] = \frac{1}{n}(-2)^n u[-n-1]$

Solution.

(a) $x[n] = (2)^n \cos(\frac{n\pi}{3})u[-n-1]$

We rewrite the signal as

$$x[n] = (2)^n \cos(\frac{n\pi}{3})u[-n-1] = (2)^n(\frac{e^{j\frac{n\pi}{3}} + e^{-j\frac{n\pi}{3}}}{2})u[-n-1]$$
$$= \frac{1}{2}(2)^n(e^{j\frac{n\pi}{3}})u[-n-1] + \frac{1}{2}(2)^n(e^{-j\frac{n\pi}{3}})u[-n-1].$$

As we know

$$(2)^n u[-n-1] \xrightarrow{Z} \frac{-1}{1-2z^{-1}}, \quad ROC: |z| < 2.$$

Based on the frequency shifting property, we have

$$(z_0)^n x[n] \xrightarrow{Z} X(\frac{z}{z_0}).$$

Thus

$$(2)^n(\frac{e^{j\frac{n\pi}{3}}}{2})u[-n-1] \xrightarrow{Z} -\frac{1}{2} \cdot \frac{1}{1-2e^{j\frac{\pi}{3}}z^{-1}}, \quad |z| < 2|e^{j\frac{\pi}{3}}|,$$

$$(2)^n(\frac{e^{-j\frac{n\pi}{3}}}{2})u[-n-1] \xrightarrow{Z} -\frac{1}{2} \cdot \frac{1}{1-2e^{-j\frac{\pi}{3}}z^{-1}}, \quad |z| < 2|e^{-j\frac{\pi}{3}}|.$$

Since $|e^{-j\frac{\pi}{3}}| = |e^{j\frac{\pi}{3}}| = 1$, the answer is obtained as follows:

$$X(z) = -\frac{1}{2}(\frac{1}{1-2e^{j\frac{\pi}{3}}z^{-1}} + \frac{1}{1-2e^{-j\frac{\pi}{3}}z^{-1}}), \quad |z| < 2.$$

(b) $x[n] = n(\frac{1}{2})^n u[n-2]$

Consider the following relationship:

$$(\frac{1}{2})^n u[n-2] = \frac{1}{4}(\frac{1}{2})^{(n-2)} u[n-2].$$

By obtaining the z-transform of $(\frac{1}{2})^n u[n]$ and using time-shifting property, we have

$$\frac{1}{4}(\frac{1}{2})^{(n-2)} u[n-2] \xrightarrow{Z} \frac{1}{4}\frac{z^{-2}}{1-\frac{1}{2}z^{-1}}, \quad ROC: |z| > \frac{1}{2}.$$

Now let $w[n] = \frac{1}{4}(\frac{1}{2})^{(n-2)} u[n-2]$ with z-transform as $W(z)$. In this case, $x[n] = nw[n]$

and with differentiation property in z-domain, we have

$$X(z) = -z\frac{dW(z)}{dz} = \frac{1}{2}z^{-2}\frac{1-\frac{1}{4}z^{-1}}{(1-\frac{1}{2}z^{-1})^2}, \quad ROC: |z| > \frac{1}{2}.$$

(c) $x[n] = \frac{1}{n}(-2)^n u[-n-1]$

Let $y[n] = (-2)^n u[-n-1]$. Then

$$y[n] = (-2)^n u[-n-1] \xrightarrow{Z} \frac{-1}{1+2z^{-1}}, \quad ROC: |z| < 2.$$

Now since $y[n] = nx[n]$, in the Z-domain, the signals are related as:

$$Y(z) = -z\frac{dX(z)}{dz}$$

then,

$$\frac{dX(z)}{dz} = \frac{z^{-1}}{1+2z^{-1}} = \frac{1}{z+2},$$

and we obtain the $X(z)$ as follows:

$$X(z) = \ln(z+2), \quad ROC: |z| < 2.$$

Problem 184. For each of the following signals, find the z-transform using convolution and differentiation properties.

(a) $x[n] = n\big[(\frac{1}{2})^n u[n] * (\frac{1}{3})^n u[n-2]\big]$

(b) $x[n] = n\sin(\frac{\pi}{2}n)u[-n]$

(c) $x[n] = (\frac{1}{2})^{|n|}\big[u[n+3] - u[n-4]\big]$

(d) $x[n] = \frac{1}{n}u[n]$

Solution.

(a) $x[n] = n\big[(\frac{1}{2})^n u[n] * (\frac{1}{3})^n u[n-2]\big]$

Let

$$a[n] = (\frac{1}{2})^n u[n]$$
$$b[n] = (\frac{1}{3})^n u[n]$$
$$c[n] = \frac{1}{9}(\frac{1}{3})^{n-2}u[n-2] = \frac{1}{9}b[n-2].$$

Thus,

$$x[n] = n\big[a[n] * c[n]\big].$$

We write

$$a[n] = (\frac{1}{2})^n u[n] \xrightarrow{Z} A(z) = \frac{1}{1 - \frac{1}{2}z^{-1}}, \quad ROC : |z| > \frac{1}{2}$$

$$b[n] = (\frac{1}{3})^n u[n] \xrightarrow{Z} B(z) = \frac{1}{1 - \frac{1}{3}z^{-1}}, \quad ROC : |z| > \frac{1}{3}$$

$$c[n] = \frac{1}{9} b[n-2] \xrightarrow{Z} C(z) = \frac{1}{9} \frac{z^{-2}}{1 - \frac{1}{3}z^{-1}}, \quad ROC : |z| > \frac{1}{3}$$

By utilizing the convolution and differentiation properties

$$x[n] = n\big[a[n] * c[n]\big] \xrightarrow{Z} -z\frac{d}{dz}\{A(z)C(z)\},$$

and thus we obtain

$$X(z) = \frac{1}{9}\Big[\frac{2z^2 - \frac{5}{6}z}{(z^2 - \frac{5}{6}z + \frac{1}{6})^2}\Big], \quad ROC : |z| > \frac{1}{2}.$$

(b) $x[n] = n\sin(\frac{\pi}{2}n)u[-n]$

Let

$$a[n] = \sin(\frac{\pi}{2}n)u[n].$$

Thus

$$a[-n] = \sin(-\frac{\pi}{2}n)u[-n] = -\sin(\frac{\pi}{2}n)u[-n].$$

From the Table. (10.2), the z-transfer of $a[n]$ is given by

$$A(z) = \frac{z^{-1}\sin(\frac{\pi}{2})}{1 - 2z^{-1}\cos(\frac{\pi}{2}) + z^{-2}} = \frac{z^{-1}}{1 + z^{-2}}, \quad ROC : |z| > 1.$$

Thus

$$-a[-n] \xrightarrow{Z} -A(z^{-1}) = -\frac{z}{1 + z^2}, \quad ROC : |z| < 1.$$

Since $x[n] = -na[-n]$, we get

$$X(z) = -z\frac{d(-A(z^{-1}))}{dz} = \frac{z - z^3}{(1 + z^2)^2}, \quad ROC : |z| < 1.$$

(c) $x[n] = (\frac{1}{2})^{|n|}\big[u[n+3] - u[n-4]\big]$

Let

$$x_1[n] = (\frac{1}{2})^n\big[u[n] - u[n-4]\big].$$

Then,

$$x[n] = x_1[n] + x_1[-n] - \delta[n],$$

and

$$X_1(z) = \sum_{n=0}^{n=3} (\frac{1}{2})^n z^{-n} = \frac{1 - (\frac{1}{2}z^{-1})^4}{1 - \frac{1}{2}z^{-1}}.$$

Now we write:

$$X(z) = X_1(z) + X_1(z^{-1}) - 1,$$

$$X(z) = \frac{1 - (\frac{1}{2}z^{-1})^4}{1 - \frac{1}{2}z^{-1}} + \frac{1 - (\frac{1}{2}z)^4}{1 - \frac{1}{2}z} - 1.$$

$$ROC = \text{The whole } z\text{-plane, except } 0 \text{ and } \infty.$$

(d) $x[n] = \frac{1}{n}u[n]$

As we know

$$u[n] \xrightarrow{Z} \frac{1}{1 - z^{-1}}, \quad ROC : |z| > 1.$$

Since $nx[n] = u[n]$, using differentiation in z-domain property, we must have

$$-z\frac{dX(z)}{dz} = \frac{1}{1 - z^{-1}}.$$

Thus $X(z)$ is obtained as

$$X(z) = \int \frac{-1}{z - 1} dz = -\ln(z - 1), \quad |z| > 1.$$

Problem 185. For each of the following signals find the z-transform based on the z-transform of $x[n]$.

(a) $y[n] = x_{(4)}[n - 2]$

(b) $y[n] = e^{-jn}x^*[n]$

(c) $y[n] = \begin{cases} -x[n] & \text{even } n \\ 2x[n] & \text{odd } n \end{cases}$

Solution.

(a) $y[n] = x_{(4)}[n - 2]$

Based on the definition:

$$x_{(4)}[n - 2] = \begin{cases} x[\frac{n-2}{4}] & (n - 2) \text{ is multiple of } 4 \\ 0 & (n - 2) \text{ is not multiple of } 4 \end{cases}$$

For the systems below, if the input is $w[n]$, we have indicated the output and based on that, we can track the changes of $x[n]$:

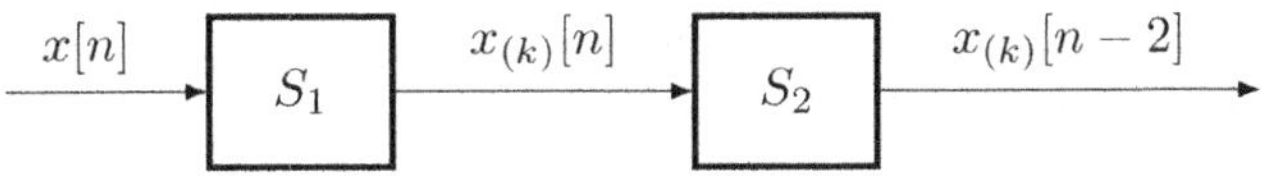

$$\begin{cases} S_1: & w[n] \to w_{(k)}[n] \\ S_2: & w[n] \to w[n-2] \end{cases}$$

In the Z-domain domain, the systems are shown in the figure below:

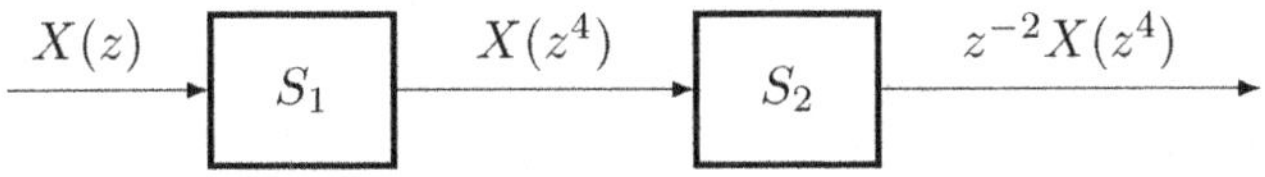

Finally, answer is obtained as

$$Y(z) = z^{-2}X(z^4)$$

(b) $y[n] = e^{-jn}x^*[n]$

For the systems below, if the input is $w[n]$, we have indicated the output and based on that, we can track the changes of $x[n]$:

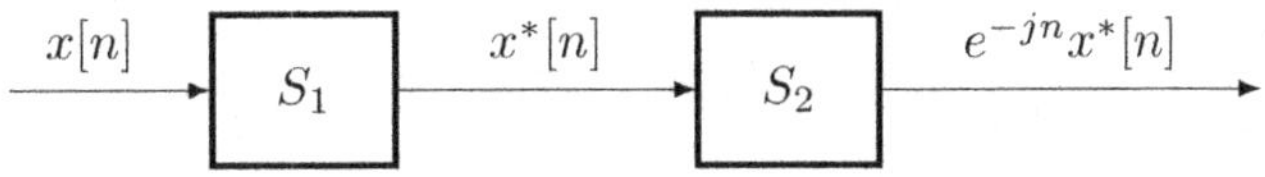

$$\begin{cases} S_1: & w[n] \to w^*[n] \\ S_2: & w[n] \to e^{-jn}w^*[n] \end{cases}$$

In the Z-domain domain, the systems are shown in the figure below:

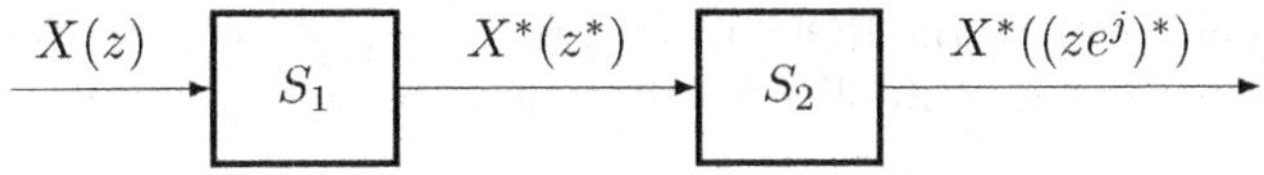

Finally, answer is obtained as

$$Y(z) = X^*(z^*e^{-j})$$

Note: The scaling in Z-domain property is as below:
if $x[n] \xrightarrow{Z} X(z)$ then $a^n x[n] \xrightarrow{Z} X(\frac{z}{a})$.

(c) $y[n] = \begin{cases} -x[n] & \text{even } n \\ 2x[n] & \text{odd } n \end{cases}$

In general, if an arbitrary signal $w[n]$ takes values A and B for n even and odd respectively, we can write it as follows:

$$w[n] = \frac{A+B}{2} + \frac{A-B}{2}(-1)^n.$$

Thus for $y[n]$ we write:

$$\begin{aligned} y[n] &= \frac{-x[n]+2x[n]}{2} + \frac{-x[n]-2x[n]}{2}(-1)^n \\ &= x[n].\Big(\frac{-1+2}{2} + \frac{-1-2}{2}(-1)^n\Big) \\ &= \frac{1}{2}x[n] - \frac{3}{2}x[n](-1)^n. \end{aligned}$$

Using the scaling in Z-domain property, we obtain

$$\begin{aligned} Y(z) &= \frac{1}{2}X(z) - \frac{3}{2}X(\frac{z}{-1}) \\ &= \frac{1}{2}X(z) - \frac{3}{2}X(-z). \end{aligned}$$

Problem 186. The transfer function of a causal LTI system is as follows

$$H(z) = \frac{1-5z^{-1}}{1+\frac{1}{2}z^{-1}}.$$

Find the system response to the input

$$x[n] = \begin{cases} 1+5^n & n \ge 0 \\ 1 & n < 0 \end{cases}.$$

Solution.

Since the system is causal, the ROC of $H(z)$ is $|z| > \frac{1}{2}$. We can use the conventional method based on convolution, but we will offer an alternative method here. We rewrite the input as

$$x[n] = \begin{cases} 1+5^n & n \ge 0 \\ 1 & n < 0 \end{cases} = 1 + 5^n u[n].$$

The input is composed of two signals $x_1[n] = 1$ and $x_2[n] = 5^n u[n]$. The output is obtained for each part. For the first part, since $z = 1$ is in the ROC, we use the definition of transfer function for LTI systems and write:

$$x_1[n] = 1 = (1)^n \Rightarrow y_1[n] = H(1)(1)^n = H(1) = -\frac{8}{3}$$

and for $x_2[n]$, we have

$$x_2[n] = 5^n u[n] \xrightarrow{Z} X_2(z) = \frac{1}{1-5z^{-1}}, \quad ROC: |z| > 5$$

$$Y_2(z) = H(z)X_2(z) = \frac{1-5z^{-1}}{1+\frac{1}{2}z^{-1}} \cdot \frac{1}{1-5z^{-1}} = \frac{1}{1+\frac{1}{2}z^{-1}}.$$

By obtaining the intersection of the ROCs of $X_2(z)$ and $H(z)$, the ROC for $Y_2(z)$ is given as $|z| > \frac{1}{2}$. Now we obtain $y_2[n]$ as

$$Y_2(z) = \frac{1}{1+\frac{1}{2}z^{-1}}, \quad |z| > \frac{1}{2} \quad \xrightarrow{Z^{-1}} \quad y_2[n] = (-\frac{1}{2})^n u[n].$$

Therefore, the system output to the input $x[n] = x_1[n] + x_2[n]$ is

$$y[n] = y_1[n] + y_2[n] = -\frac{8}{3} + (-\frac{1}{2})^n u[n].$$

Problem 187. Consider an LTI, causal and stable system with the impulse response $h[n]$. If the z-transform of the system output to the input $x[n] = h[-n]$ is

$$Y(z) = \frac{25z}{(5z-1)(5-z)},$$

find the value of following expressions.

(a) $A = \sum_{n=0}^{n=\infty} h[n]$ **(b)** $B = \sum_{n=0}^{n=\infty} h^2[n]$

Solution.

Since the $x[n] = h[-n]$, by the time reversal property, we have

$$X(z) = H(z^{-1}).$$

In an LTI system we have

$$Y(z) = X(z)H(z),$$

and thus,

$$\begin{aligned} Y(z) &= H(z^{-1})H(z) \\ &= \frac{25z}{(5z-1)(5-z)} \\ &= \frac{5}{5-z} \cdot \frac{5}{5-z^{-1}}. \end{aligned}$$

So $H(z)$ is

$$H(z) = \frac{5}{5-z^{-1}} = \frac{1}{1-\frac{1}{5}z^{-1}}.$$

As the system is stable, the ROC should the contain the unit circle. Since the system is causal, the ROC should contain $z = \infty$. Thus, the ROC is obtained as below:

$$H(z) = \frac{1}{1 - \frac{1}{5}z^{-1}}, \quad ROC : |z| > \frac{1}{5}.$$

As such

$$h[n] = (\frac{1}{5})^n u[n].$$

Now, we can evaluate the expressions:

(a)

$$A = \sum_{n=0}^{n=\infty} h[n] = \sum_{n=0}^{n=\infty} (\frac{1}{5})^n = \frac{1}{1 - \frac{1}{5}} = \frac{5}{4}.$$

(b)

$$B = \sum_{n=0}^{n=\infty} h^2[n] = \sum_{n=0}^{n=\infty} (\frac{1}{25})^n = \frac{1}{1 - \frac{1}{25}} = \frac{25}{24}.$$

Problem 188. For a causal LTI system we know that when the input is $x[n] = (\frac{1}{2})^n u[n]$, for the output $y[n]$ we have

$$y[0] = 2, \; y[1] = 4, \; y[2] = -3, \; y[3] = 1.$$

Find the value of $h[3]$.

Solution.

$$x[n] = (\frac{1}{2})^n u[n] \xrightarrow{Z} X(z) = \frac{1}{1 - \frac{1}{2}z^{-1}}, \quad ROC : |z| < \frac{1}{2}.$$

$$H(z) = \frac{Y(z)}{X(z)} = \frac{Y(z)}{\frac{1}{1-\frac{1}{2}z^{-1}}} = Y(z)(1 - \frac{1}{2}z^{-1}) = Y(z) - \frac{1}{2}z^{-1}Y(z).$$

Using the above result, the time-domain relationship between $h[n]$ and $y[n]$ is obtained as

$$h[n] = y[n] - \frac{1}{2}y[n-1].$$

Thus we have

$$h[3] = y[3] - \frac{1}{2}y[3-1] = y[3] - \frac{1}{2}y[2] = 1 - \frac{1}{2} \times (-3) = \frac{5}{2}.$$

Problem 189. The transfer function of an LTI system with the impulse response $h[n]$ is described by

$$H(z) = \frac{1 - \frac{1}{2}z^{-1}}{z^{-1}(1 - \frac{1}{3}z^{-1})(1 + 3z^{-1})}, \quad ROC : |z| > 3.$$

Comment on the following items:

(a) Causality

(b) Stability

(c) The value of $\sum_{n=-\infty}^{\infty} h[n]$

Solution.
First we rewrite $H(z)$ as

$$H(z) = \frac{z^3 - \frac{1}{2}z^2}{(z - \frac{1}{3})(z + 3)}.$$

(a) Since the degree of the numerator polynomial of $H(z)$ is larger than that of the denominator, we have $\lim_{z\to\infty} = \infty$, meaning that there is at least one pole in $z = \infty$. This implies that the ROC does not contain $z = \infty$. Consequently, the system is not causal.

(b) The ROC of this system is $|z| > 3$ and does not contain the unit circle ($|z| = 1$), so this system is not stable.

(c) From the definition of z-transform, we have

$$H(z) = \sum_{n=-\infty}^{\infty} h[n]z^{-n}.$$

Thus, $H(1) = \sum_{n=-\infty}^{\infty} h[n]$. Since $z = 1$ is not in the ROC, $H(1) = \infty$. So we have

$$\sum_{n=-\infty}^{\infty} h[n] = \infty.$$

Problem 190. Consider an LTI and causal system with the impulse response $h[n]$. The step response of this system is $s[n] = \delta[n] + ah[n-1]$, where a is a constant and positive number. Find $h[0]$ for different values of a.

Solution.
Let's write the equality in the Z-domain:

$$s[n] = \delta[n] + ah[n-1] \xrightarrow{Z} S(z) = 1 + az^{-1}H(z). \tag{10.44}$$

For an LTI system, we know that the following relationship holds between the step response and impulse response Z-transforms:

$$S(z) = \frac{H(z)}{1 - z^{-1}}. \tag{10.45}$$

Using (10.44) and (10.45), we have

$$\frac{H(z)}{1 - z^{-1}} = 1 + az^{-1}H(z)$$

$$\Rightarrow H(z) = 1 - z^{-1} + (az^{-1} - az^{-2})H(z)$$
$$\Rightarrow H(z) = \frac{1 - z^{-1}}{1 - az^{-1} + az^{-2}} = \frac{z^2 - z}{z^2 - az + a}.$$

As stated in the question, the system is causal, so the initial value theorem is applicable:

$$h[0] = \lim_{z \to \infty} H(z) = 1.$$

Problem 191. Consider the following difference equation:

$$y[n] - \frac{1}{2}y[n-1] + \frac{1}{18}y[n-2] = 2x[n].$$

(a) If the system is stable, find the response to the input $x[n] = \delta[n-1]$.

(b) Find the system response to $x[n] = e^{j\pi n}$.

Solution.

(a) We obtain $Y(z)$ by taking the z-transform from both sides of the equation:

$$y[n] - \frac{1}{2}y[n-1] + \frac{1}{18}y[n-2] = 2\delta[n-1]$$
$$\xrightarrow{Z} Y(z) - (\frac{1}{2}z^{-1})Y(z) + (\frac{1}{18}z^{-2})Y(z) = 2z^{-1}$$

and then,

$$Y(z) = \frac{2z^{-1}}{1 - \frac{1}{2}z^{-1} + \frac{1}{18}z^{-2}}$$
$$= \frac{12}{1 - \frac{1}{3}z^{-1}} - \frac{12}{1 - \frac{1}{6}z^{-1}}.$$

Now, since the system is stable, the ROC has to contain the unit circle and thus, $|z| > 1/3$. As such, the inverse z-transform of $Y(z)$ is as

$$y[n] = 12[(\frac{1}{3})^n - (\frac{1}{6})^n]u[n].$$

(b) First we obtain the impulse response of the system as follows:

$$Y(z) - (\frac{1}{2}z^{-1})Y(z) + (\frac{1}{18}z^{-2})Y(z) = 2X(z)$$

thus,

$$H(z) = \frac{Y(z)}{X(z)} = \frac{2}{1 - \frac{1}{2}z^{-1} + \frac{1}{18}z^{-2}}.$$

The input can be seen as z^n with $z = e^{j\pi}$. As such we can write

$$y[n] = H(z = e^{j\pi})e^{j\pi n} = H(1)e^{j\pi n} = \frac{18}{5}e^{j\pi n}.$$

Problem 192. The z-transform of signal $x[n]$ is $X(z) = \frac{z^2}{z^2-4}$. Find the z-transform of following signals.

(a) $y[n] = x[n-2]$

(b) $y[n] = (\frac{1}{2})^n x[n]$

(c) $y[n] = x[-n] * x[n]$

(d) $y[n] = nx[n]$

(e) $y[n] = \cos(2n)x[n]$

Solution.

(a) $y[n] = x[n-2]$

By the time shifting property

$$Y(z) = z^{-2}X(z) = \frac{1}{z^2-4}.$$

(b) $y[n] = (\frac{1}{2})^n x[n]$

From frequency shifting property

$$z_0^n x[n] \xrightarrow{Z} X(\frac{z}{z_0})$$

Thus we can write

$$y[n] = (\frac{1}{2})^n x[n] \xrightarrow{Z} Y(z) = X(\frac{z}{\frac{1}{2}}) = X(2z) = \frac{4z^2}{4z^2-4} = \frac{z^2}{z^2-1}.$$

(c) $y[n] = x[-n] * x[n]$

By considering the following properties (time reversal and convolution):

$$x[-n] =\xrightarrow{Z} X(\frac{1}{z}),$$

$$x_1[n] * x_2[n], \xrightarrow{Z} X_1(z) \times X_2(z)$$

we obtain

$$Y(z) = X(\frac{1}{z})X(z) = \frac{z^2}{17z^2 - 4z^4 - 4}.$$

(d) $y[n] = nx[n]$

We use differentiation in z-domain property as follows:

$$nx[n] \xrightarrow{Z} -z\frac{dX(z)}{dz},$$

$$Y(z) = -z\frac{dX(z)}{dz} = \frac{8z^2}{(z^2-4)^2}.$$

(e) $y[n] = \cos(2n)x[n]$

We rewrite $y[n]$ as

$$y[n] = \left(\frac{1}{2}e^{j2n} + \frac{1}{2}e^{-j2n}\right)x[n].$$

Similar to Part **b**, we write

$$Y(z) = \frac{1}{2}X(e^{-j2}z) + \frac{1}{2}X(e^{j2}z)$$

$$\Rightarrow Y(z) = \frac{1}{2}\cdot\frac{(e^{-j2}z)^2}{(e^{-j2}z)^2-4} + \frac{1}{2}\cdot\frac{(e^{j2}z)^2}{(e^{j2}z)^2-4}.$$

Problem 193. Find the inverse z-transform of each following signals using the partial fraction decomposition.

(a) $X(z) = \dfrac{4}{z^3(z+0.2)}, \quad |z| > \frac{1}{5}$

(b) $X(z) = \dfrac{4 - \frac{7}{4}z^{-1} + \frac{1}{4}z^{-2}}{1 - \frac{3}{4}z^{-1} + \frac{1}{8}z^{-2}}, \quad |z| > \frac{1}{2}$

(c) $X(z) = \dfrac{3 - 3z^{-1}}{(1 - \frac{1}{2}z^{-1})(1 - 2z^{-1})}, \quad x[n]$ is causal

(d) $X(z) = \dfrac{1}{1 + 3z^{-1} + 2z^{-2}}, \quad |z| > 2$

(e) $X(z) = \dfrac{1}{(1 - z^{-1})(1 - z^{-2})}, \quad |z| > 1$

Solution.

(a) $X(z) = \dfrac{4}{z^3(z+0.2)}, \quad |z| > \frac{1}{5}$

We write

$$X(z) = \frac{4}{z^3(z+0.2)} = z^{-4}\frac{4}{1+0.2z^{-1}}.$$

Let's define $W(z)$ as

$$W(z) = \frac{4}{1+0.2z^{-1}}.$$

Since the signal is right-sided based on its ROC, we have

$$W(z) = \frac{4}{1+0.2z^{-1}} \xrightarrow{Z^{-1}} w[n] = 4(-0.2)^n u[n].$$

Using time shifting property we obtain

$$X(z) = z^{-4}W(z) \xrightarrow{Z^{-1}} x[n] = w[n-4],$$

and

$$x[n] = 4(-0.2)^{(n-4)}u[n-4].$$

(b) $X(z) = \dfrac{4 - \frac{7}{4}z^{-1} + \frac{1}{4}z^{-2}}{1 - \frac{3}{4}z^{-1} + \frac{1}{8}z^{-2}}, \quad |z| > \frac{1}{2}$

As the polynomial degree of numerator and denominator are the same, at first the numerator is divided into the denominator and then we use the partial fraction decomposition:

$$\begin{aligned}
X(z) &= \frac{4 - \frac{7}{4}z^{-1} + \frac{1}{4}z^{-2}}{1 - \frac{3}{4}z^{-1} + \frac{1}{8}z^{-2}} \\
&= 2 + \frac{2 - \frac{1}{4}z^{-1}}{1 - \frac{3}{4}z^{-1} + \frac{1}{8}z^{-2}} \\
&= 2 + \frac{A}{1 - \frac{1}{2}z^{-1}} + \frac{B}{1 - \frac{1}{4}z^{-1}}.
\end{aligned}$$

To obtain A and B we write

$$Y(z) = \frac{2 - \frac{1}{4}z^{-1}}{1 - \frac{3}{4}z^{-1} + \frac{1}{8}z^{-2}},$$

$$A = \left[(1 - \frac{1}{2}z^{-1})Y(z)\right]_{z^{-1}=2} = \frac{2 - \frac{1}{4}(2)}{1 - \frac{1}{4}(2)} = 3,$$

$$B = \left[(1 - \frac{1}{4}z^{-1})Y(z)\right]_{z^{-1}=4} = \frac{2 - \frac{1}{4}(4)}{1 - \frac{1}{2}(4)} = -1.$$

Since $|z| > \frac{1}{2}$, the signal is right-sided and thus

$$x[n] = 2\delta[n] + 3(\frac{1}{2})^n u[n] - (\frac{1}{4})^n u[n].$$

(c) $X(z) = \dfrac{3 - 3z^{-1}}{(1 - \frac{1}{2}z^{-1})(1 - 2z^{-1})}$, $x[n]$ is causal

By the partial fraction decomposition

$$X(z) = \frac{A}{1 - \frac{1}{2}z^{-1}} + \frac{B}{1 - 2z^{-1}},$$

where, using the method in the previous part, A and B are obtained as

$$A = 1, \qquad B = 2.$$

Since the signal is causal, its ROC is $|z| > 2$ and thus the answer is obtained as

$$\Rightarrow x[n] = (\frac{1}{2})^n u[n] + 2(2)^n u[n].$$

(d) $X(z) = \dfrac{1}{1 + 3z^{-1} + 2z^{-2}}$, $|z| > 2$

Similar to the previous parts,

$$X(z) = \frac{1}{1 + 3z^{-1} + 2z^{-2}} = \frac{1}{(1 + 2z^{-1})(1 + z^{-1})}$$

$$\Rightarrow X(z) = \frac{A}{1 + 2z^{-1}} + \frac{B}{1 + z^{-1}},$$

where A and B are computed as

$$A = 2, \qquad B = -1.$$

By the given ROC we obtain

$$x[n] = 2(-2)^n u[n] - (-1)^n u[n].$$

(e) $X(z) = \dfrac{1}{(1 - z^{-1})(1 - z^{-2})}$, $|z| > 1$

Since $X(z)$ has two poles in $z = \pm 1$, its partial fractions is written as

$$X(z) = \frac{A}{1 + z^{-1}} + \frac{B_1}{1 - z^{-1}} + \frac{B_2}{(1 - z^{-1})^2}, \tag{10.46}$$

where

$$A = \Big[(1 + z^{-1})X(z)\Big]_{z^{-1}=-1} = \frac{1}{4},$$

$$B_2 = \Big[(1 - z^{-1})^2 X(z)\Big]_{z^{-1}=1} = \frac{1}{2}.$$

B_1 can be found by replacing an arbitrary value of z in (10.46). We choose $z = 2$:

$$X(2) = \frac{8}{3} = \frac{2}{3}A + 2B_1 + 4B_2 \quad \Rightarrow \quad B_1 = \frac{1}{4}.$$

Now by the given ROC the answer is

$$x[n] = \frac{1}{4}\Big[(-1)^n + 1 + \frac{1}{2}(n+1)\Big]u[n].$$

Problem 194. Let $x[n] = n^2 5^n u[n]$ with $X(z)$ as its z-transform. Find $y[n]$ based on $x[n]$ for each of the following cases.

(a) $Y(z) = X(2z)$

(b) $Y(z) = X(z^{-1})$

(c) $Y(z) = \dfrac{dX(z)}{dz}$

(d) $Y(z) = \dfrac{z^2 - z^{-2}}{2}X(z)$

(e) $Y(z) = X(z)^2$

Solution.

(a) $Y(z) = X(2z)$

We know that

$$z_0^n x[n] \xrightarrow{Z} X(\frac{z}{z_0}).$$

For $Y(z) = X(2z)$, we have $z_0 = \frac{1}{2}$ and thus

$$\Rightarrow y[n] = (\frac{1}{2})^n x[n] = (\frac{1}{2})^n n^2 5^n u[n].$$

(b) $Y(z) = X(z^{-1})$

We know that

$$x[-n] \xrightarrow{Z} X(z^{-1}).$$

For $Y(z) = X(z^{-1})$ we have

$$y[n] = x[-n] = n^2 5^{-n} u[-n].$$

(c) $Y(z) = \dfrac{dX(z)}{dz}$

We know that for any signal $a[n]$ we have:

$$np[n] \xrightarrow{Z} -z\frac{dP(z)}{dz},$$

and

$$p[n-1] \xrightarrow{Z} z^{-1}P(z).$$

Now we write $Y(z)$ as

$$Y(z) = \frac{dX(z)}{dz} = -z^{-1}\Big[-z\frac{dX(z)}{dz}\Big],$$

and thus $y[n]$ can be written as

$$\begin{aligned} y[n] &= -(n-1)x[n-1] \\ &= -(n-1)(n-1)^2 5^{(n-1)}u[n-1] \\ &= -(n-1)^3 5^{(n-1)}u[n-1]. \end{aligned}$$

(d) $Y(z) = \dfrac{z^2 - z^{-2}}{2}X(z)$

We first note that $x[n-n_0] \xrightarrow{Z} z^{-n_0}X(z)$. We write $Y(z)$ as

$$Y(z) = \frac{z^2}{2}X(z) - \frac{z^{-2}}{2}X(z).$$

Thus,

$$y[n] = \frac{1}{2}x[n+2] - \frac{1}{2}x[n-2].$$

Finally,

$$y[n] = \frac{1}{2}\Big[(n+2)^2 5^{(n+2)}u[n+2] - (n-2)^2 5^{(n-2)}u[n-2]\Big].$$

(e) $Y(z) = X(z)^2$

As we know, the convolution in time domain is equivalent to multiplication in Z-domain and vice versa. Thus we can write $y[n]$ as a convolution of $x[n]$ by itself:

$$y[n] = x[n] * x[n].$$

Thus

$$y[n] = \sum_{k=-\infty}^{+\infty} k^2 5^k u[k]\ (n-k)^2 5^{n-k}u[n-k] = \sum_{k=0}^{n} k^2 5^k (n-k)^2 5^{n-k}.$$

Therefore

$$y[n] = 5^n \sum_{k=0}^{n}\Big[k^2n^2 - 2nk^3 + k^4\Big].$$

Problem 195. The transfer function of an LTI and causal system is as below

$$H(z) = \frac{1 + 2z^{-1}}{1 - \frac{1}{2}z^{-1}}.$$

Find an input signal which results the following output:

$$y[n] = \frac{1}{3}(\frac{1}{3})^n u[n] - 3^n u[-n-1].$$

Solution.
The z-transform of the output signal is

$$\begin{aligned} Y(z) &= \frac{\frac{1}{3}}{1-\frac{1}{3}z^{-1}} + \frac{1}{1-3z^{-1}} \\ &= \frac{\frac{4}{3}(1-z^{-1})}{(1-\frac{1}{3}z^{-1})(1-3z^{-1})}, \quad ROC: \frac{1}{3} < |z| < 3. \end{aligned}$$

Since the system is LTI, the input z-transform is obtained as below

$$X(z) = \frac{Y(z)}{H(z)} = \frac{\frac{4}{3}(1-z^{-1})(1-\frac{1}{2}z^{-1})}{(1+2z^{-1})(1-\frac{1}{3}z^{-1})(1-3z^{-1})}.$$

By the partial fraction decomposition,

$$X(z) = \frac{A}{1-\frac{1}{3}z^{-1}} + \frac{B}{1-3z^{-1}} + \frac{C}{1+2z^{-1}}.$$

The values of A, B and C are computed as follows:

$$\begin{aligned} \Big[(1-\frac{1}{3}z^{-1})X(z)\Big]_{z^{-1}=3} &= -\frac{1}{42}, \\ \Big[(1-3z^{-1})X(z)\Big]_{z^{-1}=\frac{1}{3}} &= \frac{1}{2}, \\ \Big[(1+2z^{-1})X(z)\Big]_{z^{-1}=-\frac{1}{2}} &= \frac{6}{7}. \end{aligned}$$

As the system is casual, the ROC must contain ∞ and so we have the ROC $|z| > \frac{1}{2}$ for $H(z)$. Given that the ROC of $Y(z)$ is $\frac{1}{3} < |z| < 3$, the ROC for $X(z)$ has to be $\frac{1}{3} < |z| < 2$. Thus, $x[n]$ is obtained as

$$x[n] = -\frac{1}{42}(\frac{1}{3})^n u[n] - \frac{1}{2}(3)^n u[-n-1] - \frac{6}{7}(-2)^n u[-n-1].$$

Problem 196. Consider two causal systems described by the following difference equations. Let the input signal to these systems be $x[n] = (\frac{1}{2})^n u[n]$. Find the output for each system.

a) $y[n] - y[n-1] + y[n-2] = \frac{1}{2}x[n] + \frac{1}{2}x[n-1]$.

b) $y[n] + 3y[n-1] = x[n] - x[n-1]$.

Solution.

(a) At first, the z-transform is taken from the both sides of difference equation:

$$Y(z) - z^{-1}Y(z) + z^{-2}Y(z) = \frac{1}{2}X(z) + \frac{1}{2}z^{-1}X(z)$$

$$\Rightarrow Y(z)\Big(1 - z^{-1} + z^{-2}\Big) = \frac{1}{2}(1 + z^{-1})X(z),$$

and, with $X(z) = \frac{1}{1-\frac{1}{2}z^{-1}}$, we have

$$\begin{aligned} Y(z) &= \frac{\frac{1}{2} + \frac{1}{2}z^{-1}}{(1 - \frac{1}{2}z^{-1})(1 - z^{-1} + z^{-2})} \\ &= \frac{\frac{1}{2}}{1 - \frac{1}{2}z^{-1}} + \frac{z^{-1}}{1 - z^{-1} + z^{-2}}. \end{aligned}$$

We then have:

$$y[n] = \frac{1}{2}(\frac{1}{2})^n u[n] + \frac{2}{\sqrt{3}}\sin(\frac{n\pi}{3})u[n].$$

(b) Taking the z-transform from the both sides of difference equation we get:

$$Y(z) + 3z^{-1}Y(z) = X(z) - z^{-1}X(z)$$

$$\Rightarrow Y(z) = \frac{X(z)(1 - z^{-1})}{1 + 3z^{-1}}.$$

with $X(z) = \frac{1}{1-\frac{1}{2}z^{-1}}$ we get:

$$\begin{aligned} Y(z) &= \frac{1 - z^{-1}}{(1 + 3z^{-1})(1 - \frac{1}{2}z^{-1})} \\ &= \frac{A}{1 + 3z^{-1}} + \frac{B}{1 - \frac{1}{2}z^{-1}}, \end{aligned}$$

where

$$A = \Big[(1 + 3z^{-1})Y(z)\Big]_{z^{-1}=-\frac{1}{3}} = \frac{8}{7},$$

$$B = \Big[(1 - \frac{1}{2}z^{-1})Y(z)\Big]_{z^{-1}=2} = -\frac{1}{7}.$$

Then we have

$$y[n] = \Big[\frac{8}{7}(-3)^n - \frac{1}{7}(\frac{1}{2})^n\Big]u[n].$$

Problem 197. The impulse response of an LTI system is as below

$$h[n] = 2\delta[n] + \frac{5}{2}(\frac{1}{2})^n u[n] - \frac{7}{2}(-\frac{1}{4})^n u[n].$$

Is the inverse of the system causal and stable?

Solution.

If an LTI system with transfer function $H(z)$ is invertible, the transfer function of the inverse system, $H_i(z)$, is given by $H_i(z) = \frac{1}{H(z)}$. Let's find the transfer function of the system:

$$\begin{aligned} H(z) &= 2 + \frac{\frac{5}{2}}{1 - \frac{1}{2}z^{-1}} - \frac{\frac{7}{2}}{1 + \frac{1}{4}z^{-1}} \\ &= 2 + \frac{\frac{19}{8}z^{-1} - 1}{(1 - \frac{1}{2}z^{-1})(1 + \frac{1}{4}z^{-1})} \\ &= \frac{(1 - \frac{1}{8}z^{-1})(1 + 2z^{-1})}{(1 - \frac{1}{2}z^{-1})(1 + \frac{1}{4}z^{-1})}. \end{aligned}$$

Thus

$$H_i(z) = \frac{(1 - \frac{1}{2}z^{-1})(1 + \frac{1}{4}z^{-1})}{(1 - \frac{1}{8}z^{-1})(1 + 2z^{-1})}.$$

As can be seen, $H_i(z)$ has two poles in $z = \frac{1}{8}$ and $z = -2$. with one of the poles being outside of the unit circle, the inverse of system cannot be stable and causal at the same time.

Problem 198. The following information is available for a system with the impulse response $h[n]$ and transfer function $H(z)$:

1. $h[n]$ is real and is equal to 0 for $n < 0$ and $n > 5$.
2. $\sum_{n=0}^{5}(-1)^n h[n] = 0$.
3. $H(z) = 0$ at $z = \frac{1}{2}e^{\frac{-j\pi}{4}}$, $z = 2e^{\frac{-j\pi}{4}}$.
4. $h[0] = 2$.
5. $H(e^{jw})$ has a pole at the origin.

Find $H(z)$.

Solution.

Using the first assumption, we can write:

$$\begin{aligned} H(z) &= h[0] + h[1]z^{-1} + h[2]z^{-2} + h[3]z^{-3} + h[4]z^{-4} + h[5]z^{-5} \\ &= \frac{h[0]z^5 + h[1]z^4 + h[2]z^3 + h[3]z^2 + h[4]z^1 + h[5]}{z^5}, \end{aligned}$$

implying that $H(z)$ is a polynomial having five zeros. Based on the second fact, $z = e^{j\pi} = -1$ is a zero for the system. Since the system is real, conjugate of a zero is also a zero of the system. Thus using fact 3, $z = \frac{1}{2}e^{\frac{j\pi}{4}}$ and $z = 2e^{\frac{j\pi}{4}}$ are also zeros of the system. As such we have

$$\begin{aligned}
H(z) &= k(1+z^{-1})(\frac{1}{2}e^{\frac{j\pi}{4}} - z^{-1})(\frac{1}{2}e^{\frac{-j\pi}{4}} - z^{-1})(1 - \frac{1}{2}e^{\frac{j\pi}{4}}z^{-1})(1 - \frac{1}{2}e^{\frac{-j\pi}{4}}z^{-1}) \\
&= k(1+z^{-1})\left(\frac{1}{4} - z^{-1}(\frac{e^{\frac{-j\pi}{4}} + e^{\frac{j\pi}{4}}}{2}) + z^{-2}\right)\left(1 - z^{-1}(\frac{e^{\frac{-j\pi}{4}} + e^{\frac{j\pi}{4}}}{2}) + \frac{1}{4}z^{-2}\right) \\
&= k(1+z^{-1})(\frac{1}{4} - \frac{\sqrt{2}}{2}z^{-1} + z^{-2})(1 - \frac{\sqrt{2}}{2}z^{-1} + \frac{1}{4}z^{-2}).
\end{aligned}$$

By the assumption 4, $h[0] = 2$, and for $z = \infty$, we have $H(z) = h[0]$, thus

$$2 = k \times \frac{1}{4} \Rightarrow k = 8.$$

Finally,

$$\Rightarrow H(z) = 8(1+z^{-1})(\frac{1}{4} - \frac{\sqrt{2}}{2}z^{-1} + z^{-2})(1 - \frac{\sqrt{2}}{2}z^{-1} + \frac{1}{4}z^{-2}).$$

Problem 199. Consider an LTI system for which the input $x[n]$ results in the output $y[n]$, both are shown in figures below. Investigate the causality and stability of this system.

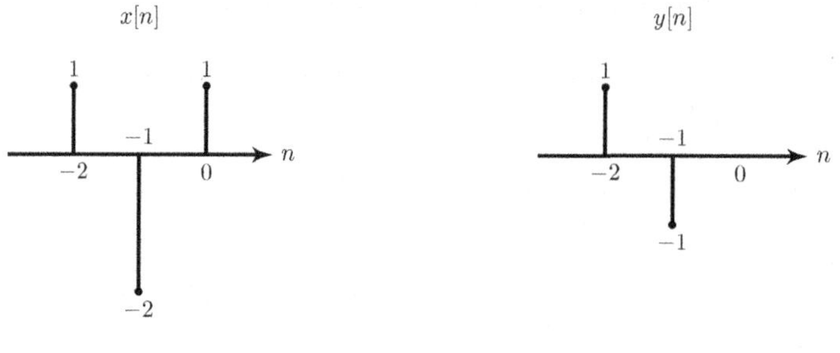

Figure 10.4: Problem 199

Solution.

At first, we obtain $x[n]$ and $y[n]$'s z-transform by writing it based on delta functions:

$$x[n] = \delta[n+2] - 2\delta[n+1] + \delta[n] \xrightarrow{Z} X(z) = z^2 - 2z + 1,$$

$$y[n] = \delta[n+2] - \delta[n+1] \xrightarrow{Z} Y(z) = z^2 - z.$$

Thus

$$H(z) = \frac{Y(z)}{x(z)} = \frac{z^2 - z}{z^2 - 2z + 1} = \frac{z(z-1)}{(z-1)^2} = \frac{z}{z-1} = \frac{1}{1-z^{-1}}.$$

As the pole of $H(z)$ is on the unit circle, this system cannot be stable. If we choose the region of convergence as $|z| > 1$, this system can be causal.

Problem 200. Consider two systems presented by the following equations. Let the impulse response of the inverse systems be denoted by $h_1[n]$ and $h_2[n]$, respectively. We let $h_3[n] = h_1[n] * h_2[n]$. Find $h_3[0]$.

$$y_1[n] = x[n] - \frac{1}{5}x[n-5],$$

$$y_2[n] = x[n] - \frac{1}{3}x[n-4].$$

Solution.

At first, we find $H_1(z)$ and $H_2(z)$ by taking the z-transform from both sides of difference equations:

$$Y_1(z) = X(z) - \frac{1}{5}z^{-5}X(z)$$
$$\Rightarrow \; H_1(z) = \frac{1}{1 - \frac{1}{5}z^{-5}},$$

and

$$Y_2(z) = X(z) - \frac{1}{3}z^{-4}X(z)$$
$$\Rightarrow \; H_2(z) = \frac{1}{1 - \frac{1}{3}z^{-4}}.$$

Using the convolution property

$$H_3(z) = H_1(z)H_2(z) = \frac{1}{(1 - \frac{1}{5}z^{-5})(1 - \frac{1}{3}z^{-4})}.$$

Now using the initial value theorem we obtain:

$$h_3[0] = \lim_{z\to\infty} H_3(z) = 1.$$

Problem 201. Transfer function of a system is given as

$$H(z) = \frac{1}{1 - \frac{1}{2}e^{\frac{j\pi}{4}}z^{-1}} + \frac{1}{1 - \frac{1}{2}e^{\frac{-j\pi}{4}}z^{-1}} - \frac{1}{1+2z^{-1}}.$$

Obtain the impulse response $h[n]$ assuming that the system is

a) Stable

b) Causal

Solution.

The system has three poles at $z = \frac{1}{2}e^{\frac{j\pi}{4}}$, $z = \frac{1}{2}e^{\frac{-j\pi}{4}}$, and $z = -2$. Since not all the poles are inside the unit circle, the system can not be both stable and causal simultaneously.

a) To have a stable system, the ROC should contain the unit circle. Thus, the ROC is $\frac{1}{2} < |z| < 2$ and $h[n]$ is obtained as

$$\begin{aligned} h[n] &= (\frac{1}{2}e^{\frac{j\pi}{4}})^n u[n] + (\frac{1}{2}e^{\frac{-j\pi}{4}})^n u[n] + (-2)^n u[-n-1] \\ &= 2 \cdot (\frac{1}{2})^n \cos(\frac{n\pi}{4}) u[n] + (-2)^n u[-n-1]. \end{aligned}$$

As can be seen, the system is not causal as $h[n] \neq 0$ for $n < 0$.

b) The ROC of a causal system should contain $z = \infty$. As such, the ROC is $|z| > 2$ and $h[n]$ is obtained as

$$\begin{aligned} h[n] &= (\frac{1}{2}e^{\frac{j\pi}{4}})^n u[n] + (\frac{1}{2}e^{\frac{-j\pi}{4}})^n u[n] - (-2)^n u[n] \\ &= 2 \cdot (\frac{1}{2})^n \cos(\frac{n\pi}{4}) u[n] - (-2)^n u[n]. \end{aligned}$$

As can be seen, the system is not stable due to the second term.

Problem 202. Find the z-transform of the following function when $|a| < 1$.

$$y[n] = \begin{cases} \sum_{k=-n}^{k=n} a^{|k|} & n \geq 0 \\ 0 & n < 0 \end{cases}$$

Solution.

Let's rewrite $y[n]$ as

$$\begin{aligned} y[n] &= \Big(\sum_{k=-n}^{k=n} a^{|k|}\Big) u[n] = \Big(1 + 2\sum_{k=1}^{k=n} a^{|k|}\Big) u[n] \\ &= \Big(1 + 2[a + a^2 + a^3 ... a^n]\Big) u[n] = \Big(1 + 2a[1 + a + a^2 ... a^{n-1}]\Big) u[n] \\ &= \Big(1 + \frac{2a(1-a^n)}{1-a}\Big) u[n] = u[n] + \frac{2a}{1-a} u[n] - \frac{2a}{1-a} a^n u[n] \\ &= \frac{1+a}{1-a} u[n] - \frac{2a}{1-a} a^n u[n]. \end{aligned}$$

Thus,

$$Y(z) = \frac{1+a}{1-a} \cdot \frac{1}{1-z^{-1}} - \frac{2a}{1-a} \cdot \frac{1}{1-az^{-1}}.$$

Problem 203. Consider an LTI and causal system which has only one pole and only one zero. In addition, the impulse response satisfies the following conditions. Find the zero and pole of this system.

a) $h[0] = \frac{1}{2}$

b) $\sum_{n=-\infty}^{n=\infty} h[n] = 2$

c) $\sum_{n=-\infty}^{n=\infty} (-1)^n h[n] = 0$

Solution.

Since the system has just one pole and zero, its transfer function can be written as below.

$$H(z) = K\frac{z - z_0}{z - p_0}.$$

As the system is causal, $h[n]$ is equal to zero for $n < 0$, and by the initial value theorem,

$$h[0] = \lim_{z\to\infty} H(z) = \lim_{z\to\infty} K\frac{z - z_0}{z - p_0} = K \xrightarrow{h[0]=\frac{1}{2}} K = \frac{1}{2}.$$

Thus

$$H(z) = \frac{1}{2} \cdot \frac{z - z_0}{z - p_0}. \tag{10.47}$$

By the definition of z-transform,

$$H(z) = \sum_{n=-\infty}^{n=\infty} h[n](z)^{-n}. \tag{10.48}$$

If we set $z = 1$ and $z = -1$ in (10.48), we have

$$H(1) = \sum_{n=-\infty}^{n=\infty} h[n](1)^{-n} = \sum_{n=-\infty}^{n=\infty} h[n],$$

$$H(-1) = \sum_{n=-\infty}^{n=\infty} h[n](-1)^{-n} = \sum_{n=-\infty}^{n=\infty} (-1)^n h[n].$$

Now using conditions **b** and **c**, and by substituting $z = 1$ and $z = -1$ into (10.47) we obtain

$$H(-1) = \frac{1}{2} \cdot \frac{-1 - z_0}{-1 - p_0} = 0 \longrightarrow z_0 = -1,$$

and

$$H(1) = \frac{1}{2} \cdot \frac{1+1}{1-p_0} = 2 \longrightarrow p_0 = \frac{1}{2}.$$

Problem 204. Transfer function of an LTI system is assumed as

$$H(z) = \frac{2}{1 - \frac{1}{2}z^{-3}}, \quad |z| > \frac{1}{2}.$$

Find the response of the system to the input $x[n] = 2 + \cos(\frac{n\pi}{3})$.

Solution.

It will not be straightforward to find the output of the system using conventional method, so we propose a shortcut solution: As unit circle ($|z| = 1$) is in the ROC, $H(z)$ is defined for $z = e^{jw}$ and so we can use the system frequency response to find the output (otherwise the output would have been unbounded). By substituting $z = e^{jw}$ into the transfer function we obtain

$$H(w) = \frac{2}{1 - \frac{1}{2}e^{-3jw}}.$$

The input can be seen as sum of two signals, $x_1[n] = 2$ and $x_2[n] = \cos(\frac{n\pi}{3})$. We now find the output for each signal separately by considering the fact that the output of an LTI system to an exponential input $e^{jw_0 n}$ is equal to $H(w_0)e^{jw_0 n}$. Thus we have

$$x_1[n] = 2 = 2e^{j0n} \Rightarrow y_1[n] = 2H(0)e^{j0n} = 2H(0) = 8.$$

$$x_2[n] = \cos(\frac{n\pi}{3}) \Rightarrow y_2[n] = |H(\frac{\pi}{3})| \cos\left(\frac{n\pi}{3} + \sphericalangle H(\frac{\pi}{3})\right),$$

where

$$H(\frac{\pi}{3}) = \frac{2}{1 - \frac{1}{2}e^{-3j\frac{\pi}{3}}} = \frac{4}{3}, \text{which means that } \sphericalangle H(\frac{\pi}{3}) = 0.$$

Therefore,

$$y[n] = y_1[n] + y_2[n] = 8 + \frac{4}{3}\cos(\frac{n\pi}{3}).$$

Problem 205. Let the z-transform of $x[n]$ be as follows:

$$X(z) = \frac{1}{1+z^2}, \quad ROC: |z| < 1.$$

Find the values of $x[0]$ and $x[-2]$.

Solution.

We know that for $|a| < 1$,

$$\frac{1}{1-a} = 1 + a + a^2 + a^3 + ...; \tag{10.49}$$

now since $|z| < 1$, then $|-z^2| < 1$. We substitute $a = -z^2$ into (10.49) and write

$$\frac{1}{1+z^2} = 1 - z^2 + z^4 - z^6 + ... \quad (10.50)$$

On the other hand, using the definition of z-transform we have

$$X(z) = \sum_{n=-\infty}^{n=\infty} x[n](z)^{-n} = ... + x[-2]z^2 + x[-1]z + x[0] + x[1]z^{-1} + x[2]z^{-2}... \quad (10.51)$$

Comparing (10.50) and (10.51), we have

$$x[0] = 1, \quad x[-2] = -1.$$

Problem 206. Consider the following difference equation which describes a stable LTI system:

$$y[n] - ay[n-1] + y[n-2] = x[n]. \quad (10.52)$$

Comment on the causality of this system for different values of a.

Solution.
We take the z-transform from both sides of the difference equation and obtain the transfer function:

$$Y(z)\left(1 - az^{-1} + z^{-2}\right) = X(z)$$

$$\Rightarrow H(z) = \frac{Y(z)}{X(z)} = \frac{1}{1 - az^{-1} + z^{-2}} = \frac{z^2}{z^2 - az + 1}.$$

Let z_1 and z_2 be the poles of the system which are in fact the roots of quadratic equation $z^2 - az + 1 = 0$. We know that in a quadratic equation, the constant term is equal to the product of the roots. Thus in our case we have $z_1 z_2 = 1$. Thus means that $|z_1||z_2| = 1$ which implies that regardless of the value of a, always one of the poles is inside the unit circle and the other one is outside the unit circle.

Let z_1 be the pole inside the unit circle, i.e., $|z_1| < 1$ and consequently, $|z_2| > 1$. Now since the system is assumed to be stable, the ROC should contain the unit circle, i.e, the ROC is $|z_1| < z < |z_2|$. This means that $z = \infty$ can not be in the ROC and thus the system can not be casual regardless of the value of a.

Made in the USA
Monee, IL
07 November 2026

67981961R00221